The Democratization of Invention

This book examines the evolution and impact of the intellectual property rights system in the United States during the "long nineteenth century." The American experience is compared to Britain and France, countries whose institutions reflected their oligarchic origins. Instead, U.S. patent and copyright institutions were carefully calibrated to "promote the general welfare." The United States created the first modern patent system, and its policies were the most liberal in the world toward inventors. When markets expanded, these inventors contributed to the proliferation of new technologies and improvements, many of which proved to be valuable in both economic and technical terms. Individuals who did not have the resources to directly exploit their inventions benefited disproportionately from the operation of efficient markets. The accumulation of such incremental efforts helped to propel the United States to the forefront of all other industrial nations. In contrast to its leadership in the area of patents, the U.S. copyright regime was among the weakest in the world, and it profited from international copyright piracy for a century. American patent and copyright institutions both promoted a process of democratization that not only furthered economic and technological progress but also provided a conduit for the creativity and achievements of relatively disadvantaged groups.

B. Zorina Khan is Associate Professor of Economics at Bowdoin College. Her research focuses on the economic history of law, technology, and institutions. She has written articles for such journals as the *Journal of Economic History* and the *Journal of Economic Perspectives*. She is on the Editorial Board of the *Journal of Economic History* and is a member of the National Bureau of Economic Research.

1. The object of the NBER is to ascertain and present to the economics profession, and to the public more generally, important economic facts and their interpretation in a scientific manner without policy recommendations. The Board of Directors is charged with the responsibility of ensuring that the work of the NBER is carried on in strict conformity with this object.

2. The President shall establish an internal review process to ensure that book manuscripts proposed for publication DO NOT contain policy recommendations. This shall apply both to the proceedings of conferences and to manuscripts by a single author or by one or more co-authors but shall not apply to authors of comments at NBER conferences who are not NBER affiliates.

3. No book manuscript reporting research shall be published by the NBER until the President has sent to each member of the Board a notice that a manuscript is recommended for publication and that in the President's opinion it is suitable for publication in accordance with the above principles of the NBER. Such notification will include a table of contents and an abstract or summary of the manuscript's content, a list of contributors if applicable, and a response form for use by Directors who desire a copy of the manuscript for review. Each manuscript shall contain a summary drawing attention to the nature and treatment of the problem studied and the main conclusions reached.

4. No volume shall be published until forty-five days have elapsed from the above notification of intention to publish it. During this period a copy shall be sent to any Director requesting it, and if any Director objects to publication on the grounds that the manuscript contains policy recommendations, the objection will be presented to the author(s) or editor(s). In case of dispute, all members of the Board shall be notified, and the President shall appoint an ad hoc committee of the Board to decide the matter; thirty days additional shall be granted for this purpose.

5. The President shall present annually to the Board a report describing the internal manuscript review process, any objections made by Directors before publication or by anyone after publication, any disputes about such matters, and how they were handled.

6. Publications of the NBER issued for informational purposes concerning the work of the Bureau, or issued to inform the public of the activities at the Bureau, including but not limited to the NBER Digest and Reporter, shall be consistent with the object stated in paragraph 1. They shall contain a specific disclaimer noting that they have not passed through the review procedures required in this resolution. The Executive Committee of the Board is charged with the review of all such publications from time to time.

7. NBER working papers and manuscripts distributed on the Bureau's web site are not deemed to be publications for the purpose of this resolution, but they shall be consistent with the object stated in paragraph 1. Working papers shall contain a specific disclaimer noting that they have not passed through the review procedures required in this resolution. The NBER's web site shall contain a similar disclaimer. The President shall establish an internal review process to ensure that the working papers and the web site do not contain policy recommendations, and shall report annually to the Board on this process and any concerns raised in connection with it.

8. Unless otherwise determined by the Board or exempted by the terms of paragraphs 6 and 7, a copy of this resolution shall be printed in each NBER publication as described in paragraph 2 above.

The Democratization of Invention

Patents and Copyrights in American
Economic Development, 1790–1920

B. ZORINA KHAN

NBER

CAMBRIDGE
UNIVERSITY PRESS

CAMBRIDGE UNIVERSITY PRESS
Cambridge, New York, Melbourne, Madrid, Cape Town, Singapore, São Paulo

Cambridge University Press
40 West 20th Street, New York, NY 10011-4211, USA

www.cambridge.org
Information on this title: www.cambridge.org/9780521811354

First published 2005

Printed in the United States of America

A catalog record for this publication is available from the British Library.

Library of Congress Cataloging in Publication Data

Khan, B. Zorina.
The democratization of invention : patents and copyrights in American economic
development, 1790–1920 / B. Zorina Khan.
p.   cm.
ISBN 0-521-81135-X (hardback)
1. Intellectual property – Economic aspects – United States – History.   2. Copyright –
Economic aspects – United States – History.   3. Patents – Economic aspects – United States –
History.   4. Inventions – Economic aspects – United States – History.   5. Democracy –
United States – History.   1. Title.
KF2979.K48   2005
346.7304′8 – dc22   2004029199

ISBN-13   978-0-521-81135-4 hardback
ISBN-10   0-521-81135-X hardback

For my parents, Charles and Dhanpat Khan,
and my sisters, Esther and Sherry

Contents

Tables

xi

Figures and Illustrations

FIGURES

ILLUSTRATIONS

Acknowledgments

Cliometrics is a development of our own times and offers a unique opportunity for younger scholars to benefit directly from leaders in the field. I would not be an economic historian today were it not for the assistance and example of colleagues such as Lance Davis, Stanley Engerman, Hank Gemery, Claudia Goldin, Naomi Lamoreaux, Peter Lindert, Joel Mokyr, Alan Olmstead, and Kenneth Sokoloff. Stanley Engerman has been exceedingly generous in offering comments on my projects, responding with insightful handwritten suggestions within the week. (Indeed, Stan takes the initiative by frequently reminding me not to delete him from my mailing list.) Claudia Goldin offered detailed comments that helped to significantly improve the first draft of this book. As coeditor of the *JEH*, Naomi Lamoreaux had a transformative effect on my papers and she has had an equally transformative effect on my perceptions of the qualities that comprise a true economic historian. Joel Mokyr, whose own work illuminates the history of technological change, has been uniformly supportive and encouraging. The All-University of California Economic History Group and its unstinting mentors, especially Alan Olmstead and Peter Lindert, provided an unparalleled resource when I was a graduate student, as they have for numerous other potential cliometricians.

Kenneth Sokoloff has consistently set a high standard for meticulous, original, and inspiring research. I entered the UCLA Ph.D. program in Economics with the intention of studying modern technology and industrial organization. Ken was looking for a research assistant at the time and offered me several options, including a "risky project on patents," which I promptly selected. His seminal work on the economic history of U.S. patenting was published in the *Journal of Economic History* in 1988 as "Inventive Activity in Early Industrial America: Evidence from Patent Records, 1790–1846." This paper is one of those rare articles that is not only enormously creative in itself but also adds value because it continues to stimulate numerous other researchers to contribute to related aspects of this "risky project." Ken supervised my thesis on the patent system and inventive activity in Britain and America and coauthored the papers that comprise the basis for Chapters 4 and 7 in this book. He reads numerous drafts of my research and never fails to offer suggestions that result in marked improvements. Thus, from the first two weeks of my life as a graduate student to the present day I have been exceptionally fortunate that Ken has served as a coauthor, role model, mentor, and friend.

One of the advantages of an academic life is that it allows you an entrée into a cooperative network of exceptional individuals whose work informs and enhances your own. Harold Demsetz and Ed Leamer of UCLA have always represented my ideal of the true economist, whose command of logic and insight clarifies widely ranging issues. Other careful scholars who have contributed to my understanding of the history of intellectual property and technology include Kristine Bruland, Larry Epstein, Catherine Fisk, Patricio Sáiz González, Christine MacLeod, Robert Merges, Liliane Hilaire-Pérez, and members of the NBER programs on Productivity and the Development of the American Economy.

James Lee, Esther Khan, Yael Elad, and Valérie Marchal (of the Institut national de la propriété industrielle in Paris) provided valuable help, as did a number of excellent undergraduate assistants at UCLA and Bowdoin College. Anonymous referees for journal articles and for this book and participants at numerous seminars also offered useful suggestions that helped to refine the exposition and arguments. Bowdoin College has provided a supportive academic and research environment. Special thanks are due to Guy Saldanha, Ginny Hopcroft, Carr Ross, and Phyllis McQuaide, some of Bowdoin's truly exceptional cadre of librarians, who have scoured national and international institutions on my behalf to secure countless obscure texts in English, French, and Spanish (and even considered giving me my own bookshelf in the Interlibrary Loan section).

I would like to express my appreciation to Martin Feldstein and to the National Bureau of Economic Research for the grant of the 2004–2005 Griliches Fellowship. Institutions that also provided generous support include the Institute for Advanced Studies at the Australian National University, the Lemelson Center for Invention and Innovation, the Department of Economics at UCLA, and the UCLA School of Law. I am also grateful to the *Journal of Economic History* and the *Journal of Interdisciplinary History* for permission to use material that was previously published.

Illustration 1. This 1792 painting by Samuel Jennings, *Liberty Displaying the Arts and Sciences* (detail), captures the view of the Framers of the U.S. Constitution that a democratic society would "promote the progress of science and useful arts." (Courtesy of The Library Company of Philadelphia.)

1

Introduction

Institutions, Inventions, and Economic Growth

"We showed the results of pure democracy upon the industry of men."
– Edward Riddle, "Report on the World's Exposition," 1851

A belief in the ability of democracy and technology to enhance the common good has defined American society since the founding of the Republic.[1] To the men who gathered in Philadelphia to "promote the general Welfare," it was self-evident that ideas, inventions, and democratic values were integrally related.[2] The intellectual property clause was included in the very first Article of the U.S. Constitution, a document that distilled the precepts of a

---

[1] According to the *American Jurist* (vol. 10, 1833, p. 121), "no government of magnitude or power, whether free or arbitrary, has hitherto been sustained without the help of the distinction of classes." Democracy is a concept that is easily recognizable in its entirety but more contentious in the details, which can be as subtle as they are multifarious. This is not a treatise in political philosophy, so I will adopt the Alice in Wonderland approach, and merely specify a list of features that indicate what I mean by the term. Democracy entails the protection of private property, freedom of choice and speech, equality of opportunity, and equal access to political and economic institutions and their benefits (but not necessarily equality of outcome), an independent judiciary that protects the rule of law, an elected government that represents the majority of the population, a system of checks and balances to prevent subversion or capture by a self-interested minority, and flexible institutions that respond to changes in social costs and benefits.

[2] A common view in the eighteenth century held "That it is impossible for the arts and sciences to arise, at first, among any people unless that people enjoy the blessing of a free government.... The first growth, therefore, of the arts and sciences can never be expected in despotic governments." ["Of the Rise and Progress of the Arts and Sciences" Volume 2 of David Hume's *Essays, Moral and Political* (1742).] A letter of James Madison to Thomas Jefferson (New York Oct. 17, 1788) distinguished between the dangers of monopolies and exclusive rights in an oligarchic society and one based on democratic principles: "With regard to monopolies they are justly classed among the greatest nusances (sic) in Government. But is it clear that as encouragements to literary works and ingenious discoveries, they are not too valuable to be wholly renounced? Would it not suffice to reserve in all cases a right to the Public to abolish the privilege at a price to be specified in the grant of it? Is there not also infinitely less danger of this abuse in our Governments, than in most others? Monopolies are sacrifices of the many to the few. Where the power is in the few it is natural for them to sacrifice the many to their own partialities and corruptions. Where the power, as with us, is in the many not in the few, the danger can not be very great that the few will be thus favored." [http://www.constitution.org/jm/17881017_tj.txt, accessed January 2005.]

democratic society. The proposal passed without any debate and with unanimous approval, because it was viewed as a prerequisite for progress.[3] The
growth of science and literature in tandem with broad-based access to an
intellectual property system was even declared to be "essential to the preservation of a free Constitution."[4] Congress was therefore given the mandate
to "promote the Progress of Science and useful Arts, by securing for limited Times to Authors and Inventors the exclusive Right to their respective
Writings and Discoveries."[5]

One of the most striking innovations of the framers of the American Constitution was their recognition of the value of contributions from the less
exceptional. American institutions were designed to ensure that rewards
accrued to the deserving based on productivity rather than on the arbitrary basis of class, patronage, or privilege. Alexis de Tocqueville, still the
shrewdest observer of the American national character, argued that "You
may be sure that the more a nation is democratic, enlightened, and free,
the greater will be the number of these interested promoters of scientific
genius, and the more will discoveries immediately applicable to productive
industry confer gain, fame, and even power on their authors. For in democracies the working class take a part in public affairs; and public honors
as well as pecuniary remuneration may be awarded to those who deserve
them."[6] The creators of supposedly heroic inventions were lauded in the
European nations; inventors and innovators of all classes were universally
celebrated in the United States. Indeed, according to Thomas Jefferson, "a
smaller [invention], applicable to our daily concerns, is infinitely more valuable than the greatest which can be used only for great objects. For these
interest the few alone, the former the many."[7]

---

[3] George Washington's First Annual speech to Congress on January 8, 1790 in Federal Hall,
New York City, stated "The advancement of agriculture, commerce, and manufactures, by
all proper means, will not, I trust, need recommendation; but I cannot forbear intimating
to you the expediency of giving effectual encouragement as well to the introduction of new
and useful inventions from abroad, as to the exertions of skill and genius in producing them
at home.... Nor am I less persuaded, that you will agree with me in opinion, that there is
nothing which can better deserve your patronage, than the promotion of science and literature. Knowledge is in every country the surest basis of publick [sic] happiness. In one, in which
the measures of government receive their impressions so immediately from the sense of the
community, as in ours, it is proportionably essential."

[4] "Literature and Science are essential to the preservation of a free Constitution: the measures
of Government should, therefore, be calculated to strengthen the confidence that is due to that
important truth," *U.S. Senate Journal*, 1st Cong. 8–10 (1790); *U.S. Annals of Congress*, 1st.
Cong. 935–36; cited in Bruce W. Bugbee, p. 137, *Genesis of American Patent and Copyright
Law*, Washington, D.C., Public Affairs Press (1967).

[5] U.S. Constitution, Art. I, § 8, cl. 8.

[6] Alexis de Tocqueville, *Democracy in America*, trans. by Henry Reeve (2 vols., London, 1889),
II, 35–42.

[7] From a letter Jefferson sent to George Fleming in 1815, excerpt from *The Jeffersonian Cyclo-
pedia*, http://etext.lib.virginia.edu.

The belief in the power of technology and industry to serve the many was not unmixed, as we know from the conflicts between Thomas Jefferson and Alexander Hamilton. However, the early Jeffersonian fear of the negative consequences of monopolies and industrialization was soon lost in the optimistic conviction that democracy was a crucible that would convert the resources of man and nature into wealth for everyone in the nation, and not just an arbitrary few. American conceptions of Utopia, such as Edward Bellamy's *Looking Backward*, were colored by rosy visions in which technological innovations conjured up a benign world of plenty that allowed the attainment of the highest social and political ideals. Ultimately, the intellectual property system would have to incorporate the more complex idea that it was necessary to construct a system that induced patentees and copyright holders to contribute to social welfare, but at the same time did not create undue obstacles to the diffusion of their creations, nor to the development of new products that built on their pioneering contributions.[8] Nevertheless, the emphasis in the nineteenth century was decidedly on the need to promote progress through security of private property rights in inventions.

The American system of intellectual property was based on the conviction that individual effort was stimulated by higher expected returns. Abraham Lincoln – who was himself a patentee – declared that the rate and direction of inventive activity were determined by "the fuel of interest." Genius was redefined as the province of the many, not the rare gift of the few, and only wanted the assurance that the inventor would be able to benefit from his investments. Supreme Court Justice Joseph Story, whose brilliant decisions are enshrined in modern patent and copyright laws, exhorted an audience of ordinary mechanics in 1829: "Ask yourselves, what would be the result of one hundred thousand minds...urged on by the daily motives of interest, to acquire new skill, or invent new improvements."[9] The answer was not long in coming, for the next few decades would lay the foundation for American industrial and cultural supremacy. Contemporary observers were dazzled by the rate of cumulative attainments, and it is worth recalling that since the days of canal-building the optimistic notion of a "new era" has persisted throughout American history.[10] At first, the British were dismissive

---

[8] See, for instance, the testimonies before Congress when patentees applied for extensions to their existing patent term. As the editors of *Scientific American* noted, "Special acts of Congress in extending patents often do injury to inventors in general; they also tend to retard the progress of invention, and for this reason we oppose the extension of patents by Congress, in cases where patentees have been sufficiently remunerated. One patentee, under a democratic government like ours, has no more right to special privileges than another." *Scientific American*, January 21, vol. 9 (19), 1854, p. 149.

[9] Speech reported in *American Jurist and Law Magazine*, vol. 1 (1829).

[10] Writing at the start of the year of 1844, Commissioner Ellsworth marvelled that "The advancement of the arts, from year to year, taxes our credulity, and seems to presage the arrival of that period when human improvement must end." Report of the Commissioner of Patents for 1843, 28th Congress, 1st Session, [Senate] [150] February 13, 1844.

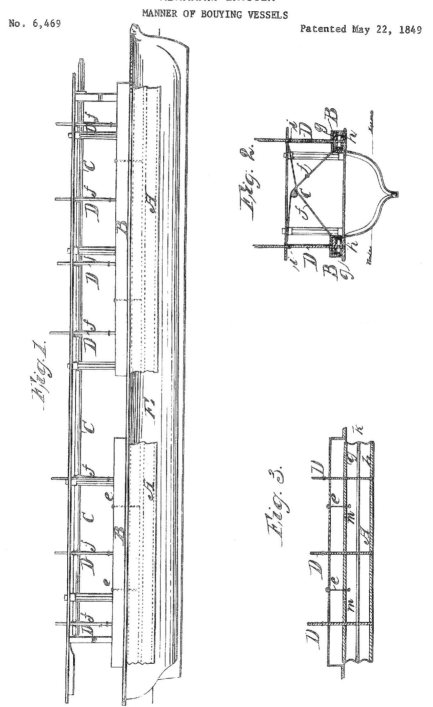

ABRAHAM LINCOLN
MANNER OF BOUYING VESSELS

No. 6,469

Patented May 22, 1849

Illustration 2. Abraham Lincoln, the only President of the United States to obtain a patent, served as legal counsel in several disputes about patent rights. (*Source*: U.S. Patent Office.)

of American efforts, and even declared it was unlikely that their former colony would ever progress beyond facile emulation of superior European technologies and culture. However, by the time of the Crystal Palace Exhibition in 1851, Europeans were surprised and alarmed to find that the United States was marshaling its resources in a way that promised to propel it to the first rank among nations.

The United States today is the most powerful nation on earth but, as Stanley Engerman and Kenneth Sokoloff remind us, early in U.S. history its standard of living was lower than the level enjoyed by many of its contemporary South American and West Indian neighbors.[11] Even on the eve of the Declaration of Independence the United States was an undistinguished developing country with an agricultural economy and few pretensions to local cultural output of any distinction. How did this former colony make the transition from follower to a leading economy in the course of one century? Numerous contemporary observers tried to uncover the reasons for the rapid trajectory in American development, and many explicitly pointed to the advantages of a democratic society for technical and cultural inventiveness.[12] For others, the answer could be found in its intellectual property system. Attribution to any single factor will obviously overstate its influence, as American economic and social progress was a function of an array of variables, including (among others) a relatively equal distribution of income, an educated and enterprising populace, enlightened legal institutions, and favorable factor endowments. Nevertheless, the reinforcing relationships between intellectual property institutions and democracy in America are worth further exploration.

---

[11] Stanley L. Engerman and Kenneth L. Sokoloff, "Factor Endowments, Institutions and Differential Paths of Growth among New World Economies," in Stephen Haber (ed.), *How Latin America Fell Behind*, Palo Alto: Stanford University Press, 1997: 260–304; Kenneth L. Sokoloff and Stanley L. Engerman, "Institutions, Factor Endowments, and Paths of Development in the New World," *Journal of Economic Perspectives*, Vol. 14, No. 3, Summer 2000: 217–32

[12] For instance, see *The American System of Manufactures*: the report of the Committee on the Machinery of the United States 1855, and the special reports of George Wallis and Joseph Whitworth 1854, Nathan Rosenberg (ed.), pp. 388–89, Edinburgh: Edinburgh University Press, 1969. They noted the favorable attitudes of American workers toward new improvements, in marked contrast to the sullen disapproval of the British working class. The commentators tried to find the causes for their pervasive inventiveness and pointed to the beneficial influence of laws in the United States, and the widespread education and literacy that characterized ordinary citizens. An American observer of the Exhibition similarly declared that the items produced and displayed by the United States contingent provided "evidence of the ingenuity, industry and capacity of a free and educated people.... We demonstrated the progressiveness of the human mind when in enjoyment of liberty." Edward Riddle, "Report on the World's Exposition," Report of the Commissioner of Patents for the Year 1851, Washington, D.C., 1852: 347–485. This is also the source of the epigraph to this chapter.

THE DEMOCRATIZATION OF INVENTION

This book examines American experience in a European mirror, and contrasts intellectual property institutions in Britain, France, and the United States. The philosophy and enforcement of intellectual property laws in Britain and France, the structure of the patent and copyright systems, and the resulting patterns of invention, were all consistent with the oligarchic nature of European society.[13] Although there is little consensus on many of these points, some have argued that early patent and copyright laws in England were conflated and tended to be explicated in terms of similar underlying principles of individual creativity and spontaneous manifestations of genius.[14] Later distinctions between patent and copyright doctrines were based on subjective estimations of the quantity and quality of mental labor involved in industrial and literary invention.[15] According to this mode of reasoning, literary and artistic inventions were more deserving of protection than pragmatic industrial inventions, and copyright piracy was regarded as a more egregious offence than patent infringement.[16] This perspective was reinforced by the grant of patents to anyone who paid the fees, regardless

[13] Modern scholars who specialize in the philosophical dimensions of intellectual property differ quite significantly in their interpretations of the implications of the nineteenth-century literature. It is not my intention to enter into this debate. The following discussion is necessarily quite general, and does not adequately document the subtleties in different perspectives, nor their changes over time. For instance, utilitarian arguments were sometimes made in Britain and France, and the appeal to natural rights were not entirely absent from American debates, at least at the rhetorical level. My outline draws on the preponderance of arguments in these jurisdictions to characterize the broad differences between U.S. and European approaches to intellectual property. However, the main emphasis here and in the following chapters is not on philosophical motivation but on policies and measurable outcomes.

[14] According to Mark Rose, in eighteenth-century Britain, "a work of literature belonged to an individual because it was, finally, an embodiment of that individual.... The basis of literary property, in other words, was not just labor but "personality," and this revealed itself in "originality." (*Authors and Owners: The Invention of Copyright*, Cambridge, Mass.: Harvard University Press, p. 114.) See also Martha Woodmansee, "The Genius and the Copyright: Economic and Legal Conditions for the Emergence of the 'Author'," *Eighteenth-Century Studies*, vol. 17 (4) 1984: 425–48.

[15] See Brad Sherman and Lloyd Bently, *The Making of Modern Intellectual Property Law: the British Experience, 1760–1911*, Cambridge: Cambridge University Press, 1999: "It is clear that during the literary property debate the quantity of mental labour which was embodied in representative objects played an important role in distinguishing between the different forms of protection then available" (p. 147) and "one of the defining features of intellectual property law in the eighteenth and the first half of the nineteenth century was its concern with mental labour and creativity" (p. 173). According to Clare Pettitt, *Patent Inventions – Intellectual Property and the Victorian Novel*, Oxford: Oxford University Press, 2004, initially the mental labor of mechanical and literary inventors was treated analogously, but "the analogy . . . between the inventor and the writer, was losing currency by the end of the century" (p. 32).

[16] See Brad Sherman and Lloyd Bently, *The Making of Modern Intellectual Property Law: the British Experience, 1760–1911*.

of whether or not they were true inventors or mere importers of inventions. Although many Europeans in the nineteenth century lobbied to repeal patent protection, the same abolitionists would have been horrified at parallel proposals to turn all literary inventions over to the public domain.[17]

European societies were organized in ways that concentrated power in the hands of elites and facilitated rent-seeking by favored producers, and the organization of invention was no exception. The hierarchical culture of Britain and France was replicated through institutions that promoted the inherent rights and genius of authors and (to a lesser extent) inventors. Intellectual property systems were derived from the grant of "privileges" or monopoly rights from the Crown, and subsequent grants reflected their provenance. In British law, patents were regarded as "pernicious monopolies," which had to be narrowly interpreted, monitored, and restricted. The legal system was biased against patents in general, and incremental improvements in particular. High transaction and monetary costs, as well as the prevailing prejudices toward nonelites, combined to create barriers to entry that excluded the poor or disadvantaged from making contributions to economic growth. Patent fees in England were so costly that they effectively (and indeed, consciously) excluded working-class inventors from patenting their discoveries. As a result, trade secrecy likely played a more prominent part in protecting new discoveries, diffusion was certainly inhibited, potential inventors faced a great deal of uncertainty, markets were thin, and the rate of technological change may have been adversely influenced.

Clearly, despite these drawbacks Britain and France still experienced early industrialization and economic growth, but it is also true that their economies were unable to sustain their initial advantage. The case of patents and copyrights suggests that their loss of competitiveness may have partially owed to policies that favored elites and deprecated the contributions of the uneducated working class. The British system restricted patent rights in ways that favored capital-intensive industries and unbalanced economic growth patterns. The elite groups who were privileged by these institutions had little inducement to adopt improvements or techniques that infringed on their rents, and in some cases had the power to suppress competing technologies. As long as their private benefits were enhanced by such a strategy, they might even have had the incentive to shift the growth path onto a lower trajectory. As an example of this, the British patent system generated surplus revenues to patent agents and administrators who lobbied against reforms.

---

[17] Sherman and Bently point out that "while the anti-patent lobby believed that the producers of literary and artistic property...were properly designated as creators, the same could not be said of inventors," p. 150. They argue that the patent controversy of the 1860s and 1870s shifted the rationale for protection away from the labor theory of value and toward a characterization of patents as part of a utilitarian social contract between inventors and the state, with a focus on the invention itself.

Recommendations such as the introduction of an examination system were
rejected in part because they threatened to erode the Royal mandate. Moreover, since creativity and genius are unlikely to vary systematically over
time, institutions that are predicated on these factors are unlikely to generate internal reforms that might induce greater inventive activity. Consequently, despite their inefficiencies, the patent rules and standards in both
France and England remained essentially unchanged for stretches of over a
hundred years. Similarly, the confused state of British copyright grants was
rationalized only in 1911 and some have even argued that their present-day
copyright laws remain "pre-modern."[18] In sum, England and France failed
to offer inducements for investments by all potential inventors regardless of
their background, and privileged the rights of elite producers in a manner
that arguably reduced social welfare.

   Instead of adhering to the European model, the United States consciously
created patent and copyright institutions that were intended to function as
the keystone of a democratic society. The Constitution specified that the
pragmatic, utilitarian objective of the intellectual property system was to
promote the public welfare through additions to knowledge and technology.
Patent rights and copyrights were clearly distinguished in separate statutes
in 1790, and developed along diametrically different lines based on a rational assessment of their costs and benefits. Policy makers in the United States
were well aware of the European experience in this and other dimensions.
They carefully considered the potential deficiencies of state grants of intellectual property rights, as well as suggestions for alternative strategies that
others considered to be superior. They did not shrink from novel approaches
that they estimated would increase social welfare, regardless of how great the
popular outcry, as witnessed by their refusal to recognize international copyrights. Thus, it is implausible to consider the early structure of U.S. patent
and copyright statutes and their implementation as haphazard or random;
rather, the innovations in these institutions were deliberate, and comprised
a critical part of a blueprint for a democratic society.

   The discussion in this book highlights the contributions of intellectual
property institutions in shaping the unique character of U.S. economic
growth in the nineteenth century. Among the leading nations of the day it
was commonplace to acknowledge that patent rights might increase the rate
of invention, but it was less conventional to propose that the background or
the identity of inventors was irrelevant to their productivity. Although the
U.S. Constitution itself fell short of true democratic ideals in many regards,

---

[18] Sherman and Bently argue that, although patent laws gradually acquired their modern form
in the second half of the nineteenth century, to some degree copyright laws in Britain still
remain "pre-modern" (p. 192). According to Catherine Seville, "the rationale for copyright remains unclear" (*Literary Copyright Reform in Early Victorian England*, Cambridge:
Cambridge University Press, 1999, p. 12).

the intellectual property system it authorized epitomized those ideals.[19] The patent system exemplified one of the most democratic institutions in early American society, offering secure property rights to true inventors, regardless of age, color, marital status, gender, or economic standing. The empirical analysis explores the extent to which outcomes accorded with these objectives. Who were the individuals contributing to the transformation of technology and society in the United States during this critical period, and what induced them to redirect their attention to creating additions to the existing stock of useful knowledge?

The patterns of patenting, when linked to biographical information, show that the expansion of markets and profit opportunities stimulated increases in inventive activity by attracting wider participation from relatively ordinary individuals. The technical skills and knowledge required for effective invention during this era were widely diffused among the general population. Rather than an elite that possessed rare technical skills or commanded large stocks of resources, the rise in patenting was associated with a democratic broadening of the ranks of patentees to include individuals, occupations, and geographic districts with little previous experience in invention. One finds among the roster of patentees not only engineers and machinists, but also candidates for the Greenback Party, schoolteachers, poets, humble factory workers, housewives, farmhands, teenagers, and even economists. *Scientific American* would later proclaim that the United States advanced "not because we are by nature more inventive than other men – every nationality becomes inventive the moment it comes under our laws – but because the poorest man here can patent his devices. . . . In the aggregate the little things – which in England or on the continent either could not be or would not be patented, owing to the excessive cost of the papers or other onerous conditions – probably add more to the wealth and wellbeing of the community, and more to the personal income of the inventor, than the great things do."[20]

The market orientation of the American intellectual property system aided the democratization of invention because it enhanced the opportunities of nonelite inventors. It is a standard libertarian claim that free markets evolve in tandem with democratic principles. However, the link between markets and democracy is often made in terms of consumer sovereignty or the freedom to choose among competing offers. The analysis here emphasizes the role that patents and copyrights played in the securitization of ideas through

---

[19] Robert Dahl points to seven undemocratic elements in the Framers' Constitution: slavery; limited suffrage; election of the president; the appointment of senators by state legislatures rather than by the people; equal representation in the Senate; judicial power; Congressional power to regulate and control the economy was constrained. (*How Democratic is the American Constitution?* New Haven, Conn.: Yale University Press [2001].) Because at least four of these elements can be disputed, it seems inevitable that the nature and extent of democracy in America will remain fuzzy and contentious to observers.

[20] *Scientific American*, October 21, 1876, vol. 35 (17), p. 256.

the creation of tradeable assets: intellectual property rights facilitated market exchange, a process that assigned value, helped to mobilize capital, and improved the allocation of resources. Access to markets and trade in inventions led to greater specialization and division of labor among inventors, and furthered the diffusion of new technologies. Extensive markets in patent rights allowed inventors to extract returns from their activities through licensing and assigning or selling their rights. The ability to transform their human inventive capital into tradeable assets disproportionately helped inventors from disadvantaged backgrounds who lacked the financial resources or contacts that would have allowed them to extract returns by commercializing their inventions on their own.

American democracy, it is sometimes proposed, benefited men at the expense of women, and many women – especially those who lived in rural areas – were excluded from the mainstream of economic progress. Patents do not capture all of the inventions that are created, but this limitation makes it all the more striking that these records indicate that nineteenth-century women were active participants in the market for technology. The diffusion of household innovations in both rural and urban regions was more pervasive than previously thought. Patents by women comprised only a small fraction of total patents, but the overall patterns of patenting and the pursuit of profit opportunities by women inventors were similar to those of male inventors. A notable departure from the parallels between male and female patenting was manifested in the higher fraction of rural women who obtained patents, relative to the patterns for men. Women in frontier regions were especially inventive, and devised ingenious mechanisms to ease the burden of an arduous existence far from the conveniences of cities and extended social networks. However, even if patent rights were well protected by the federal courts, state laws also influenced the ability to benefit from innovations. The barriers to individual initiative that state legislatures initially placed in this and other contexts illustrates the wisdom of maintaining enforcement of intellectual property rights at the federal level. For much of the nineteenth century, married women lobbied for reforms in state laws that prohibited or hindered their capacity to hold property, engage in contracts, and keep their earnings. Legal reforms in married women's property rights encouraged women to increase their investments in patenting. Their responsiveness to such institutional changes highlight the importance of specific features of other institutions, including the parameters of intellectual property rules and standards.

According to Douglass North, "The most interesting challenge to the economic historian is to account for changes in the structure and enforcement of property rights over time."[21] One way in which to do so is through the

---

[21] Douglass C. North, p. 250, "A Framework for Analyzing the State in Economic History," *EEH* vol. 16 (3) 1979: 249–59.

analysis of legal records, and I assess extensive samples of patent and copyright lawsuits. Courts confronted the continuous stream of mankind about its commonplace business of life, and from these unpropitious materials created decisions that were based on analogies drawn from historical experience, logic, and the attempt to serve the community in general. The economic history of intellectual property laws and their enforcement leads to the inevitable conclusion that the federal judiciary and the U.S. legal system played a central role in facilitating social and economic progress during the nineteenth century. In other work, I examined several thousand suits at common law that dealt with major innovations including canals, railroads, the telegraph, automobiles and medical technologies.[22] Those records likewise support the argument here that the judiciary objectively weighed costs and benefits, and ultimately the decisions that prevailed promoted social welfare rather than the interests of any single group. This is not to say that every judge was of the caliber of Joseph Story or Benjamin Cardozo, but a system of appeals assured that "the tide rises and falls, but the sands of error crumble."[23] There is little support for the notion that judges subsidized economic development by transferring resources from the working class to corporations.[24] Effective policies toward innovations required a social calculus that was far more subtle than a blind promotion of the interests of any one specific group in society. Technological advances altered the costs and benefits of transacting within a particular network of rules and standards, but open and accessible institutions proved to be sufficiently flexible to accommodate these changes.

Unlike England, where the Crown reserved the right to expropriate patent property, in the United States even federal government claims could not trump the "supreme law of the land." American judges understood that the most effective means to counter oligarchical tendencies was through secure private property rights and market competition. The judiciary in the antebellum period refuted the idea that patent grants required metaphysical inquiries into the quantity of mental labor or the degree of inventiveness. All that was required was that the invention should be new to the world. As for decisions about the utility of allegedly trivial inventions, that was to be determined by the market, not by the courts. The early judicial optimism about the coincidence between private and public welfare waned somewhat

---

[22] B. Zorina Khan, "Innovations in Law and Technology, 1790–1920," in *Cambridge History of Law in America*, Michael Grossberg and Christopher Tomlins (eds.), Cambridge: Cambridge University Press (forthcoming).

[23] Benjamin Cardozo, *The Nature of the Judicial Process*, New Haven, Conn.: Yale University Press, [c1921], p. 177.

[24] According to Morton Horwitz's influential book, *The Transformation of American Law, 1780–1860*, Cambridge, Mass.: Harvard University Press (1977), p. 101, judicial decisions were biased in favor of industrialists and were able to "dramatically . . . throw the burden of economic development on the weakest and least active elements of the population."

by the second half of the century, as equity courts mediated the efforts of patentees to protect national monopolies. A century before the introduction of formal antitrust laws, judges in the courts of equity and the Supreme Court attempted to resolve the paradox of promoting inventive rights without suppressing economic progress through the defense of monopolies. Courts also responded to changes in the nature and organization of technology, through legal innovations regarding the rights of employers, and the definition of patentable invention, among other issues. In short, since the founding of the Republic, legal institutions were modified as the scale and scope of market and society evolved, but the central policy objective of promoting the public interest remained the same. That is, after all, one of the chief virtues of a society that is bound and enabled by prescient constitutional principles.

The democratization thesis presented so far highlights the cumulative effect of ordinary patentees attempting to profit in the marketplace from improvements to existing technologies, in a manner that supports the predictions of endogenous growth theory. Some scholars argue that these nonspecialized inventors merely created minor improvements that had little impact on total value or on economic growth.[25] They reject the idea that important "macroinventions" were induced by the prospect of economic returns, and contend that important inventions were either exogenous, or else related to supply factors such as the number of technically educated individuals. Kenneth Sokoloff and I coauthored several studies based on biographies of "great inventors" to examine such issues. The record for these "great inventors" dispelled several "myths of invention." The overwhelming majority of great inventors also were patentees, and their use of the patent system made it easier for them to specialize and extract returns from inventive activity throughout their long careers. Like their less eminent counterparts, most of them had little or no formal schooling. The occupations of great inventors were similarly undistinguished as the majority were artisans, manufacturers, farmers, and others whose jobs did not require technical skills. In sum, one of the most striking features of the records for the great inventors is how similar their characteristics and patterns of patenting were to those of ordinary patentees.

The early twentieth century is usually characterized as the age of professional, science-based invention conducted by teams in research laboratories. Indeed, during this period, formal college education, human capital accumulation, and financial capital mobilization through corporate ties became more important, but independent inventors from more modest backgrounds were still able to exploit and benefit from the market for invention. At least up to the time of the Second Industrial Revolution, such relatively uneducated inventors or those from rural areas were no less likely to produce valuable

[25] The concepts of microinventions and macroinventions appear in Joel Mokyr, *The Lever of Riches: Technological Creativity and Economic Progress*, New York: Oxford University Press (1990).

inventions. The Second Industrial Revolution was a transitional period that hinted at the changes to come in the nature and organization of technology, but even in the 1920s American technology reflected the open access highlighted here. For all classes of inventors in the "long nineteenth century," technological progress in the United States involved a process of democratization in response to increases in expected benefits when markets expanded. The American patent system was a key institution in the progress of economy and technology, and it also stood out as a conduit for creativity and achievement among otherwise disadvantaged groups.

The U.S. patent system was soon acknowledged to be the most advanced in the world, and other countries drew causal connections between American achievements and its protection of inventive activity through patent property rights. Sir William Thomson, a British inventor and scientist, attended the 1876 Centennial Exhibition in Philadelphia, which featured displays for Bell's telephone, the Westinghouse airbrake, Edison's improved telegraph, sewing machines, refrigerator cars and numerous other patented discoveries. He stated, "I was much struck with the prevalence of patented inventions in the Exhibition: it seemed to me that every good thing deserving a patent was patented . . . If Europe does not amend its patent laws. . . . America will speedily become the nursery of useful inventions for the world."[26] Even the Swiss Commissioner to the Philadelphia Exhibition, Edward Bally (a noted shoe manufacturer), urged his own countrymen to introduce a patent system in order to counter the finding that "American industry has taken a lead which in a few years may cause Europe to feel its consequences in a very marked degree."[27] Toward the end of the nineteenth century Japan reorganized its patent system after a special commission visited the U.S. Patent Office. According to the Japanese envoy, Assistant Secretary of State Korehiyo Takahashi, they wished to discover " 'What is it that makes the United States such a great nation?' and we investigated and we found it was patents, and we will have patents."[28]

---

[26] *Scientific American*, October 21, 1876, vol. 35 (17), p. 256.

[27] "Another factor which favors the education of the people (of the United States) is the excellent system of patents, by means of which, at a very moderate expense, a patent is obtained; not only the inventor is protected against infringement, but the invention is made known; . . . We must introduce the patent system. All our production is more or less a simple copy. . . . It is evident that this absolute want of protection will never awaken in a people the spirit of invention. . . . America has shown us how in a few years a people, in the midst of circumstances often embarrassing, can merit by its activity, its spirit of enterprise, and its perseverance, the respect and admiration of the whole world, and acquire in many respects an incontestable superiority," cited in "Arguments before the Committee on Patents of the Senate and House of Representatives," 45th Congress, 2nd Session, Mis. Doc. No. 50, Washington, D.C.:Government Printing Office, 1878: 448–49.

[28] Cited in "Patents in relation to Manufactures," Story B. Ladd, *12th Census of the United States*, vol. 10 (4) pp. 751–66. The first Japanese patent laws were introduced in 1885, and reformed in 1888 and 1899.

However much they praised and emulated the patent policies of the United States, other countries (as well as many American citizens) failed to understand the rationale for its copyright policies. Despite the rhetoric that drew on the phrases from eighteenth-century European philosophy, U.S. rules and standards were not based on esoteric ideas of inherent rights of personhood nor creativity but, rather, on purely pragmatic and utilitarian grounds. The intellectual property clause of the U.S. Constitution was the common source of both patent and copyright policies, and the same individuals were responsible for their formulation and implementation. Yet, American copyright policies were markedly different from the procedures comprising the patent system. I contend that copyrights differed from patents precisely because the objective of both systems was, in accordance with the constitutional preamble, to "promote the general Welfare." This objective required a judicious balancing of private and public interests, the weighing of costs and benefits, and estimations of incentives and outcomes. Interests, costs, and incentives differed across technical inventions ("the useful arts") and cultural goods ("science"), and also altered over time. The system evolved or adapted endogenously to meet these changing circumstances in a way that contrasted directly with the institutional sclerosis in Europe.

Calibration of different institutional inputs in the United States resulted in significant policy variation across patents and copyrights, assignees and licensees, citizens and noncitizens, as well as producers, competitors, and consumers. Society benefited on net from the creation and commercialization of additions to the useful arts that were induced by profit incentives, despite the temporary inhibitions on diffusion, higher prices during the term of the patent, and the potential effects on cumulative inventions. Thus, to a large extent, the objectives of policy makers and the legitimate aims of patentees coincided. In the case of copyrights, the trade-offs were regarded with greater concern, for three primary reasons. First, the economic processes that give rise to cultural goods differ significantly from those that produce patentable inventions. Many copyrighted items such as academic research or religious tracts would be produced even in the absence of financial incentives because their producers could benefit from ancillary returns such as enhanced reputations or greater demand for complementary goods. Second, the risk of unwarranted monopolies (that appropriated what belonged to the public and made it private and exclusive) was higher, because cultural goods incorporated ideas that belonged to the public domain in ways that made it difficult to distinguish between the contributions of the author and those of society in general. Third, and most important, the enforcement of copyright had much more serious implications for a democratic society. Restrictions on free diffusion could result in significant social costs in terms of learning, education and free speech, in ways that promised to bolster and perpetuate the narrow redistributive claims of elites and interest groups.

It is therefore not surprising that in the United States from the earliest years patents were treated differently from copyrights. Moreover, U.S. policies departed radically from European intellectual property regimes that privileged the rights of literary elites above the rights of technicians. Although a French parliamentarian would have agreed with U.S. Senator John Ruggles that "inventors and authors stand on somewhat different ground," the ranking of the two groups would have been reversed. Ruggles noted that strong copyrights had important negative implications for the diffusion of useful knowledge "on which depends so essentially the preservation and support of our free institutions."[29] However, patentees should be accorded greater encouragement to create new inventions and also to commercialize them into valuable products. The first copyright statute granted protection to both "authors and proprietors" for the instrumental purpose of learning, whereas only the first and true inventor could claim patent rights. Similarly, for much of the nineteenth century work-for-hire doctrines led to weak employee rights in the case of copyrights, but not in the case of patents. Copyrights were administered in a registration system and were overturned if authors did not strictly comply with the rules; patents since 1836 were granted through an examination system and could not be revoked except for fraud. American patent laws prohibited compulsory licenses and unauthorized use of patent rights, unlike copyrights, in which pervasive "fair use" doctrines allowed free access if such access did not significantly reduce the author's returns, especially in the case of educational materials.[30]

Unlike the intellectual property policies of the European continent, the utilitarian orientation of American democracy supported a patent system that offered strong protections to inventors but required much weaker copyrights. The rhetoric of copyright in France and many other jurisdictions in Europe increasingly centered on the natural rights of creative individuals. Publishers in both France and Britain lobbied heavily for so-called author's rights, because these rights paradoxically redistributed income to publishing interests to a greater extent than to authors. Natural rights expanded in scope until they were enshrined in the international Berne Convention in the form of "moral rights." In contrast, U.S. copyright focused in a pragmatic fashion on the requirements of a developing society based on democratic principles. Although they were concerned with security of property rights, their major objective was not to benefit authors or publishing companies *per se*, but to increase contributions to knowledge and the dissemination of information. By rejecting the notion of copyright as an inherent and absolute right of creativity, the benefits to a privileged few were circumscribed

[29] Report to accompany Senate Bill No. 32, June 25, 1838, Senate Reports, 25th Congress, 2nd Session, No. 494, pp. 1–7.

[30] In patent doctrine, courts have allowed one extremely narrow exception to the strict holding of the patentee's right to exclude. This involved noncommercial experimental use of an invention for the purpose of amusement or for verification. The experimental use exception, which dates back to Whittemore v. Cutter, 29 F. Cas. 1120 (1813), was made even more limited in Madey v. Duke University, 307 F. 3d 1351 (2002).

in order to protect the public domain and to promote the interests of the community in lower costs of learning. To this end, the duration of copyright was among the most limited in the world. Moreover, facts, ideas, and data could not be protected by copyright and were vested in the public domain.[31]

Like other forms of intellectual property laws, the copyright system evolved to encompass improvements in technology and changes in the marketplace. Copyright law illustrated the difficulties and dilemmas that the legal system experienced in dealing with such new technologies as mimeographs, flash photography, cinematography, piano rolls, phonographs, radio, and "information technology," including the stock ticker and the telegraph. Even the preliminary decision about whether these technologies fell under the subject matter to be protected by the law created deep conflicts that were complicated by constitutional questions about freedom of speech and democracy. Copyright comprised a pervasive right against society, so judges attempted to resolve copyright disputes in ways that reduced spillovers. Thus, legal innovations expanded beyond traditional copyright doctrines to noncopyright holdings such as unfair competition, trade secrecy, and the right to privacy. These legal substitutes maintained bilateral rights against competitors and producers without imposing undue costs on society. The notion of copyright as a means of pursuing and prosecuting large segments of the consumer public was unknown and incompatible with the original copyright doctrines of the United States.

One of the most dramatic proofs of the infusion of cost-benefit analysis in early U.S. intellectual property policy appears in the treatment of international patent rights and international copyrights. A nation of artificers and innovators, both as consumers and producers, American citizens were confident of their global competitiveness in technology, and accordingly took an active role in international patent conventions that aimed to strengthen the rights of patentees. As a German judge at the Philadelphia exhibition in 1876 pointed out, "The United States of America already outstripped most of the older nations, except in matters of art, and as art required time, America would eventually not be behind other nations even in that."[32] Although

---

[31] In *Feist Publications v. Rural Tel. Serv. Co.*, 499 U.S. 340 (1991) the Supreme Court correctly rejected the argument that copyright should extent to a compilation of facts from a telephone directory, but for the wrong reasons. The Court did not endorse the view that copyright should be based on the degree of effort in compiling facts, but it incorrectly held that originality under the 1976 copyright statute meant that the work should exhibit "at least some minimal degree of creativity." Both of these doctrines, the "sweat of the brow," and the "creative spark" or degree of creativity, are drawn from the European approach to copyright, but are inconsistent with the utilitarian basis of the intellectual property clause and with the history of U.S. copyright policy. Facts cannot be protected because it would prohibitively increase the social costs of learning, which the statutes are meant to promote.

[32] "Arguments before the Committee on Patents of the Senate and House of Representatives," 45th Congress, 2nd Session, Mis. Doc. No. 50, Washington, D.C.: Government Printing Office, 1878: 445.

they excelled at pragmatic contrivances, nineteenth-century Americans were advisedly less sanguine about their efforts in the realm of music, art, literature, and drama, and this country was initially a net debtor in flows of material culture from Europe. The first copyright statute implicitly recognized this when it authorized Americans to take free advantage of the cultural output of other countries and encouraged international copyright piracy that persisted for a century. Until 1891 American policies deemed the works of foreign citizens to be in the public domain because legislators warned that reforms would not benefit the United States, and the net effects "would be, for us, on the wrong side of the leger [*sic*]." I assess the costs and benefits of copyright piracy and find that Americans likely profited from acting as "continental Brigands," so it is hardly surprising that a century of lobbying only resulted in a succession of failed legislative proposals. It was only when the balance of trade in cultural goods was more favorable to the United States that an international copyright law was finally passed. This policy was a dramatic departure from the evolution of international copyright grants in European countries such as France, which early on accorded national treatment to all countries. The significant differences in international patent and copyright laws highlight the extent to which American intellectual property policies were endogenously market-oriented.

The finding that U.S. policies toward patents and copyrights to a large extent conformed with economic conceptions does not imply that outcomes are or will be optimal. The *American Jurist* in 1833 warned that a representative democracy gives rise to the danger of "infractions of the constitution by those who have temporary objects or their own personal aggrandisement in view."[33] The history of copyright illustrates the dangers inherent in a system based on decentralization and democratic social capital, whereby public trust in institutions can perhaps all the more readily be subverted into redistributing wealth and power to a few rather than serving the common good. Copyright decisions illustrate how adjudication by analogy economized on legal inputs, and how judges introduced innovations in their interpretation of the law in order to "promote the progress of science," but they also reveal the extent to which judge-made policies were constrained by the statutes. Many of the technological innovations of the nineteenth century were sufficiently different from existing technologies as to make judicial analogies somewhat strained, and ultimately required accommodation by the legislature instead. Thus, the resolution of copyright conflicts drew on the key institutions of courts, markets, and the legislature, which ideally were intended to provide a system of checks and balances. That balance was initially effective because all parties deferred to the Constitution, but it also highlighted the potential for harm to the public domain if the legislature were captured by interest groups. Those dangers and infractions were always latent and

[33] *American Jurist and Law Magazine*, vol. 10 (1833), p. 120.

became apparent in American legislation early in the twentieth century, since when copyright doctrine has tended to be formulated through negotiations among industry representatives. It is striking that, ever since the eighteenth century, publishers have lobbied to gain copyrights in perpetuity, but they were continually defeated by defenders of the public domain including the judiciary in both Britain and America. In 2003, the Supreme Court of the United States allowed Congress to grant a virtually perpetual copyright in defiance of the Constitution's stipulation that such grants should only be "for limited Times." The majority seemed to acknowledge the validity of the economic arguments cited in Justice Breyer's dissent, but dismissed them with the declaration that "it is doubtful, however, that those architects of our Nation . . . thought in terms of the calculator."[34]

Other skeptics no doubt also would question the validity of an economic interpretation of the history of intellectual property, or the notion that early American institutions were deliberately designed to increase social welfare and varied to accommodate changes in external circumstances in accordance with this objective. This book adopts a cliometric approach because quantitative economic history ideally helps to refute the view of history as simply a Rorschach blot for one's previous convictions. Cliometrics requires the formulation of testable hypotheses, and the rejection of untenable claims that are inconsistent with the evidence. If the architects of our Nation did not think in terms of the calculator, then we should expect to find policies and outcomes that were inconsistent with economic predictions; inefficient rulings would be reflected in the common law in the form of surges in disputes and litigation; key policy statements that weigh costs and benefits such as those of John Ruggles would be atypical; and comparisons within and across countries would yield few systematic patterns.

Instead, this books finds that the rapid expansion of markets and national wealth in the United States during the long nineteenth century was supported by institutions and policies that were designed and interpreted in ways that favored broad-based economic growth. Policy in this area reflected the view that institutions mattered – indeed, institutions in the United States were carefully calibrated to promote social and economic welfare – and that appropriate rules and standards toward the protection of intellectual property were especially important in achieving these ends. Patent and copyright systems in the nineteenth century were motivated by the democratic belief that everyone, regardless of social status or economic standing, could make a

---

[34] *Eldred v. Ashcroft*, 537 U.S. 186 (2003): "JUSTICE BREYER several times places the Founding Fathers on his side. . . . It is doubtful, however, that those architects of our Nation, in framing the "limited Times" prescription, thought in terms of the calculator rather than the calendar." Other innovations today, such as appeals to the amount of creativity involved in the subject matter of copyrights, and proposals to grant property rights in databases, similarly reveal a potentially hazardous unfamiliarity with the democratic rationale for American intellectual property.

valuable contribution to social welfare. It was felt that individuals would be best induced to contribute to material progress if offered the opportunity to appropriate returns from their efforts through secure private property rights in their intangible assets. In order to achieve democratic ends patent rights were strongly enforced whereas the copyright grant was weaker and more hedged about with restrictions. Flexible and effective legal institutions played a key role in accommodating and facilitating the radical transformations that industrialization and technological change brought. Finally, the conviction that American democracy should value the contributions and well-being of ordinary citizens led to the conclusion that innovations, commercialization, and improvements in social welfare were best achieved through the decentralized mediation of markets rather than through allocations that were based on the values and actions of elites or bureaucrats. Policy makers therefore rejected the menu of policy alternatives in favor of patent and copyright systems despite their acknowledged drawbacks. In sum, American institutions during the nineteenth century created an ambience that encouraged the participation of a broad spectrum of the population, and succeeded in motivating relatively ordinary men and women to dramatically expand the existing stock of technical and cultural inventions. Patent and copyright institutions played a central role in ensuring that social and economic development were characterized by a process of "democratization."

INSTITUTIONS, INVENTION, AND ECONOMIC GROWTH

At its most general level, this book provides a historical perspective on questions that are being posed today regarding the relationship between institutions, inventions (broadly defined to include technological and cultural goods) and economic growth. Speculations about the sources of growth have always been central to economic analysis, although the focus, methods and policy recommendations of economic growth theorists have changed significantly within the past fifty years. Adam Smith proposed that expansions in market demand induced efficiency gains through specialization and the division of labor. The Smithian approach highlighted the cumulative effects of productivity gains from modest improvements in organization, learning, and techniques. Technological advances, increases in allocative efficiency and institutional changes were the result of, as well as the impetus for, market expansion. Neoclassical models featured an accounting decomposition of traditional inputs in the production function, emphasized the contributions of human and physical capital accumulation, and interpreted the residual in terms of exogenous advances in technology.[35] More recent contributions to growth theory have modified conventional mathematical

---

[35] A general discussion appears in Robert J. Barro and Xavier Sala-i-Martin, *Economic Growth*, Cambridge, Mass.: MIT Press (1998).

models to incorporate the role of ideas and knowledge in endogenous technological change.[36] The work of the new institutional economists underlines the importance of credible commitments on the part of the state, the need for an independent judiciary, and a transparent legal system that enforces contracts and private property rights. At the same time, an influential series of papers in the past decade echo the concerns of eighteenth-century social commentators: they attribute a critical role to fundamental, country-specific variables such as factor endowments, climate, and geography in determining the nature of institutions and economic growth.[37] These different approaches all agree that technological change is fundamental to sustained economic development, but how and why such changes occur and succeed are less well understood. In short, the record on growth studies indicates the need for comparative institutional analyses, especially of the way in which specific rules and standards are revised and implemented, and their consequences over time.

Another long-standing debate centers on the link between political institutions and economic outcomes. Democracy is defined in terms of widespread participatory political access or "voice," and the ability of nonelites to alter the existing rules and standards, as well as in economic terms such as respect for private property rights and equality of opportunity.[38] As such, democracy might be viewed as a "meta-institution" or a prerequisite for the formation of other appropriate institutions.[39] Although it is conventional to propose a causal relationship between democracy and higher economic growth, a number of economic studies are more ambivalent.[40] For instance, Robert Barro posited a nonlinear relationship, in which democratic institutions might reduce potential economic progress after a certain level of freedom

---

[36] For instance, see Phillippe Aghion and Peter Howitt, *Endogenous Growth Theory*, Cambridge, Mass.: MIT Press (1998); and Paul Romer, "Endogenous Technological Change," *Journal of Political Economy*, vol. 98 (October) 1990: S71–102.

[37] For an excellent overview, see Stanley L. Engerman and Kenneth L. Sokoloff, "Institutional and Non-Institutional Explanations of Economic Differences," NBER Working Paper 9989 (2003).

[38] Peter H. Lindert's, "Voice and Growth: Was Churchill Right?" (*JEH*, 63 (2), June 2003: 315–50) argues that "Using too little historical information, and mistaking formal democratic rules for true voice, has understated the gains from spreading political voice more equally."

[39] See Dani Rodrik, "Institutions for High-Quality Growth: What They are and How to Acquire Them," NBER Working Paper no. w7540 (Feb. 2000).

[40] Relative to authoritarian regimes, some have argued that democracies may raise the costs of achieving a given level of future output: they may increase the propensity to consume and support other inefficient or short-sighted policies such as protectionism, foster ethnic conflicts, reduce the responsiveness to external crises, and redistribute income from producers to consumers. Adam Przeworski and Fernando Limongi ("Political Regimes and Economic Growth," *Journal of Economic Perspectives*, Vol. 7, No. 3. (Summer, 1993), pp. 51–69) survey empirical studies of this subject, most of which were "seriously flawed." They conclude that economic models of democracy are a "house of cards" (p. 10) and "we do not know whether democracy fosters or hinders economic growth" (p. 15).

had already existed.[41] Stanley Engerman and Kenneth Sokoloff, the authors of an extensive project to explain differential growth paths in New World colonies, found that equality and democratic institutions were a function of a country's factor endowments, broadly defined.[42] They cautioned against the notion that any particular set of institutions was a prerequisite for achieving growth. Other empirical studies show that successful democracies are by nature dynamic: the inputs that democratic growth requires evolve over time. Some economic historians find that property rights and contract enforcement were important explanatory variables only up to the early part of the nineteenth century, whereas human capital formation became key after the onset of industrialization.[43] Flexibility was critical in this process, as the rules and standards that promoted economic growth needed to respond to such incentives for change as rapid technological progress. An emerging consensus seems to suggest that economic progress may require a menu of policies that are tailored to the particular needs of each society.

Such scholars as Douglass North have suggested that intellectual property systems had an important impact on the course of economic development.[44] Appropriate institutions to promote creations in the material and intellectual sphere are especially critical because ideas and information are public goods that are characterized by nonrivalry and nonexclusion. Once the initial costs are incurred, ideas can be reproduced at zero marginal cost and it may be difficult to exclude others from their use. Thus, in a competitive market public goods may suffer from underprovision or may never be created because of a lack of incentive on the part of the original provider who bears the initial costs but may not be able to appropriate the benefits. Market failure can be ameliorated in several ways, for instance, through government

[41] Robert J. Barro, *Determinants of Economic Growth: A Cross-Country Empirical Study*, Cambridge, Mass.: MIT Press, 1997.

[42] These authors have produced a significant number of insightful studies on this subject, including Stanley L. Engerman and Kenneth L. Sokoloff, "Factor Endowments, Inequality and Paths of Development Among New World Economies," *Economia* vol. 3 (Fall) 2002: 41–109; and "The Evolution of Suffrage Institutions in the Americas." NBER Working Paper 8512 (2001).

[43] See Peter H. Lindert, "Voice and Growth: Was Churchill Right?" *JEH*, 63 (2), June 2003: 315–50. The value of a historical accounting is shown by studies today that find "most indicators of institutional quality used to establish the proposition that institutions cause growth are . . . conceptually unsuitable for that purpose" (Edward Glaeser, Rafael La Porta, and Florencio Lopez de Silane, "Do Institutions Cause Growth?" from the abstract of their NBER Working Paper 10568 (2004)). The argument that human capital accumulation "is a more basic source of growth than are institutions" is somewhat debatable from a historical perspective.

[44] "The failure to develop systematic property rights in innovation up until fairly modern times was a major source of the slow pace of technological change," p. 164, Douglass C. North, *Structure and Change in Economic History*, New York: W. W. Norton (1981); as well as North and Robert P. Thomas, *The Rise of the Western World*, Cambridge: Cambridge University Press (1973).

provision, rewards, or subsidies to original creators, private patronage, and through the creation of intellectual property rights. Numerous economic studies have analyzed intellectual property rights from both a theoretical and empirical perspective. Patents and copyrights allow the initial producers a limited period during which they are able to benefit from a right of exclusion. If creativity is a function of expected profits, these grants to authors and inventors have the potential to increase social production possibilities at lower cost. Disclosure requirements promote diffusion, and the expiration of the temporary monopoly right ultimately adds to the public domain. Overall welfare is enhanced if the social benefits of diffusion outweigh the social costs of temporary exclusion. This period of exclusion may be costly for society, especially if future improvements are deterred, and if rent-seeking such as redistributive litigation results in wasted resources. Much attention also has been accorded to theoretical features of the optimal system, including the breadth, longevity, and height of patent and copyright grants.[45]

However, strongly enforced rights do not always benefit the producers and owners of intellectual property rights, especially if there is a prospect of cumulative invention in which follow-on inventors build on the first discovery. Thus, more nuanced models of patents and copyrights are ambivalent about the net welfare benefits of strong exclusive rights to inventions. Indeed, network models imply that the social welfare of even producers may gain from weak enforcement if more extensive use of the product increases the value to all users. Under these circumstances, the patent or copyright owner may benefit from the positive externalities created by piracy. In the absence of royalties, producers may appropriate returns through ancillary means, such as the sale of complementary items or improved reputation. In a variant of the durable-goods monopoly problem, it has been shown that piracy can theoretically increase the demand for products by ensuring that producers can credibly commit to uniform prices over time. Also in this vein, price and/or quality discrimination of nonprivate goods across pirates and legitimate users can result in net welfare benefits for society and for the individual firm. If the cost of imitation increases with quality, infringement also

[45] Breadth refers to the amount of related claims that are secured by the patent or copyright. Novelty requirements or "height" protect the rights holder against attempts to invent around their claim, or against minor improvements that might compete with the initial grant. The length indicates the life or term of the property right. There is some degree of substitutability across these different elements in attaining the same value for the property right. See Nancy T. Gallini, "The Economics of Patents: Lessons from Recent U.S. Patent Reform," *Journal of Economic Perspectives*, vol. 16 (2) 2002:131–54; Richard Gilbert and Carl Shapiro, "Optimal Patent Length and Breadth," *RAND Journal of Economics*, vol. 21 (Spring) 1990: 106–12; and Theon Van Dijk, "Patent Height and Competition in Product Improvements," *Journal of Industrial Economics*, vol. 44 (2) June 1996: 151–67. Michael A. Heller and Rebecca S. Eisenberg "Can Patents Deter Innovations? The Anticommons in Biomedical Research," *Science*, vol. 280 (May) 1998: 698–701, caution against the problem of patent thickets that potentially block innovation.

can benefit society if it causes firms to adopt a strategy of producing higher quality commodities.

Economic theorists who are troubled by the imperfections of intellectual property grants have proposed alternative mechanisms that lead to more satisfactory mathematical solutions. Theoretical analyses have advanced our understanding in this area, but such models by their nature cannot capture many complexities. They tend to overlook such factors as the potential for greater corruption or arbitrariness in the administration of alternatives to patents. Similarly, they fail to appreciate the role of private property rights in conveying information and facilitating markets, and their value in reducing risk and uncertainty for independent inventors with few private resources. The analysis becomes even less satisfactory when producers belong to different countries than consumers. Thus, despite the flurry of academic research on the economics of intellectual property, we have not progressed far beyond Fritz Machlup's declaration that our state of knowledge does not allow to us to either recommend the introduction or the removal of such systems.[46] Existing studies leave a wide area of ambiguity about the causes and consequences of institutional structures in general, and their evolution across time and region. This is especially true of copyrights, which are less amenable to empirical economic assessment than patents.

Although it may be impossible to draw up a uniform and universally applicable blueprint for effective institutions, a detailed examination of the historical record can provide valuable information for understanding current issues and policies. It goes without saying that such an exercise must be conducted with due caution because the modern period is different from previous eras. The intellectual property system of the twenty-first century encompasses new technologies and issues that range from the enforcement of rights to digital music in cyberspace, biodiversity, and plant protection, the privatization of genetic information and resources, the patentability of business methods and life-forms, through public health and human rights concerns. Today, the ownership of intellectual property rights is concentrated among large corporations whereas their costs are distributed among millions of users, making it more likely that policy trade-offs will be resolved in favor of producers relative to consumers. This is especially true of now-developing countries, which are subject to more compelling political-economic pressures and potential sanctions than in the nineteenth century. The majority

---

[46] Fritz Machlup, *An Economic Review of the Patent System*. Study of the Subcommittee on Patents, Trademarks, and Copyrights of the Committee on the Judiciary, U.S. Senate, study no. 15. Washington, D.C.: U.S. Government Printing Office, 1958, p. 80: "If we did not have a patent system, it would be irresponsible, on the basis of our present knowledge of its economic consequences, to recommend instituting one. But since we have had a patent system for a long time, it would be irresponsible, on the basis of our present knowledge, to recommend abolishing it." See also Edith Penrose, *The Economics of the International Patent System*, Baltimore: Johns Hopkins University Press, 1951.

of innovations originate from the largest industrial countries, and access to new technologies is even more unequally distributed both within and across nations. The costs of piracy for developing countries in the current period extend beyond incentives for innovation, to include the prospect that weak intellectual property rights might deter foreign direct investment, vital inflows of pharmaceuticals, and technology transfer.

A number of scholars are so impressed with technological advances in the twenty-first century that they argue we have reached a critical juncture where we need completely new institutions.[47] Such claims indicate a lack of historical insight. Although one readily grants the premise that the telegraph was not simply a "Victorian Internet," this does not imply that the issues of the online era are entirely novel.[48] In the realm of intellectual property, questions from four centuries ago are still current, ranging from its philosophical underpinnings, to whether patents and copyrights constitute optimal policies toward intellectual inventions, to the growing concerns of international political economy.[49] Throughout their history, patent and copyright regimes have confronted and accommodated technological innovations that were no less significant and contentious for their time. Controversies over the power of elites and their alignment to publishing interests, and the need for explicit mechanisms to protect the democratic public interest in learning, are coterminous with copyright itself. Our understanding of democratic processes

---

[47] See, for instance, David R. Johnson and David Post, "Law and Borders – The Rise of Law in Cyberspace," 48 *Stan. L. Rev.* 1367, May 1996. John Perry Barlow argues that "We will need to develop an entirely new set of methods as befits this entirely new set of circumstances," in "Selling Wine without Bottles: The Economy of Mind on the Global Net," p. 10 in P. Ludlow, *High Noon on the Electronic Frontier: Conceptutal Issues in Cyberspace*, Cambridge, Mass.: MIT Press (1996).

[48] An entertaining book by Tom Standage, *The Victorian Internet*, London: Weidenfeld & Nicolson (1998), claims that the telegraph industry in the nineteenth century comprised "online pioneers."

[49] See Brian Wright, "The Economics of Invention Incentives: Patents, Prizes, and Incentive Contracts." *American Economic Review*, vol. 73 (1983): 691–707; and Steven Shavell and Tanguy van Ypersele "Rewards versus Intellectual Property Rights," *Journal of Law & Economics*, Vol. 44, No. 2, October 2001: 525–548;Michael Kremer, "Patent Buyouts: A Mechanism for Encouraging Innovation," *Quarterly Journal of Economics*, Vol. 113 (4) 1998: 1137–1167. A number of scholars have made a putative "case against intellectual property," such as Michele Boldrin and David K. Levine, "The Case Against Intellectual Property," CEPR Discussion Paper No. 3273, March 2002. According to John Stuart Mill, "...to any other system which would vest in the state the power of deciding whether an inventor should derive any pecuniary advantage from the public benefit which he confers, the objections are evidently stronger and more fundamental than the strongest which can possibly be urged against patents" (Book V, Ch. 10, p. 25, *Principles of Political Economy with some of their Applications to Social Philosophy*, London: Longmans, Green and Co., ed. William J. Ashley, 1909. First published: 1848.) Today, even critics of patents and copyrights express concerns that state-mandated intellectual property rights might be superseded by private contracting that has the potential to bypass constitutional oversight and its balancing of costs and benefits to protect social interests.

would be enhanced by studies of the specific consequences of broadening economic opportunities to nonelites, a topic that dominated nineteenth century discourse about policies. An economist from the nineteenth century would have been equally familiar with considerations about whether uniformity in intellectual property rights across countries harmed or benefited global welfare, and whether piracy might be to the advantage of developing countries.[50] Similarly, the link between trade and intellectual property rights that informs the TRIPS agreement was quite standard two centuries ago.

Figure 1.1 shows the patterns of intellectual property litigation in U.S. federal courts, and indicates a rapid increase in recent decades in conflicts regarding patents and (to an even greater extent) copyrights. This is not entirely surprising, given the growing importance of intangible assets both in terms of industrial and strategic value, but part of the rise is also due to greater uncertainty regarding fundamental features of the intellectual property system. These patterns signal that today a reassessment of patents and copyrights from a historical perspective is timely and warranted for a number of reasons. First, and most generally, an understanding of the intellectual property system allows us to investigate the role of democratic institutions, legal systems, markets, and property rights in economic growth. Second, a number of analysts argue that the two-centuries-old American system of intellectual property needs to be revised and, in a country whose institutions must conform to the Constitution, such a revision should be based on a systematic knowledge of the origins of the existing system.

However, even "Originalists," who advocate the relevance of early intellectual property doctrines, tend to perpetuate basic errors in fact and interpretation. Decisions are being made daily that contradict the constitutional mandate and even landmark Supreme Court rulings manifest a great deal of confusion about the nature of the American intellectual property system. The attempt to accommodate technological innovations without due consideration to history and to the underlying rationale for long-standing rules and standards has led to inefficiencies in patent and copyright policies. Part of the loss of clarity owes to the capture of political fora by corporate interest groups that have reversed the historical balance of private rights and the public domain in their own favor. Moreover, harmonization of international laws has unsuccessfully melded diametrically opposed principles from the American and European systems and this has led to internal contradictions in U.S. intellectual property policies. Such inconsistencies would be more apparent if scrutinized in the context of historical experience. Finally, the United States not only supports laws to protect and enforce patents and copyrights in this

[50] Several studies of harmonization conclude that uniformity in intellectual property laws can "either enhance or reduce global welfare." See, for instance, Suzanne Scotchmer, "The Political Economy of Intellectual Property Treaties," NBER Working Paper No. w9114 (August 2002).

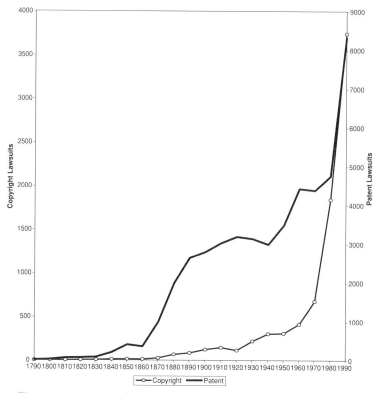

Figure 1.1. Patent and Copyright Litigation (by decade), 1790–2000

country, but also influences the policies of developing countries. Under these circumstances, insights from the period when the United States itself was a developing country, in addition to being valuable in themselves, seem to offer a useful perspective on current controversies.

This book should be of interest to economists, historians, legal scholars, and students of intellectual property, technological change, and economic development. The discussion draws on original data sets compiled from an array of sources including patent records, patent citations, city directories, manuscript censuses, biographical histories, assignment contracts, lawsuits and legal treatises, the book trade, and copyright filings. However, readers who do not share the economic historian's fascination with data can simply ignore the tables and read the text, without losing track of the general arguments. The first section discusses the economic history of patent institutions and analyzes variation in patterns of patenting. Although there are well-rehearsed reasons why one should be careful to avoid equating patent counts with inventive activity, there are just as well-rehearsed reasons why patent data can provide valuable insights into technological creativity, as I hope the following chapters will show.[51] Chapter 2 compares the evolution of patent systems in Britain, France, and the United States. Chapter 3 outlines changes in the laws toward patents and their enforcement in the United States. An empirical examination of the democratization of invention is introduced in Chapter 4 and extended to the discussion of women inventors in Chapters 5 and 6. Great inventors and their patenting (Chapter 7) underlines the extent to which even important inventions were generated through a process of democratization. The second section (Chapters 8 and 9) considers the invention of intellectual assets from the perspective of copyrights. This section highlights the evolution of copyright law and institutions, the relationship between copyright and technology, and the effects of American international copyright piracy on economic and social welfare. Chapter 10 reviews the experiences of "follower countries" in the nineteenth century, including the international harmonization of laws, and offers a summary conclusion.

[51] For a pioneering approach to the use of patent data see Jacob Schmookler, *Invention and Economic Growth*, Cambridge, Mass.: Harvard University Press, 1966. Zvi Griliches, "Patent Statistics as Economic Indicators: A Survey," *Journal of Economic Literature*, Vol. 28, No. 4 (December 1990):1661–1707, discusses the costs and benefits of analyzing patents. The major problems with patent statistics as a measure of inventive activity and technological change are that not all inventions are patented or can be patented; the propensity to patent differs across time, industries and activities; patents vary in terms of intrinsic and commercial value; patents might not be directly comparable across countries or time because of differences in institutional features and enforcement; and patents are a better gauge of inputs than productivity or output. Griliches concludes (p. 43) that "In spite of all the difficulties, patent statistics remain a unique resource for the analysis of the process of technical change. Nothing else even comes close in the quantity of available data, accessibility, and the potential industrial, organizational, and technological detail." For an excellent example of the way in which patent records can be adjusted to yield economically meaningful information, see Adam B. Jaffe and Manuel Trajtenberg, *Patents, Citations and Innovations: a Window on the Knowledge Economy*, Cambridge, Mass.: MIT Press (2002). The problem of measuring cultural goods through copyright filings is even greater since copyright protects expression, which ranges widely from rap music, poems, mathematical treatises, movies, sculpture, photographs, and commemorative porcelain plates through to computer software.

2

The Patent System in Europe and the United States

"Who, with a knowledge of facts like these, can deny protection and security to the useful arts? Such inventions do more for the nation than can be effected with armies and fleets."

– *Report on the Patent Office* (1802)

British commentators at the beginning of the nineteenth century were skeptical, if not scathing, about the degree of economic and cultural development in their former colony. They made satirical references to Thomas Paine's notion that a democratic society required a written constitution that could be carried in the pocket of each citizen, and questioned the value of contributions from those who did not belong to the elite ranks. Sir Henry Sumner Maine regarded it as self-evident that "if for four centuries there had been a very widely extended franchise and a very large electoral body in this country [Britain]. . . . The threshing machine, the power loom, the spinning jenny, and possibly the steam-engine, would have been prohibited" and "all that has made England famous, and all that has made England wealthy, has been the work of minorities, sometimes very small ones . . . the gradual establishment of the masses in power is of the blackest omen for all legislation founded on scientific opinion."[1] However, even as stringent a critic of democratic ideals as Maine conceded that the federal grant of patent rights was one of the "provisions of the Constitution of the United States which have most influenced the destinies of the American people," and was moreover responsible for the finding that the United States in 1885 was "the first in the world for the number and ingenuity of the inventors by which they have promoted the useful arts."[2]

The links among democratic institutions, intellectual property, and the wealth of nations remains much debated. Scholars such as Douglass North point to the patent system as an outstanding example of an institution that influenced the course of economic development and technological change,

---

[1] Sir Henry Sumner Maine, *Popular Government*, Indianapolis, Liberty Classics, (1976 reprint of 1885), p. 112.

[2] Sir Henry Maine, *Popular Government*, pp. 241–42. He went on to say that "on the other hand, the neglect to exercise this power for the advantage of foreign writers has condemned the whole American community to a literary servitude unparalleled in the history of thought."

whereas theorists promote the efficiency of alternatives to patents such as premiums. This chapter examines the patent institutions of France, Britain, and the United States, and proposes that the economic history of intellectual property systems provides strong support for the idea that democracy was related to social and economic progress. My assessment of the American patent system in the mirror of European experience reveals substantial differences across these countries during the early industrial period. Such differences were not merely nominal but also had a significant impact on the course of technological change and industrialization. The European systems reflected their origin in royal privilege and effectually limited access to a select class, which ultimately resulted in negative consequences for their long run competitiveness. In France the application process was convoluted, diffusion was minimal, and "the influence of men in power is great on the subject of patents as well as in other respects."[3] Similarly, in England, the industrial and patent systems generated inducements for improvements by individuals with access to wealth and influence, and skewed the nature of inventive activity toward large scale capital equipment. Institutional structures in these countries were not only biased toward the more privileged classes, they also proved to be less responsive to calls for revisions to cope with changing circumstances.

The Framers of the U.S. Constitution and statutes were certainly familiar with, and influenced by, the European intellectual property systems. Nevertheless, they made important departures in the ways in which property rights in technology were defined and awarded. Innovations in the design of the intellectual property system had a discernable impact on the course of technological and cultural change. The American path of economic and industrial development was more "Smithian" in nature: it favored modest changes in product design and the organization of production; capital goods that would reduce labor requirements and increase the productivity of labor; and the widespread participation of ordinary individuals as both producers and consumers. The United States established the world's first modern patent institution, which was consciously designed to stimulate participation in invention across a wide spectrum of the population, and to promote invention and the universal diffusion of technological information.

Democratic objectives were achieved through innovations such as reserving patent rights to the first and true inventor in the world, efficient centralized processing and examination of applications, fees that were set at a low level, and countervailing checks and balances in the legal system. The public had ready access to patent specifications, which promoted the diffusion of inventions, and the system also facilitated extensive trade in patented technologies. These provisions encouraged inventors to obtain property rights

---

[3] Thomas G. Fessenden, p. 212, *An Essay on the Law of Patents for New Inventions*, Boston: D Mallory & Co. (1810.)

in incremental inventions and small improvements in design and technique that could be applied across many industries. They further created an environment in which inventors could experiment to perfect their ideas, obtain financing for future discoveries, and sell off their rights without fear of expropriation by competitors or the state. Patent institutions and legal rules and standards altered in response to changes in external circumstances, whereas adherence to the principles of the Constitution ensured that the commitment to intellectual property rights remained credible. Overall, secure property rights in inventions contributed to the allocation of resources toward their most highly valued use, as determined by the decentralized choices of the numerous participants in markets for patent rights and for patented inventions.

BRITISH PATENT SYSTEM

The grant of exclusive property rights vested in patents developed from medieval guild practices in Europe. England and France were early leaders in the grant of royal privileges that led to monopolies, and Britain is noted for the establishment of a patent system that has been in continuous operation for a longer period than any other in the world.[4] English monarchs frequently used patents to reward favorites with privileges, such as monopolies over trade that increased the retail prices of commodities. It was not until the seventeenth century that patents were associated entirely with awards to inventors, when Section 6 of the Statute of Monopolies (1624) repealed the practice of royal monopoly grants to all except patents for inventions.[5] The Statute of Monopolies permitted patent rights of fourteen years for "the sole making or working of any manner of new manufacture within this realm to the first and true inventor."[6] Although the petition to the Crown might be

---

[4] The standard references for the economic history of the early British patent system are Christine MacLeod, *Inventing the Industrial Revolution*, Cambridge: Cambridge University Press, 1988 and Harold Dutton, *The Patent System and Inventive Activity during the Industrial Revolution, 1750–1852*, Manchester: Manchester University Press (1984.) See also Moureen Coulter, *Property in Ideas: the Patent Question in Mid-Victorian England*, Kirksville, Mo.: Thomas Jefferson Press (1991); B. Zorina Khan and Kenneth L. Sokoloff, "Two Paths to Industrial Development and Technological Change," in *Technological Revolutions in Europe, 1760–1860*, Maxine Berg and Kristine Bruland (eds.), London: Edward Elgar (1998); and Herbert Harding, *Patent Office centenary: a story of 100 years in the life and work of the Patent Office*, London: Her Majesty's Stationery Office (1953.) More generally, Joel Mokyr, *The Lever of Riches: Technological Creativity and Economic Growth*, New York: Oxford University Press (1990) provides a long term perspective on the course of technological change.

[5] The Statute of Monopolies, 21 Jac. I. C. 3, 1623, was implemented in 1624.

[6] 21 Jac. I. C. 3, 1623, Sec. 6. In Britain before this period a series of common law decisions (as opposed to statutory rules) had dealt with the requirements of patents for invention. For example, the 1602 case Darcy v. Allin held: "Where any man by his own charge and industry or by his wit or invention doth bring any new trade into the realm, or any engine tending to

viewed as a mere formality, the British patent system retained many features that reflected its origins in royal privilege. Indeed, the evidence presented later suggests that the system restricted access to property rights in inventions in ways that had the consequence of limiting the class of inventors primarily to those who were technically qualified, well-connected, or wealthy. The structure of the system served to raise the average value of patents, and favored the invention of high-valued physical capital inputs, such as textile machinery and steam engines, that were used in a narrow range of industries.

The British patent system established significant barriers in the form of prohibitively high costs that limited access to property rights in invention. Patent fees for England alone amounted to £100–£120 ($585), or approximately four times per capita income in 1860. The fee for a patent that also covered Scotland and Ireland could cost as much as £350 pounds ($1,680.)[7] Adding a co-inventor was likely to increase the costs by another £24.[8] Because there was no formal inquiry into priority or novelty the statutory phrase authorizing grants to the "first and true inventor" was merely nominal. Importers of inventions that had been created abroad could file for domestic patents, and patent agents frequently applied for patents under their own names on behalf of inventors from overseas. Patents could be extended only by a private Act of Parliament, which required political influence, and extensions could cost as much as £700.[9] These constraints favored the elite class of those with wealth, political connections, or exceptional technical qualifications, and created disincentives for inventors from humble backgrounds. Indeed, in the parliamentary debates regarding the patent system, some witnesses regarded this restrictiveness by class as one of the chief *merits* of higher fees, because they did not wish patent applications to be cluttered with trivial improvements by the "working class."[10] The Comptroller General of Patents even declared that most inventions induced by low fees were likely to be for "useless and speculative patents; in many instances taken merely for advertising purposes."[11] Patent fees provided an important source of revenues for the Crown and its employees, and created a class of administrators who had strong incentives to block proposed reforms. This group would effectively

the furtherance of trade that never was used before; and that, for the good of the Realm; that in such cases the King may grant to him a monopoly patent for some reasonable time until the subjects may learn the same."

[7] The complexity of the system is evident from the fact that nobody seems to have had a clear idea of the specific costs, and estimates ranged from £274 to £350.

[8] See Attorney General, *Hansard* (1851); and Dutton (1984.)

[9] Dutton, p. 155.

[10] Thus, in the 1829 Report of the British Committee on the Patent System, one of the questions was "Do not you think that if it became a habit among that class of people to secure patent rights for those small discoveries at low rates, it would be very inconvenient?" British Parliamentary Papers, *Reports from Select Committees on the Law Relative to Patents for Inventions, 1829–72.* 2 vols. Shannon: Irish University Press (1968.)

[11] Great Britain, p. 5, *Annual Report of the Comptroller General of Patents*, London: GB Patent Office, 1858.

have to be compensated for the loss in their rents before institutional changes could be implemented.

The complicated administrative procedures that inventors had to follow added further to the costs: patent applications for England alone had to pass through seven offices, from the Home Secretary to the Lord Chancellor, and twice required the signature of the Sovereign. If the patent were extended to Scotland and Ireland it was necessary to negotiate another five offices in each country.[12] The cumbersome process of patent applications (variously described as "mediaeval" and "fantastical") afforded ample material for satire, but obviously imposed severe constraints on the ordinary inventor who wished to obtain protection for his discovery.[13] These features testify to the much higher monetary and transactions costs, in both absolute and relative terms, of obtaining property rights to inventions in England. Such costs essentially restricted the use of the patent system to inventions of high value and to applicants who already possessed or could raise sufficient capital to apply for the patent. The complicated system also effectively inhibited the diffusion of information and made it difficult, if not impossible, for inventors outside of London to readily conduct patent searches. Patent specifications were open to public inspection on payment of a fee, but until 1852 they were not officially printed, published, or indexed.[14] Because the patent

---

[12] For descriptions of the patent procedure, see Dutton (1984), MacLeod (1988.) H. Harding, *Patent Office Centenary: A Story Of 100 Years In The Life And Work Of The Patent Office*, London: HMSO (1953), p. 5, calls the system "quite fantastic." According to Edward Holroyd, *A practical treatise of the law of patents for inventions*, London: J. Richards (1830), pp. 66–67, patent petitions had to be submitted to: the office of the Secretary of State; the Attorney General; office of the Secretary of State for King's Warrant; warrant of the Attorney General; patent office for signature of the Crown; office of the Signet; Lord Keeper of the Privy Seal; office of the Lord Chancellor; Great Seal Patent Office. Officials were not required to justify their decisions. The vast majority of patents before 1852 were filed in England – only 16 percent were extended to Scotland and Ireland as well. Hence the figures for England closely approximate the total amount of patents until 1852, when patent grants automatically covered all three kingdoms (Dutton, 1984, p. 35.)

[13] For instance, Jeremy Bentham, who favored the grant of patents, noted: "A new idea presents itself to some workman or artist.... He goes, with a joyful heart, to the public office to ask for his patent. But what does he encounter? Clerks, lawyers, and officers of state, who reap beforehand the fruits of his industry. This privilege is not given, but is, in fact *sold* for from £100 to £200 – sums greater than he ever possessed in his life. He finds himself caught in a snare which the law, or rather extortion which has obtained the force of the law, has spread for the industrious inventor. It is a tax levied upon ingenuity, and no man can set bounds to the value of the services it may have lost to the nation." From the *Works of Jeremy Bentham*, cited in Moureen Coulter, *Property in Ideas: the Patent Question in Mid-Victorian England*, Kirksville, Mo.: Thomas Jefferson Press (1991), p. 76.

[14] The following notes are based on Harold Dutton (1984) and David J. Jeremy and Darwin H. Stapleton, "Transfers between Culturally-Related Nations," in David Jeremy (ed), *International Technology Transfer: Europe, Japan and the USA, 1700–1914*, Aldershot: Edward Elgar (1991), pp. 31–50. Before 1734 tradesmen had to be instructed in the details of the invention as a condition of the patent grant, but there was no requirement for the general

could be filed in any of three offices in Chancery, searches of the prior art involved much time and inconvenience. Potential patentees were well advised to obtain the help of a patent agent to aid in negotiating the numerous steps and offices that were required for pursuit of the application in London. It is hardly surprising that the defenders of the early patent system included patent agents and patent lawyers.[15]

In the second half of the eighteenth century, nationwide lobbies of manufacturers and patentees expressed dissatisfaction with the operation of the British patent system. However, it was not until the nineteenth century that their concerns and requests for reforms were formally addressed. From 1829 onward, a series of Select Parliamentary Committees interviewed many experts and other witnesses, issued reports, and sent Bills before parliament that proposed institutional changes. Nevertheless, the reform movement in the first half of the nineteenth century was ineffective: except for a few marginal changes, the petitions of participants in this process were largely ignored. Finally, the Crystal Palace Exhibition in 1851 contributed to an official recognition of the need for legislation to meet some of these long-standing criticisms.[16] In 1852 the efforts of numerous societies, and of individual engineers, inventors, and manufacturers over many decades were finally rewarded. Parliament approved the Patent Law Amendment Act, which authorized the first major adjustment of the system in two centuries.[17]

publication or diffusion of information. After this period, and especially after a key legal decision in 1778, disclosure was conducted by means of the patent specification. However, a number of inventors successfully petitioned Parliament to keep their specifications secret and, in the absence of any examination, the adequacy of the specification was only an issue if the patent were litigated (Dutton, pp. 39–42.) Specifications could be lodged in either of three offices (the Petty Bag Office; the Rolls Chapel Office; and the Enrolment Office), and were difficult to search because records were not indexed. Only patent office employees could make copies of the records, for which a fee was charged. According to Dutton, the fee varied from 1 s. to 3s. 6d.; while Jeremy and Stapleton (1991, pp. 33–34) state that "By 1829 a copy could cost anything between two and 40 guineas." Few inventors bothered to go to the trouble of researching prior specifications (perhaps as little as 2 percent); by contrast, competitors did so in the hopes of overturning patentees' patent rights (Dutton, p. 184.)

[15] According to *The Times* of 1864, "the only persons who are benefited by [the patent system] are the Patent agents and lawyers" (cited in Coulter, p. 147.) Of course, patent officers of the Crown, whose compensation came from patent fees, also benefited; after the changes in 1852, the Irish and Scots Law Officers and clerks who were made obsolete received an annual compensation for the loss of fees that they suffered from the reforms, and in the first year alone were awarded £6,000. The Annual Report for the British Patent Office shows that in 1884 patent agents were involved in about 72 percent of all patent applications.

[16] See Nathan Rosenberg, ed., *The American System of Manufactures: The Report of the Committee on the Machinery of the United States 1855, and the Special Reports of George Wallis and Joseph Whitworth 1854*, Edinburgh: Edinburgh University Press (1969.)

[17] The 1852 law did not apply to British colonies, which were able to adopt legislation suited to their individual circumstances. (See Patent Law Amendment Act, 15 & 16 Vict., Chpt. 83, 1852.)

The new patent statutes incorporated features that drew on testimonials to the superior functioning of the American system. Significant changes in the direction of the American system included lower fees and costs, and the application procedures were rationalized into a single Office of the Commissioners of Patents for Inventions, or "Great Seal Patent Office." The 1852 patent reform bills included calls for a U.S.-style examination system but this was amended in the House of Commons and the measure was not included in the final version. Opponents were reluctant to vest examiners with the necessary discretionary power, and pragmatic observers pointed to the shortage of a cadre of officials with the required expertise.[18] The law established a renewal system that required the payment of fees in installments if the patentee wished to maintain the patent for the full term. Patentees initially paid £25 and later installments of £50 (after three years) and £100 (after seven years) to maintain the patent for a full term of fourteen years. Provision was made for the printing and publication of the patent records. The 1852 reforms undoubtedly instituted improvements over the former opaque procedures, and the lower fees had an immediate impact. Nevertheless, the system retained many of the former features that had implied that patents were in effect viewed as privileges rather than merited rights, and only temporarily abated expressions of dissatisfaction.

One source of dissatisfaction that endured until the end of the nineteenth century was the state of the common law regarding patents. At least partially in reaction to a history of abuse of patent privileges, patents were widely viewed as monopolies that restricted community rights, and thus to be carefully monitored and narrowly construed.[19] This approach also was evident in the legal system, which not only failed to protect property rights in patents but also was deemed to be actively "antipatent" and biased against inventors.[20] For instance, the law specified that patents were to be granted for inventions that were new and useful, and courts did not hesitate to enforce both of these conditions. Utility under the patent law was regarded as unrelated to the commercial success of the patented invention.[21] Moreover, "if

---

[18] One of the witnesses before the 1851 Select Committee argued against an examination system because it would "erode the royal prerogative." However, the Select Committee decided not to adopt this aspect of the American system more because they felt it would not be possible to find sufficiently skilled examiners. Law Officers did have the right to examine the technical merits of patent applications at their discretion, but there is no evidence that this right was ever exercised. See Dutton (1984), p. 61, and *British Parliamentary Papers* (1968–70.)

[19] Blanchard v. Sprague, 3 F. Cas. 648, Mass. (1839), compared the law toward patents in Britain and America.

[20] Dutton (1984), p. 35.

[21] According to *Badische Anilin und Soda Fabrik v. Levinstein*, 4 R. P. C. 462, 466: "I do not think that it is a correct test of utility to enquire whether the invented product was at the time of the patent likely to be in commercial demand or capable of being produced at a cost which would make it a profitable venture."

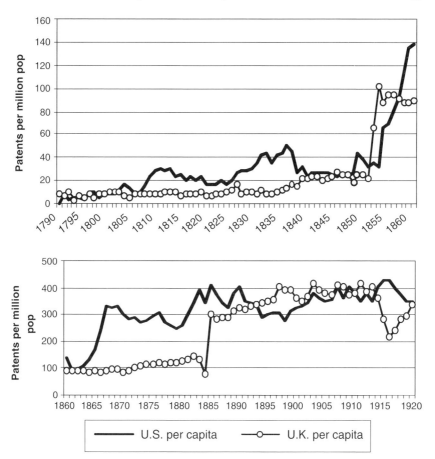

Figure 2.1. Patents Per Capita Issued in Britain and the United States, 1790–1920
Notes and Sources: For annual patent statistics, see P. J. Federico, "Historical Patent Statistics, 1791–1961," *Journal of the Patent Office Society*, vol. 46 (2) 1964: 89–170; U.S. Patent Office, *Annual Report of the Commissioner of Patents*, various years. For population statistics, see B. R. Mitchell, *International Historical Statistics: Europe, 1750–1988*, New York: Stockton Press, 1992; the annual U.S. population was interpolated from the U.S. Census for each decade.

part of an invention is found to be meritorious and part useless, the patent is likewise void."[22] The question of utility was decided by judges as well as juries, and led to decisions that were at times arbitrary.[23] Because the legal

---

[22] *United Horsenail Co. v. Stewart*, 2 R.P.C. 132.
[23] For instance, Justice Grove instructed the jury in *Young v. Rosenthal*, 1 R.P.C. 41 that if the invention "is not as good as those existing before, or no better than those existing before in any particular point, then you would say it is not useful."

system was unpredictable, patent rights could not be regarded as settled unless the patent had been contested in court with a favorable outcome.[24]

British patents were granted "by the grace of the Crown" and therefore were subject to any restrictions that the government cared to impose. According to the statutes, as a matter of national expediency, patents were to be granted if "they be not contrary to the law, nor mischievous to the State, by raising prices of commodities at home, or to the hurt of trade, or generally inconvenient."[25] The Crown possessed the ability to revoke any patents that were deemed inconvenient or contrary to public policy. After 1855, the government also could appeal to a need for official secrecy to prohibit the publication of patent specifications in order to protect national security and welfare.[26] Moreover, the state could commandeer a patentee's invention without compensation or consent, although in some cases the patentee was paid a royalty: "In England, the grant of a patent for an invention is considered as simply an exercise of the royal prerogative, and not to be construed as precluding the Crown from using the invention at its pleasure; and therefore a petition of right cannot be maintained against the Crown for using a patented invention."[27] For instance, when the British Navy freely

---

[24] According to an editorial in 1862, "there can be no doubt that a large amount of property is bound up in patent rights, and that the utmost uncertainty exists as to the legal value of that property" (*Newton's London Journal*, cited in Coulter, p. 140.)

[25] The quotation is from section 6 of the Statute of Monopolies. Also indicative of the character of the patent grant: "Though a thing be new in every particular, and the true and first inventor apply for a patent, it is in the judgment of the Crown whether it will or will not, as a matter of favor, make the grant"; and "Your petitioner therfore humbly prays Your Majesty will be graciously pleased to grant unto him ... Your Majesty's Royal Letters Patent, under the great Seal of Your Majesty's United Kingdom of Great Britain and Ireland." See Holroyd (1830), pp. 58 and 66.

[26] According to T. H. O'Dell, *Inventions and Official Secrecy: A History of Secret Patents in the United Kingdom*, Oxford: Clarendon Press (1994), the Patents for Inventions (Munitions of War) Bill was approved without debate shortly after in April 1859. See also David Vincent, *The Culture of Secrecy: Britain, 1832–1998*, New York: Oxford University Press (1998.) In the United States, official secrets statutes did not appear until the twentieth century, in connection with national security interests to withhold patent information that might be "detrimental to the public safety or defense or might assist the enemy or endanger the successful prosecution of the war." See H.R. REP. No. 1540, 96th Cong., 2d Sess. 37 (1980.) The right of the government to maintain secrecy was granted during World War I, with the U.S. Invention Secrecy Act of Oct. 6, 1917, ch. 95, 40 Stat. 394 (1917); which was reinstated during World War II, in the Act of July 1, 1940, ch. 501, 54 Stat. 710 (1940.) Inventors had the right to sue for compensation in the U.S. Court of Claims if their invention was of value to the government. See Sabing H. Lee, "Protecting the Private Inventor under the Peacetime Provisions of the Invention Secrecy Act," *Berkeley Technology Law Journal* vol. 12 (2) 1997, 345. See also Herbert N. Foerstel, *Secret Science: Federal Control of American Science and Technology*, Westport, Conn.: Praeger Publishers (1993.)

[27] According to *Samuel McKeever v. The United States*, 14 Ct. Cl. 396; 1878: "Therefore it is as historically clear as it is authoritatively settled by the decisions of the English courts that

used a patent without the consent of the patentee, the court ruled that the Crown had rightful access to any patent that had been granted.[28] In 1816, Sir William Congreve was allowed to violate an injunction that prevented him from manufacturing gunpowder barrels without the permission of the patentee, on the grounds that the infringement was in the public service on behalf of the ordnance office of the British Government.[29]

Policies toward patent assignments and trade in intellectual property rights also constrained the market for inventions. Ever vigilant to protect an unsuspecting public from fraudulent financial schemes on the scale of the South Sea Bubble, ownership of patent rights was limited to five investors (later extended to twelve.) Nevertheless, the law did not offer any relief to the purchaser of an invalid or worthless patent, so potential purchasers were well advised to engage in extensive searches before entering into contracts.[30] When coupled with the lack of assurance inherent in a registration system, the purchase of a patent right involved a substantive amount of risk and high transactions costs – all indicative of a speculative instrument. It is therefore

---

a patent in England was nothing more than a grant dependent in contemplation of law upon royal favor, and subject to the general implication of all grants wherein the contrary was not expressed, that they shall not exclude a use by the Crown." See also *Belknap v. Schild*, 161 U.S. 10; 1896.

[28] *Feather v. The Queen*, 6 B. & S. 257; *Dixon v. London Small Arms Co.*, L.R. 10 Q.B. 130. See James Johnson and J. Henry Johnson, *The Patentee's Manual: A Treatise on the Law and Practice of Patents for Invention*, 6th ed., London: Longmans, Green & Co. (1890): "It was decided in the case of Feather v. The Queen (6 B. & S. 257) that, as all grants made by the Crown must be construed favorably to the grantor and adversely to the grantee, there was nothing in the ordinary form of Letters Patent in use up to January 1, 1884, to debar the Crown from using the invention comprised therein, and consequently that the Crown had a right, either by itself or by its officers, agents, or servants, to exercise such invention and manufacture the patented object without paying royalties or compensation ... However, in Dixon v. The London Small Arms Company (L.R. 1 App. Ca. 632), where a manufacturing company had entered into a contract with the Government for the supply of a patented article for the public service at a fixed price, and in pursuance thereof had manufactured the article without the patentee's licence, and delivered the same and received payment, it was held by the House of Lords that the manufacturers were not in the position of servants or agents of the Crown, although they had received an indemnity from the Government against the claims of the patentee; and, therefore, that they were not entitled to the privilege of the Crown against a patentee, but were liable to an action for an infringement of the patent. See also *Vavasseur v. Krupp* (L.R. 9 Ch. D. 351.)" (pp. 266–267.)

[29] See *Walker v. Congreve*, 1 Carp. Pat. Cas. 356. Lord Eldon allowed that the patentee should be granted a hearing on the validity of the patent, and the government was requested to keep an account of its infringement until the trial settled the matter. However, there is no record of such a trial.

[30] The case law on licenses was more convoluted. For instance, in *Lawes v. Purser*, 6 Ell. and Bl. 930, a licensee refused to continue payments on the grounds that the patent was void. It was held that the licensee could not make such a defense as long as the contract for the invalid patent had been executed without fraud.

not surprising that the market for assignments and licences seems to have been quite limited, and even in the year after the 1852 reforms only 273 assignments were recorded as the law required.[31]

In 1883 new legislation introduced procedures that were somewhat simpler, with fewer steps. The fees fell to £4 for the initial term of four years, and the remaining £150 could be paid in annual increments.[32] For the first time, applications could be forwarded to the Patent Office through the post office. This statute introduced opposition proceedings, which enabled interested parties to contest the proposed patent within two months of the filing of the patent specifications.[33] Compulsory licences were introduced in 1883 (and strengthened in 1919 as "licences of right") for fear that foreign inventors might injure British industry by refusing to grant other manufacturers the right to use their patent. The 1883 act provided for the employment of "examiners" but their activity was limited to ensuring that the material was patentable and properly described. Indeed, it was not until 1902 that the British system included an examination for novelty, and even then the process was not regarded as stringent as in other countries. Many new provisions were designed to thwart foreign competition. Until 1907 patentees who manufactured abroad were required to also make the patented product in Britain. Between 1919 and 1949 chemical products were excluded from patent protection to counter the threat posed by the superior German chemical industry. Until 1977, licences of right enabled British manufacturers to compel foreign patentees to permit the use of their patents on pharmaceuticals and food products.

In sum, changes in the British patent system were initially unforthcoming despite numerous calls for change. Ultimately, the realization that England's early industrial and technological supremacy was threatened by the United States and other nations in Europe led to a slow process of revisions that lasted well into the twentieth century. The debate about patent rights in this period was far ranging, and (like today) explicitly linked questions of trade, comparative advantage and intellectual property. Proposals had ranged from the creation of a national fund to reward inventors

---

[31] See the first report of the British Commissioner of Patents in 1853. The patent agency of Munn & Co. noted with some complacency: "From January 1, 1865 to the 1st of December, the whole number of applications for patents to the British Patent Office will not have exceeded three thousand. Within the same period the applications made by Munn & Co. to the United States Patent Office number at least three thousand five hundred; thus showing that our professional business considerably exceeds the entire business of the British Patent Office." *Scientific American* v. 13, 23 Dec. 1865, p. 415.

[32] Despite the relatively low number of patents granted in England, between 1852 and 1880 the patent office had made a profit of over £2 million (*Report of the Commissioners of Patents for 1880.*)

[33] The patent would be refused if the idea had been stolen, if it had previously been patented in Britain, or if the patent specification was different from the description in the provisional patent. See 46 and 47 Vic. C. 57, 1883.

through the abolition of any property rights in inventions, but policies that emerged from this era of activism were far from optimal. Legal advances in the nineteenth century were inevitably piecemeal and incomplete, consisting as they did of compromises between those with vested interests in maintaining rents under the former system, inventors (especially those of limited means) who stood to benefit from improvements, and manufacturers and politicians who wished to deter short-run foreign competition even if at high costs in the long run. One commentator summed up the series of developments by declaring that the British patent system at the time of writing (1967) remained essentially "a modified version of a pre-industrial economic institution."[34]

FRENCH PATENT SYSTEM

Early French policies toward inventions and innovations in the eighteenth century are worth a close examination. They were based on an extensive but somewhat arbitrary array of rewards and incentives, and illustrate the relative benefits and costs of alternative routes to statutory grants of intellectual property rights.[35] During this period inventors or introducers of inventions could benefit from titles, pensions that sometimes extended to spouses and offspring, loans (some interest-free), lump-sum grants, bounties or subsidies for production, exemptions from taxes, or monopoly grants in the form of exclusive privileges. Exclusive rights could extend to a specific region or throughout the entire kingdom, and their term varied from five years to perpetuity. Privileges were awarded to many who were not inventors, and challengers to a favored inventor could face serious consequences, including imprisonment.[36] Jacques de Vaucanson claimed to be the first to produce a mechanical duck; when a physicist, Dominique de Châteaublanc, contested

[34] "The whole subsequent history of the principal measures of patent reform has indeed been a story of caution and circumspection . . . the current patent system is in very many essentials the system developed . . . during the second half of the sixteenth century and the first half of the seventeenth." Klaus Boehm in collaboration with Aubrey Silberston, *The British Patent System: I. Adminstration*, Cambridge: Cambridge University Press (1967), p. 32.

[35] Excellent assessments of such issues during the Enlightenment include Liliane Hilaire-Pérez's thesis, *Inventions et Inventeurs en France et en Angleterre au XVIIIè siècle*, Lille: Université de Lille (1994); and her book *L'invention technique au siècle des Lumières*, Paris: Albin Michel (2000.) Also see Shelby T. McCloy, *French Inventions of the Eighteenth Century*, Lexington: University of Kentucky Press (1952.) For the nineteenth century, see B. Zorina Khan and Kenneth L. Sokoloff, "The Innovation of Patent Systems in the Nineteenth Century: A Comparative Perspective," Unpublished manuscript (2001.)

[36] "Unfortunately the government had no uniform policy for all inventors. Some it aided financially; others it did not. More flagrantly inconsistent was its policy of granting monopolistic rights in many lines to court favorites, so that the greater number enjoying exclusive privileges were not inventors" (McCloy, p. 171.)

Vaucanson's priority in court and lost the case, Châteaublanc was imprisoned for two and a half years for libel.[37]

Alternatives to formal privileges illustrate the advantages and disadvantages of prizes and financial awards that were administered by the state on a
case-by-case basis. It is true that, on several occasions, prior examination by
a committee of qualified individuals proved to be an effective means of ensuring that productive applicants received awards and of benefiting the public.
In 1790, M. Devilliers demonstrated two inventions: a device to measure
longitude and one to purify the town water supply. He claimed a total of
4,200 livres in expenses, and offered to donate a quarter of the award to the
paupers in the town. The Committee on Agriculture and Commerce thought
it was obviously a good project and saw no need for deliberations.[38] In other
instances, funding was doled out in stages depending on the results in previous stages. When Abbé D'Arnal invented a windmill that could be used to
pull boats, he asked for a ten-year annuity of 33,000 livres. His invention
was favorably reviewed by a member of the Paris Academy of Sciences, but
he was given only a part of the money, with the prospect of further payments
if the invention proved to be useful.[39] Nevertheless, the overall system
of grants and privileges for the most part tended to be arbitrary and based
on noneconomic criteria.[40]

The researcher who peruses the massive files on individual inventors that
survive in the National Archives in Paris cannot fail to be struck by the
tremendous administrative costs associated with each application for funds,
the potential for arbitrary decisions, and the possibilities for facilitating corruption. Eighteenth-century correspondence and records provide numerous
examples of premiums that were based on court connections.[41] At the other
end of the spectrum large sums were awarded to the "deserving" on the basis
of age, conduct, or family need.[42] Members of the scientific community who

---

[37] McCloy, p. 106.

[38] Archives Nationales [AN] holdings F12/992, No. 1106 (16 June 1790.)

[39] AN F12/992, No. 155.

[40] A law of October 16, 1791 created the Bureau of Consultation of Arts and the Trades, which
consisted of 30 members drawn from various academies. They were to examine and report
on the inventions, making recommendations about the rewards to offer to inventors. In
1797 this committee was replaced by the National Institute of the Sciences and Arts. The
Minister of the Interior was also authorized to propose to the National Assemby any major
discoveries which had been made either in France or imported to France "particulièrement
lorsque ces descouvertes feront dues a des travaux pénibles, ou a voyages longs et perilleux."
[Sec. V of Law of 1791]. Under this law Coste d'Arnobat received 5,000 livres on the 29th
of December for the importation of rhubarb into France. F/12/2424 "Encouragement donné
aux artistes et aux inventeurs de 1786 à 1793."

[41] AN F12/992, No. 239 (Oct. 1781.) M. le Chevalier de Gruyere requested a privilege for the
manufacture of a vegetable-based cosmetic rouge. He was willing to pay 1.2 million livres
for the grant. His application was supported by influential women at court.

[42] Archives Nationales, F12/992, No. 3376. M. Chevalier (painter and gilder of the buildings
belonging to the King's brother) asked for a pension for two machines he invented to safely

examined applications were not necessarily qualified to assess their novelty or potential commercial value, and others had vested interests to protect.[43] Even if the privilege was commercially successful, active trade in the rights was inhibited because permission had to be obtained first.[44] Moreover, the opportunity costs of such a system were nontrivial on the part of both supplicants and the state bureaucracy.[45]

Applicants were well aware of the political dimension of invention.[46] They also were aware that promises made as inducements were not necessarily enforceable once the inventor had made fixed investments. The plaintive correspondence of the famous Lancashire textile inventor, John Kay (b. 1704), illustrates the asymmetries involved in individual bargains struck with state

pulverize colors. With the old method of manufacture twelve hundred men died each year from lead poisoning. His letter begins: "Chevalié, père de douze enfants vivants, quatre filles et huit garcons." He is sixty-six years old, married at age thirty-six. An administrative Report of February 1783 notes that "Les douze enfants paroissent bien élevés, le plus jeune a 13 ans, et le père et la mère ont une bonne conduite et paroissent aises dans leur ménage."

[43] The Abbé de Mandre invented a motor that could pull a train of thirty boats on the river. The Academy of Sciences found the motor to be "new and ingenious" but opined that the value was not large enough to warrant a significant reward. The motor however proved to be a useful invention. The Abbé made nothing from the invention and died in obscurity (McCloy 115.) J R Harris, *Industrial Espionage and Technology Transfer*, Aldershot: Ashgate (1998), p. 562, argues that "The place of the Academy in the central establishment of the State meant that innovators had almost invariably to prove their innovations to Academicians in Paris, whether that was the sensible place or not. All these problems had their effect. There was by the Revolution a large cohort of the frustrated and aggrieved among inventors and innovators, the last including some transferers of technology, who believed that their efforts had been seriously affected by theoretically biased, impractical Academicians.... it was an important reason for the dissolution of the Academy in 1793."

[44] See AN F12/992 (1787.) Les Sieurs Défevres et Cie had to request permission to purchase a fifteen-year privilege from Dubusques.

[45] For instance, AN F/12/4824 includes about three inches of documents relating to help accorded to a single individual, Albert Charles, an English machinist who introduced new methods of textile manufactures, including cotton carding machines, that he learned in Manchester. Albert Charles was given a pension of 500 francs per year from 1840 until his death in 1852. After his death, his widow was given annual sums in recognition of the "services signales rendus a l'industrie par Monsieur son mari." The files include in tabular form the biography of Charles each year from his birth in 1764. The table notes the facts of his contributions as well as the evidence to support each fact. Also included are the annual letters that the widow sent to claim her pension, which was increased from 300 francs in 1867 to 400 francs in 1868.

[46] Liliane Hilaire-Pérez refers to "La forte liaison qui existe en France au XVIIIe siècle entre technique et politique," (30) and argues that inventors were "plaidoyers (accumulant les preuves), car la technique n'est pas neutre, elle est porteuse des rêves, de revendications, d'ambitions calculées, d'utopies refondatrices et de politiques réalistes" (p. 34.) The accuracy of this observation is readily borne out by a perusal of correspondence such as F12/992 (8th October 1777.) Les Sieurs De la Fosses invented an improvement in yeast making, and submitted a request for a privilege for thirty years, from "votre sujet, amateur des sciences, qui n'avoir rien de plus précieux q'à s'occuper pour accomplir ses souhaits quant travaillant a tout ce qui pouvoir avoir rapport à votre gloire."

authorities.[47] Kay immigrated to France in 1747 because of official promises of generous subsidies, and he substantially aided his adopted country in the diffusion of textile machinery. The Bureau of Commerce gave him an annual pension for training French artisans to use the invention. However, Kay claimed that when he left Paris he was owed 15,000 livres as well as the cost of the machines. The Society for the Encouragement of Arts and Manufacturing in England offered him a generous award to return there, but they reneged once he was back in London and even abused him. When Kay met with the officers, they "began on me about going abroad and endeavour'd for to make it appear that I was a rogue for so doing," and threw him out because he had instructed the French in English methods. The Chairman called him back and placated him, saying that the Chair could not be held accountable for every member of the committee, but it seems that the trip was in vain. Kay wrote early in 1761 to Prudaine de Montigny, Conseiller d'Etat in London, to explore the possibility of receiving French financial aid if he again immigrated to Paris rather than dealing with the English Society. "I have borrowed a guinea of your ambassador but he will not advance any more without your order," he added. Later that same year, we find Kay pleading with M. de Brou, Intendant de Rouen, for he was still not receiving the pension he had negotiated.

This complex network of state policies toward inventors and their inventions was revised but not revoked after the outbreak of the French Revolution. The modern French patent system was established according to the laws of 1791 (amended in 1800) and 1844. The Revolutionary Assembly intended to avoid the excesses involved in previous grants of privileges, and proclaimed that it had drafted the outlines of a system that created a distinct break with the past.[48] The decree of 1790 (passed early the following year) declared the natural right of the inventor to obtain property rights in patents since "every discovery or invention, in every type of industry, is the property of its creator; the law therefore guarantees him its full and entire enjoyment."[49] But in effect, as Alexis de Tocqueville pointed out, many features of the institutions of the *ancien régime* survived the revolution, and this was no less evident in the workings of the intellectual property system.[50]

[47] The information for this paragraph is drawn from Kay's correspondence files [F12/992] in the National Archives in Paris, France.
[48] The French were somewhat reluctant to adopt a system of patents that they associated with the English, but the Chevalier de Boufflers pointed out to the members of the Assembly: "Ils n'ont point eu ce vain scrupule, ces fiers et sages Américains, ces dignes amis de toute liberté, qui, dans leur nouvelle constitution, on adopté la législation de l'industrie anglaise comme le plus sur moyen d'assurer aussi l'affranchissement et la prosperité de leur industrie."
[49] See the Decret du 30 Decembre 1790, in the Code des Pensions, 30 Decembre 1790, p. 45.
[50] Alexis de Tocqueville, *L'Ancien Régime et La Révolution*, Oxford: Basil Blackwell (1952.)

Patentees filed through a simple registration system without any need to specify what was new about their claim, and could persist in obtaining the grant even if warned that the patent was likely to be legally invalid. On each patent document the following caveat was printed: "The government, in granting a patent without prior examination, does not in any manner guarantee either the priority, merit or success of an invention."[51] The inventor decided whether to obtain a patent for a period of five, ten, or fifteen years, and the term could only be extended through legislative action.[52] Protection extended to all methods and manufactured articles, but excluded theoretical or scientific discoveries without practical application, financial methods, medicines, and items that could be covered by copyright. The 1791 statute stipulated patent fees that were costly, ranging from 300 livres through 1500 livres, based on the declared term of the patent. The 1844 statute maintained this policy because fees were set at 500 francs ($100) for a five-year patent, 1,000 francs for a ten-year patent and 1,500 for a patent of fifteen years, payable in annual installments.[53] The high price of protection led to difficulties among ordinary inventors, whose correspondence included pleas for extensions or for a waiver of the tariffs, or resigned acknowledgments that they were forced to let the patent right expire for want of funds.[54] Nante, a master locksmith who obtained a ten-year patent on a lock, was obviously better connected with influential friends. His file includes a letter of recommendation from a count, as well as a letter Nante addressed to the king, which the monarch forwarded to the patent officials. Nante stated that he

---

[51] AN F/12/1028 (1817): Printed on patent documents according to the amendments of the statute of 1800. This disclaimer was inserted at the initiative of Lucien Bonaparte, who protested because, according to the law of registration, he could not prevent the issue of a patent for an invention of an "invisible woman." See Antoine Perpigna, *The French Law and Practice of Patents for Inventions, Improvements, and Importations*, Philadelphia: J. S. Littell (1852), p. 5. See also Perpigna's *Manuel des inventeurs et des brevetés*, 8. éd. Paris: Chez l'auteur (1852.)

[52] Extensions were rare occurrences: of some five thousand patents obtained in the first forty years of the system, only twenty were extended. "What makes the government so averse to prolongations, is that they are never demanded but for successful inventions, and such as society at large is most anxious to enjoy. They are detrimental to trade and damp the spirit of enterprise . . . " Antoine Perpigna, *The French Law and Practice of Patents for Inventions, Improvements, and Importations*, p. 32.

[53] Early fees were 300 livres for five years, 800 for ten, years and 1,500 for fifteen years. Anyone who wished to consult a description paid 12 livres and those who merely wished to consult the index paid 3 livres.

[54] AN F/12/1025 (1816). Jean Bozon sent a letter regarding the difficulties he was having finding the 150 francs that was due to satisfy the patent fees (five-year patent for shoes.) He asked them to pity "un honnête père de famille." Francois Gury asked on November 4, 1816 for an extension on the payment of the patent fees for his hat invention; six months later he assigned the five-year patent to M. Cousteau, a manufacturer, and it might be speculated whether the sale was partially caused by his difficulties in meeting the annual payments.

could not meet the patent fees and asked to be given the patent for free. The Bureau of Arts and Manufactures subsequently paid the 800 francs on his behalf, ostensibly because locks enhanced security and this was beneficial to social welfare.[55]

Although the legal rhetoric implied that the primary intent of the legislation was to recognize the natural rights of inventors, the actual clauses led to results that were different and reflected former mercantilist policies. In an obvious attempt to limit international diffusion of French discoveries, until 1844 patents were voided if the inventor attempted to obtain a patent overseas on the same invention.[56] By contrast, the first introducer of an invention covered by a foreign patent would enjoy the same "natural rights" as the patentee of an original invention or improvement, although the term would expire at the same time as any foreign patent on the item. In order to qualify for a patent of importation, the applicant was required to gain practical knowledge through personal risk and effort, although he was not obliged to prove that the invention had been patented elsewhere nor to even state its country of origin.[57] The statutes placed a limit on the rights of inventors in the form of working requirements because "it would be injurious to society at large, to allow any one individual to cramp the efforts and attempts of more industrious inventors by obtaining a patent upon which he did not intend to work."[58] Patentees therefore had to put the invention into practice within two years from the initial grant, or face a tribunal that had the power

---

[55] AN F/12/1017A. At least one patentee thought that his connections were relevant: Nicholas Arnoult Humblot-Conte includes a letter in his application for a ten-year patent in 1807, in which he points out that his wife (a member of a noble family, with all five of her first names given) is a wealthy heiress.

[56] "The legislators feared the prosperity of their country might be impaired, if foreign countries were allowed to use every new invention as well as France, and thus were enabled to compete with French manufacturers: or they thought the French patentee would be more likely to carry his invention into extensive use in France, if he was...thus obliged to direct all his means and attention to the success of the French patent"(Perpigna, p. 28.) According to Perpigna, "this provision of the law can be evaded with impunity, it is quite useless..." so it was repealed in the 1844 revision of the statutes.

[57] "It is necessary to obtain a practical knowledge of its way of working, and for that purpose, to travel and reside some time in the country where it has been invented; to enter, often with risk and never without expense, into different manufactories, and see the machine at work: to study it in its results, and ascertain by inquiries and experiments the most beneficial mode of establishing and using it. All this requires great expense and loss of time, which the importer must incur, before he can qualify himself to introduce successfully an invention in another country" (Perpigna, p. 12.) In a dispute the burden of proof for regarding any element of the patent was on the accuser not the patentee.

[58] See Perpigna, p. 29. In 1762, the king abolished perpetual privileges and limited them to 15 years, and they could only be transferred with royal permission. They would expire if they had not been put to use within one year of the grant. (Harold T. Parker, *The Bureau of Commerce in 1781 and Its Policies with Respect to French Industry*, Durham, N.C.: Carolina Academic Press (1979), p. 57.)

to repeal the patent, unless the patentee could point to unforseen events that had prevented his complying with the provisions of the law. The rights of patentees also were restricted if the invention related to items that were controlled by the French government, such as printing presses and firearms.[59]

The French patent statutes included a statement regarding the right of the public to view patent specifications, which echoed the "bargain theory" of patents that underlay American and British grants. In return for the limited monopoly right, the patentee was expected to describe the invention in such terms that a workman skilled in the arts could replicate the invention and this information was expected to be made public. However, as no provision was made for the publication or diffusion of these descriptions, this statutory clause was a dead letter. At least until the law of April 7, 1902, specifications were only available in manuscript form in the office in which they had originally been lodged, and printed information was limited to brief titles in patent indexes.[60] The attempt to obtain information on the prior art was also inhibited by restrictions placed on access: viewers had to state their motives; foreigners had to be assisted by French attorneys; and no extract from the manuscript could be copied until the patent had expired.

The state remained involved in the discretionary promotion of invention and innovation through policies beyond the granting of patents. In the first place, the patent statutes did not limit their offer of potential appropriation of returns only to property rights vested in patents. According to Article V of the decree of 1790, "when the inventor prefers to deal directly with the government, he is free to petition either the administrative assemblies or the legislature, if appropriate, to turn over his discovery, after demonstrating its merits and soliciting a reward."[61] In other words, the inventor of a discovery of proven utility could choose between a patent or making a gift of the invention to the nation in exchange for an award from funds that were set aside for the encouragement of industry. Second, institutions such as the Société d'encouragement pour l'industrie nationale were established.[62] The society

---

[59] In France printers were required to obtain licenses from the government, and weaponry could not be manufactured without permission. Thus, the patentee who wanted to benefit from his invention in these areas could only do so if he obtained further authority from the government. See Perpigna, p. 23.

[60] The law of 1844 only allowed for the publication of the full text of patents that were judged to be important. "C'est donc bien avec la loi de 1902 que le brevet a définitivement perdu son charactère de document d'archives." Ministère de l'industrie et du commerce, *Brevets d'invention français, 1791–1902: un siècle de progrès technique*, Paris: Le Ministère (1958), p. 12.

[61] Code des Pensions, 30 Decembre 1790.

[62] These policies were viewed as necessary because "pour seconder l'industrie dans son développement, pour lui donner tout l'essor dont elle est capable, trois sortes de secours sont nécessaires: les lumières de l'instruction, des encouragements sagement conçus et appliqués et l'influence générale de l'esprit public" (cited in Pietro Redondi, "Nation et entreprise: la Société d'Encouragement pour l'Industrie Nationale, 1801–1815," in P. Redondi and R. Fox

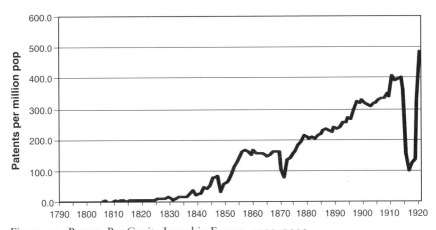

Figure 2.2. Patents Per Capita Issued in France, 1790–1920

Notes and Sources: For annual patent statistics, see P. J. Federico, "Historical Patent Statistics, 1791–1961," *Journal of the Patent Office Society*, vol. 46 (2) 1964: 89–170; U.S. Patent Office, *Annual Report of the Commissioner of Patents*, various years. For population statistics, see B. R. Mitchell, *International Historical Statistics: Europe, 1750–1988*, New York: Stockton Press, 1992.

consisted of eminent scientists and industrialists who awarded a number of medals each year to stimulate new discoveries in areas they considered to be worth pursuing, and also to reward deserving inventors and manufacturers.[63] In 1856 the society gave out sixty-five prizes, including twelve gold medals, six platinum, and twenty-four silver medals. It also offered cash awards to those who addressed questions that the Society proposed each year, such as 10,000 francs to combat diseases threatening vineyards. The recipients ranged from horticulturalists to manufacturers of cutlery and the head physician in a military hospital.[64] Third, the award of assistance and pensions to inventors and their families continued well into the nineteenth century. Fourth, at times the society purchased patent rights and turned the invention over into the public domain.

As a result, inventors had an incentive to direct their attention to rent-seeking activities in addition to productive efforts to commercialize their

(eds.), *History and Technology* vol. 5 (2–4) (1988) Special Issue, *French Institutions from the Revolution to the Restoration.*, p. 201.)

[63] AN F/12/4824 includes a thick folder of documents regarding Bauwens, who was rewarded because he had rendered "les plus grands services à l'industrie française par l'introduction de la filature du coton." After his death his widow begged for assistance and received 600 francs in 1822. Several other members of the family did so, including a daughter in 1864, 1865 and 1866. The Ministry asked a professor at the Conservatoire des Arts et Métiers to assess whether Bauwens was indeed responsible for the introduction of the jenny, and he pointed out that Bauwens was only one of many early introducers.

[64] See the society's report in Louis Figuier, *L'année scientifique et industrielle*, Paris: Hachette (1857.)

discoveries. The "privilege mentality" could be detected in the records for Felix Lemaistre of Paris, who invented a shoe that could be made in one piece without sewing, and tried to get the state to purchase the invention. His file includes a letter from the undersecretary of state rejecting Lemaistre's proposal to have the government take over the invention, and underlining that "It is your own responsibility to manage the exploitation of your invention, or else you should interest a few investors in advancing you the necessary capital."[65] Premiums from the state did not preclude inventors from also pursuing profits through other means, including patent protection. For instance, Napoleon III offered a prize for the invention of a cheap substitute for butter that allegedly induced Hippolyte Mège to make significant improvements in margarine production. In assessing the efficacy of this prize it should be noted that many inventors worldwide were already pursuing the idea of a cheap and longer-lasting substitute for butter. Mège not only won the prize but also obtained patent protection for fifteen years in France in 1869, and patented the original invention and several improvements in England, Austria, Bavaria, and the United States.

The basic principles of the modern French patent system were evident in the early French statutes and were retained in later revisions.[66] Because France during the *ancien régime* was likely the first country to introduce systematic examinations of applications for privileges, it is somewhat ironic that commentators point to the retention of registration without prior examination as the defining feature of the "French system."[67] In 1910 fees remained high, although somewhat lower in real terms, at 100 francs per year. Working requirements were still in place, and patentees were not allowed to satisfy the requirement by importing the article even if the patentee had manufactured it in another European country. However, the requirement was waived if the patentee could persuade the tribunal that the patent was not worked because of unavoidable circumstances. The list of acceptable reasons that could be presented to the courts to justify inaction included a lack of capital, political or commercial crises, the availability of superior inventions that rendered the patent unprofitable, high prices of raw materials, or competition

[65] [AN F/12/1025 (1816)] Lemaistre sold the rights in October of the following year to a businessman in Paris.
[66] "French patent law remained for nearly 150 years practically unchanged and unaffected by modern ideas in legislation," Jan Vojacek, *A Survey of the Principal National Patent Systems*, New York: Prentice-Hall (1936), p. 139.
[67] In 1968 a partial examination system was adopted which was similar to the early British reforms along these lines, because it did not include a search for novelty, merely a test for accordance with the law: "[il] se situe à mi-chemin entre la libre délivrance et l'examen préalable... en effet, l'administration n'avait pas les moyens de pratiquer un tel examen" (p. 21, Yves Marcellin, *La Procédure Française de Délivrance des Brevets d'Invention*, Rosny-Sous-Bois: Editions Cédat, 1983.) The changes were made to give value to patents and to protect the interests of third parties. It was only in 1978 that an examination for novelty was introduced.

Illustration 3. The nineteenth-century Patent Office at 8th and F Streets, Washington, D.C., was regarded as a "temple to technology." President Andrew Jackson laid the cornerstone in 1836, and its architects included patentee Ithiel Town. The third federal building completed in Washington, the Patent Office building (now the National Portrait Gallery) is held to be "one of the finest examples of Greek Revival architecture in the United States." (*Source*: Library of Congress.)

from infringers. Thus, with a modicum of ingenuity this particular restriction could be evaded, but the transaction costs and uncertainty could not be avoided.

Similar problems were evident in the market for patent rights. Contracts for patent assignments were filed in the office of the Prefect for the district, but because there was no central source of information it was difficult to trace the records for specific inventions. The annual fees for the entire term of the patent had to be paid in advance if the patent was assigned to a second party. Like patents themselves, assignments and licences were issued with a caveat emptor clause. Indeed, according to one nineteenth-century source, they evinced a "remarkably hazardous and uncertain nature."[68] This was partially because of the nature of patent property under a registration system, and partially to the uncertainties of legal jurisprudence in this area. The case law suggested that the burden of proof of validity was on the purchaser of a patent in the case of "vices apparents" (overt flaws) such as a lack of novelty. The purchaser could be protected if the exchange involved "vices cachés" (hidden flaws.) However, it was not evident which specific circumstances

[68] This section is drawn from Eugene Pouillet, *Traité Théorique et Pratique des Brevets d'Invention*, Paris: Marchal et Billard (1879.) The phrase is a translation of "comporte un charactere aléatoire tout à fait remarquable," p. 219. "Pour couper court à toute difficulté, le breveté agira sagement en déclarant, dans l'acte, qu'il cède sans garantie; cette clause a pour effet d'exprimer nettement ce qui, selon nous, est sous-entendu dans tout contrat de cession" (p. 225.)

would qualify as overt or covert, especially as the jurisprudence contained conflicting decisions. Legal authorities advised the patentee to draw up a contract explicitly stating what was implicit, that the trade was conducted without any guarantees. For both buyer and seller, the uncertainties associated with the exchange likely reduced the net expected value of trade.

THE AMERICAN PATENT SYSTEM IN EUROPEAN PERSPECTIVE

The drafters of the American Constitution and of its patent laws were familiar with European practice, so it might reasonably be inferred that departures from British precedent were self-conscious and deliberate attempts to establish a different system and pursue an alternative path of development.[69] In particular, the Framers of the U.S. system believed that inventors benefited society, and minor improvements might be even more important than grand discoveries, if they affected a larger market. The environment was markedly more favorable for all inventors, whether their contributions were regarded as "heroic" or marginal, than was the case in Europe. Policy makers recognized the importance of the "fuel of interest," and acknowledged that inventive efforts were significantly influenced by security of property rights and by the prospects for material gain. An examination of the legislative history of the first U.S. Patent Act supports the view that early policy makers attempted to establish a system that would enhance rewards to inventors – whatever their social status – and accordingly stimulate inventive activity and economic growth.

George Washington urged delegates early in 1790 to give "effectual encouragement, as well to the introduction of new and useful inventions from abroad, as to the exertions of skill and genius in producing them at home."[70] Congress responded with a bill, HR-41, that was designed to meet these objectives, but before the bill was passed it was subject to several amendments. The most minor of these amendments is suggestive: the phrase "great seal of the United States," which emulated the British custom, was altered to read simply "seal of the United States."[71] Indeed, the patent bill

---

[69] See the discussion in Bugbee (1967), p. 145, of the resemblance between the copyright law of 1790 and the Statute of Anne (1710.) Note also his conclusion that (p. 157), "The national patent and copyright systems created in 1790 under the Constitution were founded not only upon English precedents...but also...upon a century and a half of distinctive provincial tradition."

[70] Bugbee (1967), p. 137.

[71] Another amendment insisted that, rather than in the name of the President, "Patents, licenses, commissions, etc., shall be in the name of the *people* of the United States and certified in the name of the President," p. 346, *Documentary History of the Ratification of the Constitution*, vol. 18 [*Commentaries on the Constitution, Public and Private*, vol. 6, John P. Kaminski et al. (eds.)] Madison: State Historical Society of Wisconsin (1995.) Emphasis in original.

that was approved by Washington in April 1790 differed significantly from the British system, in ways that favored the rights of inventors. First, the House deleted Section 6, which had imitated the English policy of granting patents for imported inventions. Second, the Senate extended the initial definition of novelty – which referred to inventions that were new within the United States – to cover only discoveries that were new to the world. The laws still employed the language of the English statute in granting patents to "the first and true inventor." Nevertheless, unlike in England, the phrase was used literally, to grant patents for inventions that were original in the world, not simply within U.S. borders. Third, a section regarding interferences (or conflicting applications) was replaced with a clause that ensured information about prior inventions was readily available to potential patentees. Fourth, the Senate suggested forcing patentees to work the patent or else license others to do so, but the House rejected this as an infringement of the patentee's rights. Also, small reductions were made to the fee schedule, which was modest to begin with, so the total fees amounted to no more than $4 or $5.[72]

The careful calibration of these different features resulted in a striking contrast to European patent institutions. Whereas British institutions retained a bias toward power, wealth, and privilege that limited access and participation, the United States early on established a set of socioeconomic and political institutions that generated benefits to a large proportion of the population. For the first time in the world, the exclusive rights of inventors were protected by a commitment in a national constitution, with the declared conviction that this would serve "to promote the progress of science and useful arts." Moreover, secure patent rights were considered to be necessary to stimulate the efforts of inventors. When the centenary of the patent system was celebrated in 1891, the President of the United States proclaimed that "it cannot be doubted by any, I think, that the securing of property in inventions is essential and highly promotive of the advance of our country.... Nothing more stimulates effort than security in the result of effort."[73] Rather than monopolists, patentees were viewed as beneficent contributors

---

[72] See Linda Grant De Pauw, *Documentary history of the First Federal Congress of the United States of America, March 4, 1789–March 3, 1791*, Baltimore: Johns Hopkins University Press (1977), pp. 1631–1637. Also see Bugbee (1967), p. 147, "It will be observed that certain other provisions which had appeared at one time or another in American intellectual property enactments were omitted from the Federal patent and copyright statutes of 1790 – in favor of the inventor or author.... During the term of a patent or copyright legally acquired from the United States, the property of the creative individual in his production was to be absolute."

[73] Benjamin Harrison, in an 1891 speech to commemorate the Centenary of the patent system. Dutton (1984), p. 18, notes that the idea of a natural property right in invention was not generally supported in Britain, which is perhaps inevitable in a system that permits patent protection for mere importation.

to progress, and the consistent goal of those who shaped the system was to encourage domestic ingenuity, whatever the social class of the inventor. By refusing to protect imported inventions, the laws discouraged rent-seeking by monopolists.

Americans had recently fought a revolution to escape the confines of a world view based on inherited privileges to a few, so it is hardly surprising that U.S. doctrines emphatically repudiated the notion that the rights of patentees were subject to the arbitrary dictates of government. The state itself was not privileged above the patentee, for "The title of a patentee is subject to no superior right of the government. The grant of letters patent is not, as in England, a matter of grace or favor, so that conditions may be annexed at the pleasure of the executive....It should be premised that our law differs from that of England as to the right of the government to use, without compensation, an invention for which it has granted letters patent."[74] U.S. courts generally held that the U.S. government had no higher claim than would a private employer. Indeed, some patentees who confronted the federal government were arguably in a stronger position than if the case had been brought against private firms. For instance, private employers were acknowledged to possess shop-rights (royalty-free use of the inventions), but a 1903 statute offered government employees the additional inducement of a waiver of the patent application fees if they agreed to allow the government to use the invention without payment of royalties. In *United States v. Dubilier Condenser Corp.*, 289 U.S. 178 (1933) the federal government claimed ownership of radio-communication inventions that were patented by government researchers who had been expressly employed to conduct research in this field. The court did grant "shop-rights" but rejected the government's petition that the patents should be assigned to the lab, even though the patentees had been paid to research this question. The refusal to apply takings doctrines or eminent domain in the realm of patents underlined the credible commitment by the state toward security in inventive property rights.

In keeping with the consciously utilitarian purpose of early economic and social policy in America, the patent application process was straightforward, and involved impersonal, routine administrative procedures. For the first few years after the Patent Act of 1790 was passed, patent applications were examined by a tribunal comprising the Secretaries of State (Thomas Jefferson) and War (Henry Knox), and the Attorney General (Edmund Randolph.) This practice proved unwieldy and was replaced by a registration system

---

[74] *United States v. Dubilier Condenser Corp.*, 289 U.S. 178 (1933.) According to *Samuel McKeever v. The United States*, 14 Ct. Cl. 396 (1878), "The course of the legislature and the practice of the executive alike forbid the assumption that the United States government has ever sought to appropriate the property of the inventor, or that it has ever asserted an inherent right to do so analogous to that reserved in Great Britain by the Crown."

Illustration 4. Senator John Ruggles of Maine (1789–1874) is credited with being the "Father of the Modern United States Patent Office." (Portrait courtesy of the Thomaston Historical Society, Thomaston, Maine.)

in 1793. Concerns about conflicting rights under such a system caused a committee headed by Senator John Ruggles of Maine to conduct an inquiry into the patent system. As a result, the Patent Act of 1836 instituted significant changes, and the Senator obtained the first patent awarded under the new system. The most significant outcome of the reform was the reinstatement of the examination requirement. Under the new system, each application was scrutinized by technically trained examiners to ensure that the invention conformed to the law, and constituted an original advance in the state of the art. The French had opposed examination in part because they were reluctant to create positions of power that could be abused by officeholders, but the characteristic U.S. response to such potential problems was to institute a policy of checks and balances. Employees of the Patent Office were not permitted to obtain patent rights. In order to constrain the ability of examiners to engage in arbitrary actions, the applicant was given the right to file a bill in equity to contest the decisions of the Patent Office with the further right of appeal to the Supreme Court of the United States.

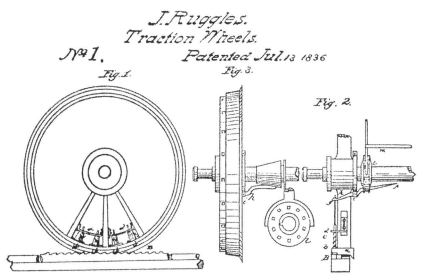

Illustration 5. Patent specification of the first numbered patent issued under the reformed patent system of 1836, made out to Senator John Ruggles, for an improvement in locomotive wheels. (*Source*: U.S. Patent Office.)

The patent system that the United States established in 1836 was the world's first modern patent institution. First, its objective was not to grant or limit monopoly rights and privileges, nor to raise revenues for the state. Instead, it served to promote invention and domestic ingenuity, and to ensure the diffusion of information and innovation. Second, it largely ignored social class and privilege, and routinely granted all inventors the right to property in their discoveries. Later, when the patenting process became more stringent, the criteria for granting the patent depended on the merits of the application and not of the applicant. Third, the administration was straightforward and uncomplicated, and employees of the Patent Office – especially after 1836 – were skilled, specialized professionals. For instance, the Patent Office in 1892 numbered over six hundred employees, including some two hundred specialized technical examiners. As the Commissioner of Patents pointed out in his Annual Report for that year, "there is no similar number of men in the world, gathered into one body, performing duties as delicate and difficult as those performed by the examining corps of the Patent Office," and other countries were deterred from adopting such a system because of a lack of comparable personnel. Fourth, the parameters of the patent institution were established by statute, rather than individual discretion, and this established rational and predictable rules.[75] Finally, it facilitated the operation

[75] Note that Britain passed no major patent statutes between 1624 and 1852 (two minor changes were recorded in 5 & 6 Will. 4, (1835) and 2 & 3 Vict. (1839).) Patent policies were therefore

of an extensive and effective market in the sale and licensing of patent rights.

The legislature ensured a democratic orientation through institutions that were decentralized and market-based. Similarly, by creating a modern patent system, the United States encouraged the participation of a wide range of inventors, including technically unqualified individuals. This feature of the patent system possibly enhanced the value of incremental inventions and small improvements in design and technique. Thus, when markets expanded in the United States during early industrialization, a broad spectrum of the population was in a position to take advantage of the opportunities that were emerging throughout the economy. The emphasis on democratic access is evident in specific features that were most likely to have influenced the path of technological progress, such as the cost of securing a patent grant, criteria for patentability, the availability of specifications of patented inventions, and restrictions on the transfer of patent rights and commercial development. The first of these features, the patent fee, highlights one of the most striking contrasts between the British and American systems. After 1793 American applicants paid a fee of $30 ($35 after 1861) that would remain among the lowest in the world in relation to per capita income.

According to many contemporary observers, "The cheap patent law of the United States has been and still is the secret of the great success of that country."[76] How sensitive were potential patentees to the costs of securing property rights to inventions? Figure 2.1 suggested that British inventors were significantly influenced by the administrative and monetary costs. Indeed, according to the Attorney General of the United Kingdom, one of the objectives of changing to the renewal system in 1852 was to encourage poorer inventors to obtain property rights at a relatively low cost, until the

---

unpredictable, effected on an ad hoc basis by the courts, Law Officers, and the Lord Chancellor (Harding, 1953, p. 4.) As US Supreme Court Justice Baldwin declared in *Whitney v. Emmett*, 29 F. Cas. 1074 (1831): "The silence of the [English] law left a wide field open to the discretion of the courts . . . But in this country the law is more explicit. The Constitution . . . is a declaration of the supreme law of the land . . . which leaves no discretion to the judges to assign or presume any other." In keeping with the consciousness that patents were critical to national policy, reforms in the antebellum period were accomplished and incorporated into U.S. law fairly rapidly, by means of the Patent Law Acts of 1790, 1793, 1800, 1819, 1832, 1836, 1839 and 1842. The 1836 statute set in place the basic system that operated in the twentieth century.

[76] *Scientific American*, vol. 43 (ns) no. 17, p. 256, October 23, 1880, "The Relations of Cheap Patents to Industrial Prosperity"; "The question was taken up at the August meeting of the London Association, and while the inaction of British inventors was admitted and deplored, the blame was traced to the working of the British patent system. Said the essayist of the occasion, Mr. John Standfield: "The chief cause of our commercial suffering and stagnation is a barbarous law. . . . The cheap patent law of the United States has been and still is the secret of the great success of that country." In the subsequent discussion this point was dwelt upon at great length."

value of the invention became more evident and they were better able to interest potential financiers.[77] These changes reduced the costs of obtaining a patent, not least because consolidating the reviews into one office simplified the process considerably. As a result, the number of patents issued jumped from 455 in 1851 to 1,384 in 1852 and 2187 in the following year. In 1883, the initial fee was lowered to £4, although the cost was approximately the same if the patent were taken to term, and applications again surged from 5,993 in 1883 to 17,100 in 1884.[78] Thus, the evidence supports the view that inventors were sensitive to the cost of obtaining patent protection.

From the patentee's perspective, the high fees and procedural costs of the British system may have been offset by the lack of any further examination of applications for novelty or utility. It is possible that, after the introduction of the examination system in 1836, U.S. patentees incurred higher costs in order to prepare their patent application for the examination. What is essential, however, is that although a registration system with high patent fees and an examination system both apply filters to the population of potentially patentable inventions, the samples of inventions and inventors that survive the filter are likely to differ in important respects. In particular, the U.S. examination system selected on the basis of novelty and contribution to knowledge, whereas the English registration system favored applicants who already had access to the substantial capital outlay required simply to obtain the patent. In markets with complete information an inventor with a valuable idea would be able to raise sufficient capital, but in general it might be expected that uncertainty about the value of a new invention, coupled with asymmetric information, were likely to pose significant obstacles. The bias also affected the distribution of British patents, which was more skewed toward high-value, capital-intensive inventions. The effective result was that a broader segment of the population in the United States was able to secure property rights in a wider array of inventions, which thus generated greater incentives for inventive activity.

The two patent systems also differed in their interpretation of the requirements for novelty in the patent grant and in their policies toward imported inventions. English judges pronounced that patents should be granted for "new manufactures within this realm ... whether learned by travel or by

---

[77] Under the renewal system, registration fees for a British patent fell to £25 in the first year; patentees who extended the term of the patent in the third year paid an additional £50, and £100 more to maintain the patent past the seventh year. According to the Attorney General, a renewal system would "give poor men an opportunity which they did not at present possess, of protecting inventions till such time as they might be able to derive advantage from them, and at a cost adapted to their means." See *Hansard* (1851), p. 1537.

[78] Fees were £4 for the first application; £50 to maintain the patent after the fourth year; and £100 due before the end of seven years. See Patents, Designs and Trademarks Act, 46 & 47 Vict. Chpt. 57 (1883.)

study, it is the same thing."[79] Patents were granted to importers of inventions from overseas as well as to original inventors, a policy which certainly must have favored members of the commercial and elite classes, who were more likely to be familiar with developments in other countries. Moreover, this practice implied that individuals with sufficient capital to obtain a patent for foreign technology could legally constrain competitors. Patents were not examined for novelty and it was difficult for specifications to be searched to ascertain the prior arts, so it is likely that a number of patents were not strictly valid. American laws employed the language of the English statute in granting patents to "the first and true inventor," but Congress deliberately and repeatedly rejected the option of patents for importations.[80] Moreover, unlike in England, the phrase was used literally, to grant patents only for inventions that were original to the world, not simply within U.S. borders.[81] This policy of a global definition of novelty had at least two important outcomes. First, in cases where foreign inventors did not seek U.S. patents,

[79] *Edgeberry v. Stevens*, 1 Web. P.C. 35 (1691.) Colonial American patent grants, like British patents, included imported inventions. See Bugbee (1967.)

[80] An amendment ordered on December 9, 1790 [HR-121]. Received and read February 7, 1791. Vol. vi: *Legislative Histories*: text of patents bills 41 and 121, Patents Bill [HR-41], February 16, 1790:

> "Sec. 6: *And be it further enacted*, That any person, who shall after the passing of this act, first import into the United States from any foreign country, any art, machine, engine, device or invention, or any improvement thereon, not before used or known in the said States, such person, his executors, administrators and assigns, shall have the full benefit of this act, as if he were the original inventor or improver within the said States. [p. 1631] [fn 42, p. 1631: "The House struck out this section."]

> Throughout the nineteenth century, admirers of British institutions submitted bills to grant importers patent rights, which consistently failed to pass. See for instance, Journal of the Senate of the United States of America, 1789–1873, February 27, 1795: "The memorial of Rd. Claiborne, was presented and read, praying that "a law may pass, authorizing the importation of inventions, and allowing to original importers a certain privilege in proportion to that allowed by the patent law to inventors." Ordered, That this memorial lie on the table."

[81] This question was settled early on: "The inventor must be the original inventor as to all the world, to be entitled to a patent." See *Reutgen v. Kanowrs*, 1 Wash. 188 (Pa) 1804; *Dawson v. Follen*, 2. Wash. 311 (Pa) 1808; *Lowell v. Lewis*, 1 Mass. 190 (Mass.) 1817. To be practicable, this meant (and still means) that the invention had not been described in a printed publication or patented overseas. If it had not been published or patented, the deciding criterion was whether the inventor knew about the foreign invention or had come up with the idea himself. If the former, the patent was void; if the latter, the patent was valid. According to *Gayler v. Wilder*, 51 U.S. 477 (1850), "if the foreign discovery is not patented, nor described in any printed publication, it might be known and used in remote places for ages, and the people of this country be unable to profit by it. The means of obtaining knowledge would not be within their reach; and, as far as their interest is concerned, it would be the same thing as if the improvement had never been discovered. It is the inventor here that brings it to them, and places it in their possession. And as he does this by the effort of his own genius, the law regards him as the first and original inventor, and protects his patent, although the improvement had in fact been invented before, and used by others."

Americans had free access to the benefits from foreign technology. Second, American inventors had a greater incentive to create useful improvements in borrowed technology (for which a patent could be granted) rather than simply acquiring monopolies over existing inventions from overseas.

The earliest U.S. statutes restricted rights in patent property to citizens or to residents who declared their intention to become citizens. The first patent statute of April 1790 made no distinction regarding citizenship, but in 1793 patents were limited to citizens of the United States. In 1800 patent rights were extended to foreigners living in the United States for two years who swore that the invention was new to world. The gradual easing of restrictions on citizenship continued with the 1832 Patent Act, which allowed patents to resident aliens who intended to become citizens, provided that they introduced the invention into public use within one year, a period that was changed to eighteen months with the 1836 reforms. In 1836, the stipulations on citizenship or residency were removed, but were replaced with discriminatory patent fees that retaliated for the significantly higher fees charged in other countries: nonresident foreigners could obtain a patent in the United States for a fee of $300, or $500 if they were British. After 1861, patent laws that stipulated discriminatory treatment of foreign nonresidents were repealed, and utility patent rights (with the exception of caveats) were available to all applicants on the same basis without regard to nationality. Although in the antebellum period the United States varied in its treatment of foreign inventors, its provisions were much more favorable toward aliens than was true of other countries. Morever, it should be noted that, despite the statutes, numerous inventors who lived in other countries applied to Congress for exemptions, and were granted the right to apply for U.S. patent rights.[82] This liberality to foreign residents was obtained at low cost, as for

---

[82] For a small sample of such foreign inventors, see: Journal of the Senate of the United States of America, 1789–1873, March 28, 1816: The bill entitled "An act authorizing and requiring the Secretary of State to issue letters patent to Andrew Kurtz;" An act authorizing the Secretary of State to issue letters patent to Henry Burden, Journal of the House of Representatives of the United States, 1819–1820, April 12, 1820; Bill S. 38, 16th Congress, 1st Session, February 17, 1820: A Bill Authorizing the Secretary of State to issue letters patent to Richard Willcox; H.R. 57, 20th Congress, 1st Session, January 11, 1828, "An Act for the relief of Simeon Broadmeadow;" a petition of Richard Holden Approved by the President, Journal of the Senate of the United States of America, 1789–1873, May 8, 1822;18th Congress, 1st Session, H.R. 230, May 26, 1824, Congress passed a law, authorizing the Department of State to issue a patent for a similar invention, to a certain Nathaniel Sylvester; Christopher Bechtler, praying that letters patent may be granted him for invention of two new and useful machines for the purpose of washing gold ores, without requiring the two years' residence as is by law now required. Approved by the President of the United States on March 3, 1831, Journal of the Senate of the United States of America, 1789–1873; Anthony Hermange, an alien, residing in the City of Washington, and Paul Steenstrup, of Kongsberg, in the Kingdom of Norway praying that letters patent for an invention may be granted, Approved by President May 1, 1828. However, not all petitions were granted; see for instance, Peter Hirgoyen's petition to be able to apply for a patent for a distilling apparatus (Journal of the House of

most of the nineteenth century patenting in the United States by overseas inventors was trivial relative to the total.

American inventors were more likely to file for patents themselves because the application procedure was straightforward. The complexity of the British system enhanced the role of middlemen who specialized in activities that reduced transactions costs for potential patentees. The overwhelming majority of British patentees employed patent agents who provided information and advice, and channeled the invention through the bureaucracy for a charge of £40 to £100 in addition to the patent fees.[83] This industry was oligopolistic in structure, for a small number of agents – such as the Newton family, Moses Poole, and Pierre Fontainmoreau – dominated the patent agency business. Some of the more active patent agents were at the same time employed by the Patent Office, including Newton and Co., who had 192 patents to their credit, and Moses Poole, who obtained over 100 patents in his own name, and allegedly granted his own clients preferential treatment. Between 1816 and 1852, English patent agents obtained 537 patents in their own names, mostly on behalf of their clients. As stated before, patent agents and Patent Office officials were among the most vociferous and determined opponents of the 1852 patent reforms.[84]

Patent agents were more critical to British patent applicants in part because of their access to information. Before the middle of the nineteenth century, inventors could not readily obtain copies or see the descriptions and

Representatives of the United States, 1817–1818, January 21, 1818.) See also "The memorial of J. Burrows Hyde, praying that all foreign inventors may be placed on an equal footing as it regards the obtaining letters patent for their inventions in the United States," which was referred to the Committee on Patents: Journal of the House of Representatives of the United States, 1847–1848, April 17, 1848; Journal of the House of Representatives of the United States, 1853–1854, March 7, 1854.

[83] "The great majority of patents are still however procured through the instrumentality of Patent Agents, ... The proportion obtained in the year 1852–3, by Patentees or Solicitors, to that by Patent Agents, was one to fifty," according to John Coryton, *A treatise on the law of letters-patent: for the sole use of inventions in the United Kingdom of Great Britain and Ireland*, Philadelphia: T. & J. W. Johnson (1855), p. 147. Most British patent agents were located in London, in closer proximity to the patent offices (Dutton, 1984.) In the United States, the Patent Office itself disseminated information and forms. Copies of prior patents could be obtained by writing to the Patent Office, which charged 10¢ per hundred words. The Secretary of State published lists of expired patents in the Washington newspapers. Patent agents could be found in most major cities in the United States, from San Francisco to Boston. Competition among numerous patent agents kept fees low: for instance, in the 1860s Munn and Co., the largest patent agency, charged $5 for a patent search and $1 to copy patent claims. Rather than negotiating with a bureaucracy, American patent agents offered advice on the technical merits of the invention, drafted technical drawings, and served as mediators between buyers and sellers of technology. For, as *Scientific American* [(1842) p. 62] – published by Munn & Co – noted, "we advise every inventor who is able, to make application for himself, and thereby save some expense. There are forms and rules that will require study, but you can soon master them."

[84] See Harding (1953); Dutton (1984.)

specification of patents that were previously granted.[85] The patent documents were stored haphazardly in three separate offices, to which an admittance fee was charged. The fee for simply reading a patent was two and a half shillings, and the cost for a copy of the patent varied between 2 and 40 guineas in 1829.[86] Specifications were initially shielded partly to prevent foreign competitors from acquiring British technology, and partly because the initial objective of administrators was to prevent monopolies through disclosure.[87] Undoubtedly, technical diffusion also was constrained by the lack of access to information. Lower diffusion would tend to increase the value of property rights belonging to those who were able to obtain a patent, and reduce the likelihood of further developments in the area by other inventors. In contrast, American legislators were concerned with ensuring that information was readily available, and even considered ruling that "Copies of such Specification together with similar Models to be made at the public Expence, and lodged in . . . each State."[88]

The ease of access to information was likely to influence the geographical distribution of inventive activity, which is typically viewed as localized in urban areas because of such advantages. However, rural inventors in the United States could apply for patents without significant obstacles, because applications could be submitted by mail free of postage. The U.S. Patent Office also maintained repositories throughout the country, where inventors could forward their patent models at the expense of the Patent Office. As such, it is not surprising that much of the initial surge in patenting during early American industrialization occurred in rural areas. The English pattern was quite different, for patents were awarded predominantly to residents of cities, particularly to patentees with addresses in London.[89] Between 1617 and 1852, London patentees consistently received the majority of patents, followed by Manchester as a distant second; at the other end of the spectrum were some thirteen hundred towns and villages with only one patent

---

[85] Specifications were not generally published in England until 1852. The various patent offices concerned prohibited extracts of key aspects of the invention (Select Committee, 1829, *Parliamentary Papers.*) Trade journals such as the *Repertory of Arts, Manufactures and Agriculture* did publish patent descriptions. However, these published lists tended to be incomplete and the length of time between patent grant and publication varied between five months and five years (David J. Jeremy, *Transatlantic industrial revolution: the diffusion of textile technologies between Britain and America, 1790–1830s*, Oxford: Basil Blackwell, c. 1981, pp. 47–49.)

[86] Jeremy (1981), pp. 46–47. Jeremy also notes that descriptions were filed haphazardly and that searches could take hours. In the 1820s an employee of the Patent Office and a patent agent on his own account, had compiled an index, but it remained "under Poole's personal control," and probably contributed to the value of his services as a patent agent.

[87] See MacLeod (1988), p. 51 for a discussion of the development of the specification requirements for patentability.

[88] See HR-41 Bill in de Pauw (1977) for details.

[89] See MacLeod (1988) and Dutton (1984.)

each.[90] The extreme geographical concentration partially reflected the difficulty of negotiating from outlying areas the complex procedures required to apply for a patent. It may be expected that the wealthy, who could afford to travel to supervise the process, or to retain agents to do so, were less disadvantaged by their distance from the capital. This provides yet another instance in which the English system not only increased costs in general but ultimately distributed them unequally across different segments of the population.

Policies that affected the means of appropriating returns from the patent constitute another important feature of the patent institution whose impact might vary across social class, and influenced the overall incentives for inventive effort. After obtaining a patent, the patentee typically attempted to pursue profit opportunities. Alternative strategies for doing so included assigning (or selling) the property rights in total or in part, licensing the technology, exploiting the patent directly as a sole proprietor or partner of an enterprise that used the invention, or some combination of these approaches. But the patentee's ability to derive returns from these strategies also depended on the patent laws and other aspects of the legal system. These, too, varied between the two countries. The U.S. examination system reduced uncertainty about the validity of patents, and economies of scale in certification provided the inventor with a lower-cost signal of potential value than was true of a registration system with private investments in certification. Centralized examination helped to resolve problems of asymmetrical information and, by helping to create tradeable assets from inventions, enabled financially disadvantaged inventors to appropriate returns through the market for invention.

English law initially limited the number of individuals who could share in the patent rights to a rather small group, whose size could only be increased by a private Act of Parliament The legal system also prohibited the patentee from subdividing the right across geographical regions. Such strictures made it more difficult for inventors to raise capital from outside investors to cover the patent application costs, or to commercially exploit the invention, and one would suppose that it constrained the working class to a greater extent.[91] In contrast, U.S. inventors routinely traded patent rights that were divided and subdivided among numerous assignees.[92] By the 1870s the number of

---

[90] Harding (1953), p. 13. See MacLeod (1988) for an analysis of the geographical concentration before 1800.

[91] In the 1860s, assignments in England varied between six hundred and seven hundred per annum. The law limited the number of assignees to five (twelve after 1832.) At least until 1830, it was not clear whether this limit extended to licensees (Edward Holroyd, *A practical treatise of the law of patents for inventions*, London: Printed by A. Strahan, for J. Richards (1830), p. 145.) This law in effect also restricted financing for inventions, as *Duvergier v. Fellows*, 10 C.B. 826 (1830) indicates.

[92] See William Newton's printed bill that warned potential infringers in 1815: "As attorney in fact for Michael Withers, the patentee of cast iron wing gudgeons; I do hereby give all

assignments averaged over nine thousand assignments per year, and this increased in the next decade to over twelve thousand transactions recorded annually. Patentees were thus able to raise capital to support their inventive activity and were better positioned to extract material returns from their efforts.[93] This flourishing market for patented inventions provided an incentive for specialization and for further inventive activity by inventors who were able to appropriate the returns from their efforts. The ability for other parties to purchase the rights to own or use inventions also enhanced the link between patents and productivity growth.

The American legal system further encouraged the evolution of trade in patent rights by protecting assignees and investors against fraudulent patents – unlike England, where payments for invalid patents could not be recovered.[94] The ability to profit thus depended on legal institutions and the attitude of the courts, which influenced investors as well as current and potential infringers. The development of the British system was shaped by judges, rather than by statute, and the courts appear to have been at best inconsistent, if not hostile, toward patentees.[95] Partly because of the registration system, judges claimed broad scope over the determination of the validity of patents, and this may have increased uncertainty regarding property rights, and made it more costly to enforce the patent.[96] By contrast, as the next chapter shows, American judges consistently attempted to implement the spirit of the Commerce Clause of the Constitution and the subsequent patent statutes, in protecting the rights of "meritorious patentees." In short, in this regard as in others, American patentees submitted their inventions to

persons public notice who use Michael Wither's invention or improvement in the said patent gudgeons, that the price is thirty dollars per pair, with interest from the date of the patent until paid, by suit or otherwise, without license had for using the same, from under my signature, or other agents authorised by me or said Withers. – See laws of Congress, vol. 2, page 200, an act to promote the progress of the useful arts; and to repeal the act heretofore made for that purpose." Library of Congress, Printed Ephemera Collection; Portfolio 244, Folder 30.

[93] In 1845 the US Patent Office recorded some 2,108 assignments. For an extensive study of the market for inventions during the Second Industrial Revolution, see work by Naomi Lamoreaux and Kenneth L. Sokoloff, including "Inventors, Firms, and the Market for Technology in the Late Nineteenth and Early Twentieth Centuries," in Naomi R. Lamoreaux, Daniel M. G. Raff, and Peter Temin, eds., *Learning By Doing in Markets, Firms, and Countries*, Chicago: University of Chicago Press (1999): 19–57.

[94] If a British patent was useless, the purchaser could not recover his payment: *Hall v. Conder*, 2 C.B., N.S. 22 (1857.) Assignees or licensees of American patents that lacked utility were entitled to refunds of their payments.

[95] See MacLeod (1988); Dutton (1984.)

[96] According to Lord Coke, an invention should "not be mischievous to the State" nor "hurt trade," (Coryton, 1855, pp. 60–61.) Patents could be invalidated because of lack of utility, and according to Hill v. Thompson, Web. P.C. 237 (1817) "the utility of an invention is a question for the jury." Importers of inventions were at times deemed to be of lower status than original inventors, as In re. Soames' Patent, 1 Web. P.C. 733 (1843.)

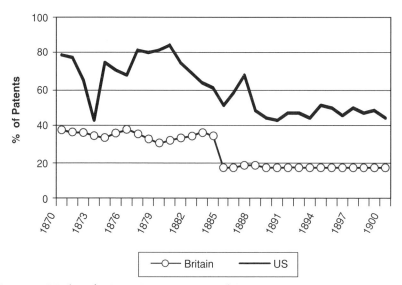

Figure 2.3 Markets for Invention in Britain and America, 1870–1900
(Ratio of Assignments to Patents in the United States, and the Ratio of All Assign-
ments and Licences to Patents in Britain)
Notes and Sources: United States Assignments estimated from U.S. Patent Office, *Annual
Report of the Commissioner of Patents,* Washington, D.C.: GPO, various years. Britain: The
series represents the ratio of assignments AND licences filed as a fraction of total patents sealed
in the same year. Great Britain Patent Office, *Annual Report of the Commissioners of Patents,*
London: HMSO, various years (Comptroller-General after 1883.)

an arena in which all participants were provided with relatively equal, low
cost, access.

The evidence supports the idea that patent institutions made a difference.
First, Figure 2.1 compared the rates of patenting per capita in the two countries, and showed that patenting per capita was markedly and consistently
higher in the new Republic. The three short intervals during which England
rivaled the United States in patenting were just after the former had lowered
fees or the latter introduced its examination system. Second, the distribution
of patent grants per person provides another perspective on the impact of
the more restrictive granting of patents in England.[97] Patent-holding was
much more concentrated in England than in the United States. For example,
between 1812 and 1829, only 42.9 percent of English patents were granted
to individuals who received a single patent over their career, compared to
57.5 percent in the United States.[98] Third, the same qualitative pattern

[97] See B. Zorina Khan and Kenneth L. Sokoloff, "Two Paths to Industrial Development and
    Technological Change," in *Technological Revolutions in Europe, 1760–1860,* Maxine Berg
    and Kristine Bruland (eds.), London: Edward Elgar (1998.)
[98] See the discussion in Chapter 4.

Table 2.1. *Sectoral Distribution of Patents in Britain and America*

	Britain		America	
	No.	%	No.	%
Agriculture	190	4.8	1009	22.3
Construction	399	10.0	753	16.7
Manufactures	2083	52.4	1812	40.1
Transportation	830	20.9	580	12.8
Other	469	11.8	361	8.0
Total	3972		4515	

Notes and Sources: The British sample was randomly selected from Bennet Woodcroft, *Alphabetical index of patentees of inventions*, [from March 2, 1617 (14 James I) to October 1, 1852 (16 Victoriae) New ed.]; London: Evelyn, Adams & MacKay, 1969, and comprises 26.7 percent of all patents issued between 1790 and 1852. The American sample accounts for 29.6 percent of all patents issued between 1790 and 1846. It is described in Kenneth L. Sokoloff, "Inventive Activity in Early Industrial America: Evidence From Patent Records, 1790–1846," *Journal of Economic History*, Vol. 48 (December) 1988: 813–850.

emerges from an analysis of the occupation of patentees over this period.[99] English and French patentees were more likely to belong to the relatively elite classes ("gentlemen," merchants, and professionals) and the occupational composition of patentees and patenting rates changed quite slowly over time. Differences in effective access to the use of the patent system and to commercial opportunities generally must play a large role in explaining the substantial share of English patents going to "gentlemen" (generally between 20 and 30 percent over the period) and to titled Frenchmen.

Table 2.1 illustrates the degree of sectoral concentration among patents in Britain relative to the United States. Patenting in Britain appears to have been more directed at capital-intensive industries such as transportation than was the case in the United States. The difference in patenting parallels the contrast between the two countries in the degree of balance across manufacturing industries in productivity growth. American manufacturing experienced productivity gains that were diffused across a wide array of industries, whereas British improvements were concentrated among a few key industries.[100] Inventions that were incremental, aimed at labor-intensive

---

[99] Occupation data are available for most English patentees, because it was required on applications.

[100] See Kenneth L. Sokoloff, "Invention, Innovation, and Manufacturing Productivity Growth in the Antebellum Northeast," in Robert Gallman and John Wallis, eds, *Growth and Standards of Living Before the Civil War*, Chicago: University of Chicago Press (1992):345–378.

industries, and discovered by ordinary workers who were confronting problems that were related to their occupation, were more accordingly more likely to be patented in the United States than in England.[101] The difference in institutions meant, at the minimum, that commercial opportunities at invention and innovation were less available to undistinguished individuals in England than in the United States, with possible implications for relative degrees of equality and social mobility. The English patent system may have led to lower rates of invention (or different types of inventions) and not just of patenting, if the more restricted provision of property rights meant that the expected returns to inventive activity were lower for a significant group of potential inventors, or for certain types of inventions. Yet other reasons why the European policies were less effective in generating broad-based technological progress include the difficulty in revising institutions if strong interest groups have a vested interest in preventing changes that redistribute resources away from themselves, however great the overall social benefit that are likely to result from reforms.

Economic theorists have supported such European alternatives to the U.S. patent system as prizes and state-sponsored rewards.[102] Their models typically pay little attention to the pragmatic or distributional consequences of nonmarket alternatives to the impersonal market-oriented grant of property rights that the United States introduced. Such consequences included higher transactions costs, greater uncertainty, greater potential for corruption, and less flexibility in responding to environmental changes. In France, some inventors were diverted into unproductive rent-seeking because of the array of state-mandated rewards and prizes, whose allocation was based on arbitrary criteria and created opportunities for corruption or favoritism. The state at times duplicated returns that inventors obtained elsewhere, and possibly led to overinvestment in certain areas of research. European patent systems encapsulated the same privilege mentality, for they favored the wealthy or well connected, and supported interest groups that had an incentive to block progressive changes. British inventors focused on technologies that were highly capital intensive or otherwise perceived ex ante to be more valuable, arguably leading to unbalanced growth. The high cost of French and British patent rights meant that their ownership was more concentrated (unequally distributed) than in the United States. The Europeans tended to

---

[101] Indeed, in eighteenth-century England "patenting of mere improvements was frowned upon," as MacLeod (1988, p. 13) points out.

[102] The prize for determining longitude at sea is a favorite example that is often adduced to support theoretical models of prizes, which is somewhat ironic. The harrowing details of John Harrison's trials in trying to collect his just returns provides proof of the pragmatic inefficiency of nonpatent rewards, the potential for corruption, and the likely distributional consequences. Part of Harrison's travails owed to the bias of the elite members of the Longitude Board against an uneducated working-class clockmaker from the provinces. See Dava Sobel, *Longitude*, New York: Walker and Company (1995.)

regard patent policies as part of an overall strategy of mercantilism, and the French actually prohibited inventors from patenting overseas to ensure that foreigners were less able to compete with French manufacturers. The British and French would later exert direct influence on the patenting systems of their colonies and other "follower countries" such as Spain. Within this global institutional environment, the American adherence to a flexible, decentralized, and democratic approach was exceptional, and worth considering more closely.

3

Patent Laws and Litigation

"I knew that a country without a patent office and good patent laws was just a crab, and couldn't travel any way but sideways or backwards."
– Mark Twain, *A Connecticut Yankee* (1889)

Thomas Paine felt that in the United States "the law is king" and Alexis de Tocqueville similarly argued that American courts wielded enormous political power, because "scarcely any question arises in the United States which does not become, sooner or later, a subject of judicial debate."[1] American legal institutions exerted a significant influence on social and economic development, and technological change was no exception. The primacy of legal institutions in American technology policy was deliberate. The statutes from the earliest years ensured that the "progress of science and useful arts" was to be achieved through a complementary relationship between law and the market in the form of a patent system. Numerous proposals were repeatedly submitted to Congress throughout the nineteenth century to replace the patent system with more centralized systems of national prizes, awards, or subsidies by the government.[2] Such proposals failed to persuade, because the process of

[1] Alexis de Tocqueville, *Democracy in America*, ed. J. P. Mayer, trans. George Lawrence, 1835; New York: Harper & Row (1969), Vol. I, Chap. 8, p. 311.
[2] For instance, Charles B. Lore of Delaware submitted H.R. 5,925 in 1886 to set up an alternative system of rewards for inventors, to be administered by an "Expert Committee." As the editors of *Scientific American* pointed out [*Scientific American*, v 54 (14), p 208, 3 April 1886], "The Expert Committee would have a very delicate duty to perform in fixing the cash valuations, and they would constantly be subjected to risks and probabilities of making egregious errors. For instance, if they were to allow $10,000 as the value of the patent for the thread placed in the crease of an envelope to facilitate opening the same, how much ought they to allow for the second patent, that was granted for the little knot that was tied on the end of the thread, so that the finger nail could easily hold the thread? Then, again, how much ought the committee allow for a simple device like the patent umbrella thimble slide, a single bit of brass tubing that costs a cent and a quarter to make? Probably the committee would think that one thousand dollars would be a most generous allowance, while two hundred thousand dollars – the limit of the bill – would, of course, be regarded as a monstrous and dishonest valuation. But the real truth is, the patent for this device is actually worth nearer one million dollars than two hundred thousand. The inventor, Dr. John J. Higgins, of this city, has already received over one hundred thousand dollars cash in royalties for his patents, and probably will receive three times that sum before they expire; while the licensees, the

democratization was most likely to be attained through decentralized decision making by inventors themselves, and through enforcement by judges confronting individual conflicts on a case by case basis. The wish to further technological innovation through private initiative created a paradox: in order to promote diffusion and enhance social welfare, it would first be necessary to limit diffusion and to protect exclusive rights. Thus, part of the debate about law, patents, and technology has always centered on the boundaries of the private domain relative to the public domain.

Patents and copyrights, as the subject of federal law, exhibited greater uniformity than if under state jurisdiction, and thus facilitated the development of a national market. Intellectual property law had a direct effect on the rate and direction of inventive activity and cultural innovations. As the creators of the patent system recognized, inventors would be motivated to address important needs of society if they were able to appropriate returns from their efforts. Patent laws ensured the security of private property rights in invention and facilitated the development of extensive and deep markets in such rights. However, it also was necessary to balance the just claims of inventors against the dangers of exclusive monopolies that would restrict the scope of current and future cumulative innovation. As the next chapter shows, legal rules and doctrines influenced the identities of inventors and the nature of their inventions. For instance, relatively low patent fees served to encourage ordinary citizens to invest in creating new discoveries, whereas an examination system increased the average technical value of patents, promoted a market in inventions, and encouraged the diffusion of information. Technology also was shaped by other areas of property law, as well as by rules regarding contract, torts, crime, and constitutional issues.

The relationship between law and technology was reciprocal for, just as law shaped technology, technical innovations significantly influenced legal innovations. How and why the common law changed constitutes a standard debate in political and legal histories. A classic source of dissension relates to the arguments of scholars who agree that American legal institutions were flexible but contend that the judiciary was captured by the interests of a small group in society. Substantial debate similarly occurs over the nature and efficacy of judicial influence in conflicts about patent rights. The attitudes of the judiciary were relevant, because if courts were viewed as "antipatent" this would tend to reduce the expected value of patent protection. A common view is that the early legal system was hostile toward patent grants and judges routinely overturned patent rights, especially those involving important inventions. Popular histories highlight the allegedly tragic experiences

---

umbrella makers, are supposed to have already realized a million dollars' profits directly or indirectly arising from the control of this little article." The cited inventor, John J. Higgins of New York, obtained two patents for improvements in umbrella runners (No. 189859 (1877) and No. 157915 (1874)) as well as several others for photographic equipment.

of Eli Whitney, Oliver Evans, and early steamboat inventors.[3] Patent laws might have been framed to be favorable to inventors, but if patentees could not enforce their rights in court, then patents would comprise little more than worthless pieces of paper. Serious inventors would opt for alternative methods of protection, such as secrecy, or exploiting strategies based on lead-time and the learning curve. Under such circumstances, an index of patented inventions would be of little relevance to charting and understanding the progress of technology, because the value of a patent right to the individual inventor is determined partly by the value of the invention and partly by the "legal tender" vested in the patent grant. Thus, the market value of patented inventions depended on the attitudes of the judiciary, as well as on the ability of the legal system to defend the right of patentees to use, and exclude others from using, their property.

How enforceable were property rights in patents in the early years of the patent system? My approach is to examine all 795 reported antebellum patent cases, in order to assess the legal doctrine and standards that prevailed at the inception of the United States patent system, and the strength of property rights in patents.[4] Patent litigation records can give us some insight into

---

[3] On examination, the claim that these "great inventors" were unable to enforce their patents because of an unjust system is disputable. Eli Whitney had attempted to extend his monopoly of the patented invention to the output market for cotton: "The unfortunate arrangement of Whitney and Miller, toward the close of the year, to erect gins throughout the cotton district, and engross the business of ginning for a toll of one third, instead of selling the machines and patent rights, stimulated the spirit of infringement," according to J. L. Bishop, *A History of American Manufactures from 1608–1860*, Philadelphia: E. Young (1868), p. 49. Oliver Evans's patent – extended by Congress for a full term of thirty-two years – claimed that he invented the automated grist mill, although his invention was merely for an *improvement* of already existing automated grist mills. Similarly, early steamboat patents failed to distinguish between the state of the art and the inventor's contribution. Both Eli Whitney and Oliver Evans, contrary to anecdotal evidence, prospered from their business deals. Whatever the merits of these examples, although individual histories yield interesting and useful insights, a few exceptions, however notorious, provide insufficient evidence to prove or disprove a theory.

[4] The data set was compiled from: *The National Reporter System, Federal Cases in the Circuit and District Courts of the United States, v. 1–31, 1789–1880* St. Paul, 1894; Samuel Fisher, *Fisher's Patent Cases*, Cincinnati, 1868; James Robb, *Robb's Patent Cases* (vols. 1 and 2), Boston, 1854; *Brodix's American and English Patent Cases*, vol. 4 (1754–1847); and Charles Whitman, *Whitman's Patent Cases Determined in the United States Supreme Court*, Washington, D.C., 1875–1878; Elgar Simond, *Simond's Digest of Patent Cases in the Federal and State Courts, 1789–1888*, New York, 1888; William G. Meyer, *Federal Decisions: Cases Argued and Determined in the Supreme, Circuit and District Courts of the United States*, St. Louis, Mo., 1886; and Stephen Law, *Digest of American cases relating to Patents for Inventions and Copyrights, including Numerous Manuscript Cases, 1789–1862*, New York, 1877.
These data are indicative of general trends, but are necessarily limited to reported decisions and the few manuscript cases available or mentioned in court citations. However, in most states the law stipulated that decisions of the higher courts were to be reported, and in 1839

the effectiveness of the patent system if their limitations are recognized.[5] A careful reading of all lawsuits shows that, at this critical juncture of American development, the courts recognized the value of individual property in ideas, and helped to create a patenting process that was sustained by the belief that property rights in inventions would be protected by legal unformity and certainty. Patent rights were abridged in England because they were associated with undeserved privileges, whereas in this country patentees were deemed to be meritorious because true inventors alone were rewarded with patents. Consequently, one of the major concerns of the United States legal system was to identify and enforce the rights of the first and true inventor, rather than to protect public welfare by deterring private monopolies. Judicial doctrines were flexible and responded to changes in the organization and progress of technology. Effective policies toward innovations, whether by statute or common law, required a balancing of costs and benefits that was far more subtle than a monolithic promotion of the interests of any one specific group in society.

PATTERNS OF PATENT LITIGATION

Legal institutions formed an integral part of American life, and were deliberately designed to facilitate the efficient operation of the patent system and markets in invention. Jurisdiction of patent law and litigation was entrusted

the *American Jurist* opined that "within the last 50 years, no country has done so much in this department of juridical literature, as the United States." Individual reporters, rather than the state system, published most of these reports; consequently, the cases of more general interest tended to be published, whereas some cases that were privately circulated and cited existed only in manuscript form. (In the current chapter, these unreported decisions are represented in aggregate statistics, but not in the tables that involve details about the lawsuit.) The decisions of the Supreme Court itself were written since its inception by presiding justices, but variations do exist in the quantity and quality of reporting in the different circuits – ranging from exceptional in the first circuit, to chaotic in the newly acquired territories. The circuit courts (1802 to 1866) comprised the following: first circuit – Maine, New Hampshire, Massachusetts, Rhode Island; second circuit – Vermont, Connecticut, New York; third – Pennsylvania, New Jersey; fourth – Delaware, Maryland; fifth – Virginia, North Carolina; sixth – South Carolina, Georgia; seventh – Kentucky, Tennessee, Ohio (from 1837 to 1866 the seventh circuit consisted of Ohio, Michigan, Indiana, and Illinois.) The terms "plaintiff" and "patentee" are used interchangeably in the text, although they are not identical: the patentee was defendant in 4.3 percent of the 580 cases for which information was available. The data take this into account in assessing quantitative outcomes.

[5] Reported lawsuits may be unrepresentative of the entire population of conflicts about patent rights. Typically only a small proportion of all disputes ever reach the courts, because it is cheaper to settle cases before going to trial. Reported cases tend to deal with issues where the greatest uncertainty exists about the outcome of the trial. Models of law and economics show that quantitative evidence about the proportion of favorable plaintiff decisions cannot be used to make inferences about the attitudes of the courts. Still, lawsuits provide a rich store of information for historians who are willing to make the rather significant investments that are necessary to accommodate the flaws in litigation as a data source.

to the federal courts because of the prevalent concern with fostering inter-state commerce and national markets.[6] Consistent regional decisions would serve to increase the value of holding a patent, by expanding the coverage of the patent to a much wider market and eliminating the uncertainty and costs of enforcement if litigation were governed by the laws of individual states. In the earliest years of the Republic, the federal courts were little used, except for admiralty and maritime cases.[7] The first patent case on record was for the district of New York, when a patent granted to Benjamin Folger for the production of candles was repealed in August 1792. This did not deter Folger from subsequently obtaining a patent for pumps in 1804. The case is a precedent for a similar pattern in later years, when unsuccessful litigants still proceeded to file further patents. From 1790 to 1860 a total of 795 patent cases were reported or cited in federal judicial decisions.[8]

Between 1793 and 1836, American inventors could obtain a patent simply by paying a one-time filing fee, with contested claims settled through litigation. By the 1830s, the Ruggles Senate Committee claimed in a much cited report that the registration system was untenable, because "Out of this interference and collision of patents and privileges, a great number of lawsuits arise, which are daily increasing in an alarming degree, onerous to the courts, ruinous to the parties, and injurious to society."[9] Americans have always been inclined to view themselves as an exceptionally litigious society, but the data on the proportion of total patents litigated before the federal courts is similar to the pattern for England, as Table 3.1 shows. In both countries there is a jump in the number of lawsuits in the fourth decade, but the

---

[6] The Supreme Court noted in *Allen v. Blunt et al., 1 F. Cas. 450* (1846): "One of these ends undoubtedly is, uniformity in the construction of our great system of Patent Laws throughout the U.S., and because questions connected with the Patent Laws themselves, when decided, govern numerous other cases and much larger amounts than are disclosed in any one verdict." This division was discussed in *Livingston v. Van Ingen*, 9 John, 507, N.Y. (1812); it was ruled that under federal law the right to exclude related to intangible property, whereas the states' function was to supervise tangible property such as the product of the patent, and to protect its residents against fraudulent patent claims and licenses (a license does not partake of the patent right; it is merely a permit to use the invention.) At the federal level, after 1819, "The jurisdiction of the circuit courts embrace all cases, both at law and in equity, arising under the patent laws... without regard to the citizenship of the parties or the amount in controversy," *Day v. Newark Rubber Co.*, 1 Blatch. 628 (1850.)

[7] Litigants in civil suits had the right to a trial by jury once the value at issue exceeded $20, but in the first decade the courts dealt with an annual average of only 380 suits for the entire nation.

[8] The phenomenon of multiple litigation was a notable feature of the American legal system, with 76 patents accounting for some 585 cases. More than a third of the latter cases involved patents that had been extended beyond the usual term of fourteen years, such as William Woodworth's planing machine, and the Blanchard lathe patent, which was active for forty-two years.

[9] Although the comments of the promoters of the 1836 Act are widely cited by scholars today, the polemical nature of any lobby should be taken into account.

Table 3.1. *Litigation of Patented Inventions, 1800–1860*

Decade	Patent Cases	Number of Patents Litigated	Total Patents Granted	Cases as Percent of All Patents
1800–1809	6	6	911	0.6
1810–1819	37	20	1998	1.8
1820–1829	36	27	2697	1.3
1830–1839	37	14	5077	0.7
1840–1849	198	95	5516	3.6
1850–1859	415	171	19661	2.1
1860	64	18	4363	1.5

Notes and Sources: See text, footnote 4 and notes to Table 3.6 for sources. Statistics for the percent of patents litigated are not reported, because in 115 cases no information was available about the patent at issue. The final column shows the number of patent cases within a decade as a percentage of all patents filed within that decade. These figures include common law and Supreme Court cases, equity cases, interferences, and appeals from decisions by the Commissioner of Patents. Interference and appellate cases from the Commissioner of Patents were introduced with the reforms of the patent system in 1836.

United States exhibits a greater tendency for multiple suits to be filed for a single patent, implying that a particular patent in Britain had a higher probability of being litigated. However, these overall similarities conceal marked differences in the attitudes of the courts, and in the ability of patentees to enforce their rights. Before explicitly considering these questions, it is informative to examine the portfolio of patent cases that appeared before the United States courts between 1790 and 1860 (Table 3.1.)

Litigation patterns reflect the use of inventions, because legal jurisdiction was limited to the circuit in which an alleged violation of the patent right occurred. Table 3.2 shows that the proportion of lawsuits relative to the number of patents awarded was lowest in less developed regions such as Northern New England and the Southern Mid-Atlantic states. Patents were often litigated in regions other than where they were invented, so these percentages may provide only limited information. However, patenting and conflicts about patent rights were clearly both concentrated disproportionately relative to the population in the emergent industrial markets of New York and Massachusetts, followed by Pennsylvania and Ohio. When regional markets such as the Midwest expanded, patenting and litigation likewise increased: after the 1830s the Midwest accounted for a rapidly growing share of patent cases, totalling 11 percent of antebellum litigation. These patterns suggest a growing concern about property rights and the extraction of returns to inventions as markets developed.

If general enforceability and weak property rights were at issue, one might expect the sectoral distributions of patents and cases to vary together. Yet, most patent cases related to inventions in the manufacturing sector, especially

Table 3.2. *Percentage Distribution of Patents and Cases by Region and Sector,*
1790–1860 (row percentages)

	Agriculture	Building	Manufacturing	Transport	Other	Total
Northern New England						
Cases	20.0	20.0	50.0	10.0	0.0	1.7
Patents	27.3	21.3	38.8	7.8	4.9	7.7
Southern New England						
Cases	13.0	17.6	55.6	6.5	7.4	18.5
Patents	13.2	16.4	53.4	9.5	7.5	21.0
New York						
Cases	10.8	22.8	45.6	15.2	5.7	27.0
Patents	22.9	17.9	38.1	13.4	7.8	31.7
Pennsylvania						
Cases	11.1	16.7	33.3	22.2	16.7	12.3
Patents	19.0	14.4	41.6	14.9	10.1	13.6
Southern Mid-Atlantic						
Cases	11.8	17.6	58.8	11.8	0.0	2.9
Patents	24.9	2.3	35.1	18.3	9.3	7.4
Midwest						
Cases	13.1	31.2	42.6	11.5	1.6	10.4
Patents	33.3	16.3	31.6	12.5	6.3	6.4
District of Columbia						
Cases	15.9	19.6	38.4	15.9	10.1	23.6
Patents	12.5	25.0	34.0	21.9	6.0	1.4
Other						
Cases	28.6	28.6	19.1	9.5	14.3	3.6
Patents	34.8	15.8	27.4	11.8	7.0	9.8
Total Cases	79	124	254	81	47	585
	13.5	21.2	43.4	13.9	8.0	100
Total Patents	1009	753	812	580	361	4515
	22.4	16.7	40.1	12.9	8.0	100

Notes and Sources: See the text, footnote 4 and notes to Table 3.6 for the source of litigation figures. The table excludes *ex parte* appeals from the Commissioner of Patents. The District of Columbia data are not representative of local litigation, because all interference cases were tried in that region; cases before the Supreme Court are also included in D.C. litigation. Litigation in the building sector is inflated by an outlier: one patent accounts for seventy-eight cases in this sector. If that patent is removed, manufacturing litigation by region amounts to 50 percent, 63, 53, 38, 62, 55, 41, and 27, respectively, with a 49 percent share of litigation overall. Patent data are from a random sample of 4,515 patents categorized by sector of final use. The litigation data cover the years 1790 to 1860, whereas the patent data are for the period from 1790 to 1846.

in the more commercialized or industrialized regions of New York, southern New England, and indeed in the Northeast overall. Even in the Midwest, litigation was highest in manufacturing (43 percent), followed by construction (31 percent), although agriculture dominated both local patenting and economic activity. Manufacturing litigation is just as pronounced when we consider the distribution of contested patents (as opposed to patent cases), for almost one half of all disputed patents related to manufacturing. Patent disputes, patenting, and manufacturing are so integrally related that it is difficult to fully disentangle the explanations behind the number of litigated patents and cases.[10] Mean awards to plaintiffs in agriculture ($7,360) were at least three times as high as those for manufacturing ($2,463), implying that higher average commercial value was not the full explanation.[11] The potential for conflict increased with market exchange, and manufacturing patent rights were more likely to be sold or licensed. Firms in the manufacturing sector were in more direct competition, and thus exhibited a lower tolerance for infringement and a higher propensity to litigate. For instance, 50 percent of all interference cases (which involve patent applications that include similar claims) related to manufacturing inventions. The interference records suggest a higher likelihood of conflict because inventors of manufacturing patents were directing their attentions to similar problems, and also attempting to invent around existing patents.[12] In short, litigation patterns tended to be associated more with business strategy rather than questionable property rights.

Both Tables 3.1 and 3.3 indicate that the overall propensity to litigate (gauged by cases relative to patents) decreased over time, from the first major surge in patenting late in the first decade of the nineteenth century, through

---

[10] "A presentation of the industries which are based upon patented inventions – either inventions which have created new industries or inventions which have revolutionized old industries – would include almost all of the manufacturing industries of the present day," reported Story B Ladd, in "Patents in Relation to Manufactures," 12th *Census of the United States*, vol. x (iv), Washington, D.C.: GPO (1900): 751–766. A rich source of details is J. Leander Bishop's *History of American Manufactures*. The second volume contains detailed information on commercially successful patents, and makes it clear that businesses of the day tended to seek patent protection, rather than relying on secrecy.

[11] Damages were decided by the jury, who were instructed to consider foregone profits, including interest sacrificed, which the judge could treble at his discretion, with costs. In *Whitney et al. v. Emmett et al.*, 29 F. Cas. 1074 (1831), the plaintiffs "contended that, as an item in the estimation of actual damages, the jury may examine and determine the loss sustained by the reduction of the price of the articles manufactured by the patented machine, in consequence of the competition brought into the market against them, when the patentee had a right to a monopoly; and going yet further, they say, that the injury done to the reputation of the manufacture, by the inferior skill and workmanship of the offender, may be fairly and legally brought into the calculation of actual damages." Story advised the jury in *Lowell v. Lewis*, 15 F. Cas. 1018 (1817): "Let the damages be estimated as high as they can be ... if the plaintiff's patent has been violated; that wrongdoers may not reap the fruits of the labor and genius of other men." Only one in ten cases reports the damages awarded, ranging from 3 cents (*Kneass v. Schuylkill Bank*, 14 F. Cas 746, 1820), to $23,220 with costs (*Parkhurst v. Kinsman*, 18 F. Cas, 1848.) Costs typically amounted to around $300, but some cases involved expenses as high as $20,000 (*Seymour v. McCormick*, 57 U.S. 480, 1850.)

[12] Interference proceedings are initiated when two or more applications claim priority to the same invention, either wholly or in part. Litigation in manufacturing patents was the least affected by changes in the patent system in 1836, as shown by the jump in its share of patent cases, from 40 percent in the 1820s to 78.6 percent in the 1830s.

The Democratization of Invention

Table 3.3. *The Regional Propensity to Litigate: Cases, Patents and Litigation Rates in the United States, 1790–1860*

Region	1790–1799	1800–1829	1830–1839	1840–1849	1850–1859	1860
Northern New England						
Cases	0	4	6	6	9	0
Patents	15	465	589	281	932	238
(%)	(0.0)	(0.8)	(1.0)	(2.1)	(1.0)	(0.0)
Southern New England						
Cases	0	20	9	49	47	7
Patents	71	1408	1152	1096	4219	786
(%)	(0.0)	(1.4)	(0.8)	(4.5)	(1.1)	(0.9)
Southern Mid-Atlantic						
Cases	0	3	1	2	10	2
Patents	28	445	392	419	1113	204
(%)	(0.0)	(0.4)	(0.3)	(0.5)	(0.9)	(1.0)
New York						
Cases	1	13	7	47	117	6
Patents	39	1680	1594	1759	5791	1278
(%)	(2.6)	(0.8)	(0.4)	(2.7)	(2.0)	(0.5)
Pennsylvania						
Cases	4	23	4	27	31	2
Patents	67	752	709	776	2499	513
(%)	(6.0)	(3.1)	(0.6)	(3.5)	(1.2)	(0.4)
Midwest						
Cases	0	0	6	29	45	6
Patents	0	167	324	630	3026	730
(%)	(0.0)	(0.0)	(1.8)	(4.6)	(1.5)	(0.8)
Other						
Cases	2	2	0	5	4	1
Patents	27	534	574	531	2456	614
(%)	(7.4)	(0.4)	(0.0)	(0.9)	(0.2)	(0.2)
Total U.S.						
Cases[a]	8	79	37	198	415	64
Patents	247	5451	5077	5516	19661	4363
(%)[a]	(3.2)	(1.4)	(0.7)	(3.6)	(2.1)	(1.5)

Notes:
[a] The U.S. total includes cases in the District of Columbia, which is atypical because all appeals from the Commissioner of Patents and interferences are filed in D.C. If the D.C. cases are excluded, the final three percentages fall to 3.2 percent (1840–1849), 1.5 percent (1850–1859), and 0.6 percent (1860.) The Midwest category consists overwhelmingly of cases for the district of Ohio. The regional shares of patent cases over the entire period are as follows: Northern New England 3.1 percent; Southern New England 16.5 percent; New York 23.8 percent; Pennsylvania 11.4 percent; Southern Mid-Atlantic 2.2 percent; Midwest 10.7 percent; Other (including D.C.) 33.5 percent.

Sources: See the notes to Table 3.6 and footnote 4 for general sources. The totals for regional patents are from the *Commissioner of Patents Annual Report for 1891*, Washington, D.C., 1892.

the Act of 1836. The extremely low average propensity to litigate evident in all regions during the 1830s appears anomalous, given the alleged dissatisfaction with the patent system. However, the litigation rate prior to the 1836 change may reflect the degree of uniformity in judicial patent policy, due to an appeal system sanctioned by the Constitution, which ultimately ensured a high degree of certainty.[13] The structural change in 1836 was associated with a jump in litigation, perhaps because of uncertainty about the new system. Subsequent policies proved relatively consistent and predictable, however, leading to the decline in later decades.[14] The marked increase in the number of plaintiffs seeking injunctions before courts of equity also appears to have been a factor.[15]

Table 3.4 provides further evidence of a relationship between markets and litigation. The rows of the table show the residence of the inventor of the primary patent at issue in the lawsuit, whereas the columns show the location of the alleged violation of the patentee's right. Over the period up to 1860 Massachusetts accounted for 15 percent of total patents granted, and New York patents amounted to 30 percent of all patents. Clearly, these two areas were the focus of a higher propensity to litigate; more than two thirds of all 387 lawsuits involved patents filed in Massachusetts and New York (20.7 and 47.8 percent respectively.) New York patentees were involved in legal disputes about their inventions across the country, and the column percentages for the District of Columbia (representing appellate cases) indicate that inventors located in New York were also more likely to appeal decisions.

---

[13] The point is not that the law was immutable and always consistent. Rather, the right of appeal and the rule of precedent ensured that dubious decisions would be filtered out to attain a consistent equilibrium. For instance, Philos Blake's patent for bed-casters was upheld in Connecticut, but overturned by a New York jury. The judge ordered a new trial in New York in order to obtain uniform decisions in both regions (*Blake v. Sperry*, 2 N.Y. Leg. Obs. 251, 1843.) Second, although the law changed, the constitutional belief in the "sacred rights of genius and property" was unchanged, as all participants recognized. "Such has been the uniform construction of the law in the circuit courts, that a patent can be declared void for no other defect in the specification than fraudulent concealment or addition." *Whitney v. Emmett*, 1831. See also *Gray v. James*, 1 Robb 120 (1817); *Reutgen v. Kanowrs*, 1 Wash. 168 (1804); *Park v. Little*, 18 F. Cas. 1107 (1813); *Lowell v. Lewis; Whittemore v. Cutter*, 1 Robb 28 (1813); *Evans v. Eaton*, 16 U.S. 454 (1818). Unquestionably, judges did vary at times in their interpretation of the statutes, especially at the district level. However, circuit decisions were the responsibility of Supreme Court justices, who lived in residence together when the Supreme Court was in session. This network of close communication was evident in frequent citations to decisions from other circuits, and contributed to the formulation of a coherent national policy toward patents and patenting.

[14] For a formal model, see George Priest, "Measuring Legal Change," *Journal of Law, Economics and Organization*, vol. 3(2), 1987:193–226.

[15] The significance of equity courts is discussed later. Courts of equity were little used before the 1840s. In 1840–1844, 24.1 percent of all disputes were in equity; 1845–1849, 42.1 percent; 1850–1854, 30.1 percent; 1855–1860, 21.4 percent. Again, the increase in equity litigation was possibly because of heightened industrial competitiveness.

Table 3.4. *The Location of Patentees Relative to Litigation of their Patents,*
1800–1860

Location of Patentee	Location of Patent Litigation						
	DC	MA	NE	New York	PA	SMA	Total
District of Columbia							
Number	1	1	2	0	0	0	4
Row %	25.0	25.0	50.0	0.0	0.0	0.0	100.0
Col. %	1.4	1.1	8.0	0.0	0.0	0.0	1.0
Massachusetts							
Number	4	54	10	6	6	0	80
Row %	5.0	68.0	12.5	7.5	7.5	0.0	100.0
Col. %	5.4	59.4	40.0	4.8	10.7	0.0	20.7
New England							
Number	7	4	7	16	0	0	34
Row %	20.6	11.8	20.6	47.1	0.0	0.0	100.0
Col. %	9.5	4.4	28.0	12.8	0.0	0.0	8.8
New York							
Number	40	25	4	85	19	12	185
Row %	21.6	13.5	2.2	46.0	10.3	6.5	100.0
Col. %	54.1	27.5	16.0	68.0	33.9	75.0	47.8
Pennsylvania							
Number	12	4	1	8	18	1	44
Row %	27.3	9.1	2.3	18.2	40.9	2.3	100.0
Col. %	16.2	4.4	4.0	6.4	32.1	6.3	11.4
Southern Mid-Atlantic							
Number	10	3	1	10	13	3	40
Row %	25.0	7.5	2.5	25.0	32.5	7.5	100.0
Col. %	13.5	3.3	4.0	8.0	23.2	18.8	10.3
Total							
Number	74	91	25	125	56	16	387
Row %	19.0	23.5	6.5	32.3	14.5	4.1	100.0

Notes and Sources: The column for D.C. litigation represents appellate cases. The Southern Mid-Atlantic region includes New Jersey, Delaware, and Maryland. The total represents all cases for which information about the inventor and his residence is currently available. The data were obtained by matching Meyer's index of lawsuits in *Federal Decisions*, with entries in the *Annual Report* of the Commissioner of Patents for various years.

Although these figures raise a number of questions about the underlying causal mechanisms, they do seem to confirm that litigation was more related to markets and competition than to problems in enforcement. *Sickels v. Rodman* (1843) illustrates the use of litigation as a competitive strategy. Frederick Sickels, a "talented and very ingenious young mechanic" from

New York, invented a cut off valve for steam engines that resulted in large cost savings. Within a few months of obtaining a patent in May 1842, he successfully prosecuted a manufacturer of steam engines "to establish the originality and validity of his patent," and so send a signal to deter potential infringers.

It also should be noted that multiple litigation was characteristic of the American legal system, with 76 patents accounting for some 585 cases. More than a third of those cases involved a small number of patents that had been extended by Congress beyond the usual term of fourteen years, such as William Woodworth's planing machine, and the Blanchard lathe patent. Extended or renewed patents inevitably generated a great deal of controversy, pitting assignees and licensees, and other members of the public, against the patentee. Members of the Supreme Court themselves were divided about the division of rights between patentees and owners of machines protected by patents that would normally have expired after fourteen years but had been extended for a total of as much as forty-two years. Thus, a significant fraction of patent disputes related to a few atypical inventions, rather than to the issue of the general enforcement of patent rights.[16]

Some scholars contend that the probability of a decision upholding the patent right increased in the 1830s because judges recanted on their former legalistic doctrines. This claim is considered below in greater detail, but such arguments ignore the drastic revision of the patent system in 1836, which devolved some of the functions of the courts to the Patent Office. In the new regime, patent applications and interference proceedings were examined by employees of the Patent Office for novelty and conformity with the patent statutes. Previously, the burden of prosecuting infringement claims and interferences was on the patentee and the courts. Plaintiffs in interference cases before the Commissioner of Patents faced a 65 percent rejection rate; the Patent Office thus filtered out potential patent disputes that previously would have appeared before the courts. In short, the composition of patent cases had changed – it is therefore hardly surprising that the proportion of decisions for plaintiffs increased after 1836.

In direct contrast, litigation rates in Britain increased over time. British cases were decided on an ad hoc basis, by means of "judge-made law," leading to uncertainty for both plaintiffs and defendants. Researchers argued that the eighteenth-century English patent system was ineffectual, for the "odds were stacked against patentees."[17] According to this perspective, the value

---

[16] See *Sickels v. Rodman*, 22 F. Cas. 26 (1843); and *Wilson v. Rousseau*, 45 U.S. 646 (1846.)

[17] Christine MacLeod, *Inventing the Industrial Revolution, 1660–1800*, Cambridge: Cambridge University Press (1988.) See also Harold Dutton, *The Patent System and Inventive Activity during the Industrial Revolution*, Manchester: Manchester University Press (1984.)

of property in patented inventions was tenuous or at best highly uncertain. A patent was of little commercial value until it had been successfully litigated; yet the process of litigation was costly, the courts inconsistent and "notoriously full of pitfalls for patentees." The sparse literature that exists for the United States similarly proposes the thesis that in the early period legal and judicial policies debased the value of patent rights. In the next section I consider the attitudes of the courts, and conclude that the British experience was decidedly different from the United States. Unlike Britain, the judiciary in this country attempted to defend and enforce patent rights in accord with a Constitution that favored and identified inventive rights with patent property.

THE ATTITUDES OF THE COURTS TOWARD PATENT RIGHTS

Some researchers designate 1794–1831 as the "Embarrassing Era," when antipatent judges invalidated patents because of trivial technicalities or arbitrary dicta.[18] A superficial examination of the available records appears to support the view that the proliferation of patent grants and changing judicial attitudes adversely affected patent values. Table 3.5, for instance, indicates that 75 percent of the verdicts for the 1820s were decided against the patentee, whereas the outcomes for the second half of the period seem to be more "fair." Because the 1793 act left the validation of patents to the courts, the situation appears analogous to the English experience of arbitrary judge-made law. This was decidedly not the case, however.

Those who question the attitudes of the courts toward patentees have tended to base their conclusions on the proportion of outcomes decided in favor of the plaintiff. According to this criterion, the judiciary was "antipatent" prior to 1836, implying that inventors were unable to enforce their rights, and patents amounted to little more than documents. The number of verdicts for the plaintiff increases after this period, hence the inference is made that judicial attitudes shifted toward the support of patent rights. However, these data should be interpreted with caution. The use of data on decisions in patent cases to make inferences about the attitudes of the courts implicitly assumes that cases before the courts are representative of

---

[18] See Frank Prager, "Trends and Developments in American Patent Law from Jefferson to Clifford (1790–1870)", *American Journal of Legal History*, vol. 6, 1962: 45–62. Lubar asserts that "In the first three decades of the 19th century, Congress was not favorably disposed to patentees, and courts dismissed many patent-enforcement cases on narrow technical grounds." Steven Lubar, "The Transformation of Antebellum Patent Law," *Technology and Culture*, vol. 32(4), October 1991: 932–59. Kent Newmeyr also supports this view, *Supreme Court Justice Story: a Statesman of the Old Republic*, Chapel Hill: UNC Press (1985.) Both Lubar and Newmeyr cite Prager's work as the evidence for their arguments.

Table 3.5. *Reported Decisions in Patent Cases, 1800–1860*
(includes appeals from the Commissioner of Patents)

Decade	No. For Patentee	(%)	No. Against Patentee	(%)	ND*
1800–1809	3	50	3	50	–
1810–1819	11	39	15	54	2
1820–1829	8	25	24	75	–
1830–1839	7	54	5	39	1
1840–1849	95	56	65	39	9
1850–1859	146	51	131	46	10
1860	25	56	20	44	–
Total	295	51	263	45	22

Reported Decisions, 1840–1860
(Excludes appeals from the Commissioner of Patents)

Decade	No. For Patentee	(%)	No. Against Patentee	(%)	ND*
1840–1849	94	61	52	34	8
1850–1859	113	53	93	44	8
1860	11	58	8	42	–
Total (1800–60)	247	53	200	43	19

Notes and Sources: See the text, footnote 4 and notes to Table 3.6 for sources. The patentee of the invention at issue was the defendant in twenty-five of these cases. The data refer to decisions where the patentee won the lawsuit ("for"); or lost the case ("against"). ND* represents cases in which no decision was reached either for or against the patentee, such as in the event of a mistrial, or where the case was remanded to another court for a lack of jurisdiction. Appeals from the Commissioner of Patents were introduced with the patent system reforms of 1836.

the underlying distribution of all patent disputes.[19] If all disputes were litigated, or if litigated cases were selected randomly, then we could be certain that a change in the patentee recovery rate (from 37 percent prior to 1836, to 55 percent after this period) reflected a change in judicial attitudes. However, litigated cases (those that actually reach the courts) are not drawn randomly from the population of disputes. Given this problem of selection bias, the rate of plaintiff victories cannot be used to gauge judicial attitudes nor changes in those attitudes. Accordingly, "It is impossible to use reports of litigated cases to generate anything more than a qualitative judgment of the direction of legal change."[20]

[19] For a more detailed model, see George Priest and Benjamin Klein, "The Selection of Disputes for Litigation," *Journal of Legal Studies*, vol. 13 (1)(1984: 1–55; and George Priest, "Selective Characteristics of Litigation," *Journal of Legal Studies*, vol. 9 (1980): 399. George Priest, "The Common Law Process and the Selection of Efficient Rules," *Journal of Legal Studies*, vol. 6 (1977) suggests that, in the limit, the judiciary is effective only in forming patentees' expectations by adhering to predictable policies.

[20] George Priest, "Measuring Legal Change," *Journal of Law, Economics and Organization*, vol. 3(2), 1987:193–226.

Research that relies solely on data about outcomes to make inferences about patent property rights is overly simplistic, for it assumes that plaintiffs pursue only one objective, that of obtaining a settlement. Even if the case is lost, the wording of the decision may uphold the patent right. Consequently, as Justice Baldwin emphasized, a verdict against the patentee did not necessarily imply that the patent property was void, merely that he could not recover his claims regarding that specific lawsuit. The stakes for the patentee were much greater than the damages awarded in any specific lawsuit, because the decision to uphold his property rights supported his claims over all users of his invention. For instance, Cyrus McCormick – who made a fortune from royalties and manufacturing profits for his agricultural machines – lost the verdict against an alleged infringer. Before dismissing the charge, the judge noted in his summing up: "Having arrived at the result, that there is no infringement of the plaintiff's patent by the defendant, as charged in the bill, it is announced with greater satisfaction, as it in no respect impairs the right of the plaintiff. He is left in full possession of his invention, which has so justly secured to him, at home and in foreign countries, a renown honorable to him and to his country – a renown which can never fade from the memory, so long as the harvest home shall be gathered."[21] Although the competing harvester invention was ruled to be different and thus noninfringing, McCormick was still able to enforce his rights against unauthorized users of his own patented invention.

The frequency with which later courts cited decisions from the previous period provides another perspective on the issue of whether early decisions were arbitrary. If early decisions were indeed idiosyncratic and prejudiced against patentees, one might expect that subsequent – allegedly more liberal – courts would reject the former legal standard. But as Figure 3.1 shows, pre-1831 cases have been cited as frequently as later decisions.[22] Indeed, decisions for early suits are still cited in legal decisions today, suggesting that these cases provided a lasting foundation for policy toward patent disputes. As emphasized earlier, patent law was conducted at the federal level, and decisions in the United States were ultimately governed by the Constitution, which recognized a natural right in inventive property, rather than by the individual court. Extensive reading of antebellum patent cases indicate that American courts from their inception attempted to establish a store of doctrine that fulfilled the intent of the Constitution.[23]

---

[21] *McCormick v. Manny et al.*, 15 F. Cas. 1314 (1856.)

[22] A thesis of structural change in the legal system would more plausibly refer to *Hotchkiss v. Greenwood*, 52 U.S. 248 (1851), which altered the standard of patentability and required patents to fulfill a technical standard of "nonobviousness"; however, the time series of citations presented here shows that the Hotchkiss decision was not implemented in the courts until after the Civil War.

[23] Thus, Justice Story pointed out, *Blanchard v. Sprague*, 3 F. Cas 648 (1839), the English courts tended to be hostile toward patent grants, but "In America, this liberal view of the subject has

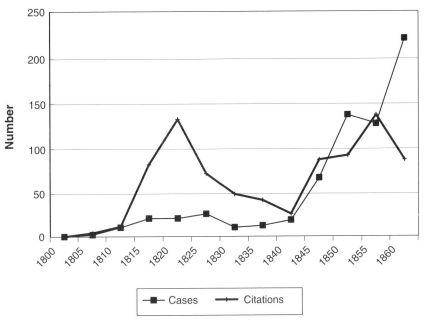

Figure 3.1. Patent Litigation: Citations and Cases, 1790–1860
Notes and Sources: The citations were obtained from the *Federal Reporter* and *LexisNexis®* and include total counts of subsequent citations for cases tried in that year. Citations for the landmark case *Hotchkiss v. Greenwood*, 52 U.S. 248 (1850–1851) were excluded, because the decision was first implemented in the 1870s.

Numerous reported decisions before the early courts clearly and repeatedly declared that patent rights were "sacred" and to be regarded as the just recompense to inventive ingenuity. Supreme Court Justice Joseph Story, the acknowledged patent expert of the antebellum courts, was responsible for forty-eight reported patent decisions during his tenure. He indicated in *Lowell v. Lewis* (1817) that "the proper duty of the court" was to ensure "that wrongdoers may not reap the fruits of the labor and genius of other men." Story's Supreme Court decision in *Ex parte Wood and Brundage* (1824) delineated the policy underlying the 1793 act, which allowed district courts the right to repeal patents if the patent had been fraudulently

always been taken, and indeed, it is a natural, if not a necessary result, from the very language and intent of the power given to congress by the constitution on this subject.... Patents, then, are clearly entitled to a liberal construction, since they are not granted as restrictions upon the rights of the community, but are granted to 'promote science and useful arts'" (my emphasis.) According to Justice Baldwin (*Whitney v. Emmett*, 1831), "The silence of the [English] law left a wide field open to the discretion of courts ... But in this country the law is more explicit. The Constitution ... is a declaration of the supreme law of the land ... which leaves no discretion to the judges to assign or presume any other."

obtained, and if the claim to repeal was brought before the courts within three years of issuance of the patent. The relevant tenth section of the act was ambiguous, but Story's reference point was the Constitution itself. His argument was a logical extension of the care with which rights of patentees were secured by constitutional edict. "It would be somewhat surprising if, after such anxious legislation," he wrote, "there should exist in the act a clause which, in a summary manner, enables any person to repeal [the patentee's] patent, and thus sweep away his exclusive property." The Act of 1793 therefore did not imply execution, but was in the nature of a *scire facias* or a process to call for the repeal of a patent, which involved a trial by jury if the defendant (the patentee) should choose to contest the claim. In all cases, the burden of proof rested with those who wished to challenge the patent, and the challenger was liable for costs if judgment went against him.[24] For, "The inventor has a property in his invention; a property which is often of very great value, and of which the law intended to give him the absolute enjoyment and possession ... involving some of the dearest and most valuable rights which society acknowledges, and the constitution itself means to favor."[25]

The attitudes of the judiciary toward patent conflicts are primarily shaped by their interpretation of the monopoly aspect of the patent grant. In *Whitney v. Emmett* (1831), Justice Baldwin contrasted the policies in Britain and America toward the patent contract. English Courts, he pointed out, interpret the patent grant as a privileged exception from the general ban on monopolies.[26] Apart from this proviso, the judiciary had total discretion in interpreting and deciding the ends that would promote public welfare. The patent was seen as a trade-off, a bargain between the inventor and the public with a negotiable outcome. In contrast, in the United States the patentee was not recognized as a monopolist *per se*: "In England a patent is granted as a favor, on such terms as the King thinks proper to impose. ... Here a patent is a matter of right, on complying with the conditions prescribed by

---

[24] This system is called the English rule. See "Legal Fees: A Comparison of the American and English Rules," Charles Plott, *Journal of Law, Economics and Organization*, vol. 3(2) 1987: 185–97 for a model suggesting that the English rule conveys perverse incentives that may escalate litigation, bounded only by bankruptcy of litigants. This may partly explain the behavior of inventors such as Goodyear, Whitney, Evans, and Sickels, who dissipated profits in litigation.

[25] Full citations are *Ex parte Wood and Brundage*, 22 U.S. 603 (1824); *Lowell v. Lewis*, 15 F. Cas. 1018 (1817.)

[26] *Whitney v. Emmett*, 29 F. Cas. 1074 (1831.) The Supreme Court emphasized in *Pennock v. Dialogue*, 27 U.S. 1 (1829): "In the courts of the United States, a more just view had been taken of the rights of inventors. The laws of the United States were intended to protect those rights, and to confer benefits; while the provisions in the statute of England, under which patents are issued, are exceptions to the law prohibiting monopolies. Hence, the construction of the British statute had been exceedingly straight and narrow, and different from the more liberal interpretation of our laws."

Illustration 6. Supreme Court Justice Joseph Story (1779–1845) played a key role in the formulation of U.S. intellectual property doctrines. (*Source*: 1 *Green Bag*, 10–25 (1889): 12.)

the law."[27] Baldwin argued that conflicts between private and social benefit did not exist under the United States statutes, governed as they were by acts

---

[27] "Patentees are not monopolists.... A monopolist is one who, by the exercise of the sovereign power, takes from the public that which belongs to it, and gives to the grantee and his assigns an exclusive use. On this ground monopolies are justly odious.... Under the patent law this can never be done. No exclusive right can be granted for anything which the patentee has not invented or discovered. If he claim anything which was before known, his patent is void, so that the law repudiates a monopoly. The right of the patentee entirely rests on his invention or discovery of that which is useful, and which was not known before. And the law gives him the exclusive use of the thing invented or discovered, for a few years, as a compensation for 'his ingenuity, labor, and expense in producing it.' This, then, in no sense partakes of

of Congress that clearly recognized the property in ideas and inventions to be private and exclusive. Rather, the explicit intention of the patent law was to benefit the inventor, in the belief that maximizing individual welfare leads to maximum social welfare.[28]

Baldwin's statement conveys an exaggerated assessment of the matter, because its logical extension would be to grant the inventor a monopoly in perpetuity. Still, the general tenor does echo the attitude of the time. Some have claimed that the courts and the patent system alike were indifferent about whether the patentee was able to appropriate benefits from his efforts.[29] However, a major concern of these institutions was to ensure that inventors were amply rewarded. Policies toward the term of the patent mirror this concern, for it was recognized that the term of the patent affected the profitability of the invention. The patentee gained exclusive rights for fourteen years, but this period could prove inadequate for inventors to capture the returns from discoveries of great commercial value. Congress therefore reserved the right, in such cases, to extend the life of the patent. For instance, Amos Whittemore's cotton card machine was covered by a patent due to expire in 1811. Although efforts to infringe on his claim proved fruitless, he failed to recover his expenses and applied to Congress in 1809 for an extension. The discovery was acknowledged to be valuable to the country, and there was some discussion of a perpetual grant, but Congress finally authorized a renewal of fourteen years from the expiration of the original patent. At the end of this period, the Whittemore enterprise sold several of its machines in anticipation of a rapid decline in the business, because the monopoly could no longer be retained. Varying the term of the patent in this manner enabled patentees of important inventions to extract a higher potential income from their discoveries than was possible for "mere gadgets."[30]

---

the character of monopoly." *Allen v. Hunter*, 6 McLean 303 (McLean, 1855.) "Probably of all species of property, this property in patent rights should be most carefully guarded and protected, because it is so easily assailed ... Now, patents are not monopolies ... a patent is that which brings out from the realm of the mind something that never existed before, and gives it to the country," *Singer v. Walmsley*, 1 Fisher 558 (Md. 1859.)

[28] Baldwin approvingly cited an English decision, that "nothing could be more essentially mischievous, than that questions of property between A and B, should ever be permitted to be decided upon considerations of public convenience or expediency." See, too, *Whittemore v. Cutter*, 29 F. Cas 1120 (1813); and *Lowell v. Lewis*, 15 F. Cas 1018 (1817), in which Story discusses the common view that the public benefits from the ownership of the idea once the patent expires. However, according to Sec. 6 of the 1793 Act, it was immaterial whether the patent was described fully enough to enable a skilled mechanic to recreate the invention, unless the defective description were intended to deceive the public. Story enforced this section of the statutes even though, as he pointed out, an accidental omission negated the benefit to the public after the expiration of the patent as effectively as a defective description due to fraud: "We must administer the law as we find it."

[29] Steven Lubar supports this viewpoint in "New, Useful and Non-Obvious," *American Heritage of Invention and Technology*, vol. 6(1) 1990: 8–16.

[30] The policy of varying the term of the patent according to value was continued in the statutes of 1832, and reiterated in the Patent Act of 1836. The patentee was required to show a

If the early courts were indeed conscious of the importance of defending private property in inventions, and of securing the profits of patentees from infringement, one might ask why patentees would ever receive an adverse ruling. First of all, property rights are delimited by and must conform to the law, which defines what constitutes a true invention. Second, although judges supported inventive property rights, they also were conscious that their decisions involved a wider class of patentees and patent rights than merely those of the litigants before the court. Third, it is simplistic to assume that a verdict against the plaintiff indicated an antipatent stance, when in many cases both plaintiffs and defendants owned valid patents. For, in such cases, as Justice Woodbury stated in *Colt v. Mass. Arms Co.* (1851), "They both have a right to have these patents protected, so far as they can be, without conflicting with each other."[31]

The major criterion of the judiciary was the identification of the first individual to have made a workable version of a device that was likely to have been the object of attention for a large number of talented individuals. Policies toward slaveholders illustrate the degree of importance attached to the principle that only the first and true inventor was entitled to a patent. Even though state laws regarded slaves and their output as the property of their owners, a slaveholder could not obtain a patent in his own name for a device that his slave had created. In his *Annual Report* for 1857, the Commissioner of Patents noted that several white men had applied for patents to protect inventions that had been created by their slaves, a situation that never had occurred before. One of these, a slaveholder from Mississippi, Oscar J. E. Stuart, wished to be given a patent in his own name for a cotton scraper plough invented by one of his late wife's slaves. Both the Patent Office and the Attorney General on appeal rejected Stuart's claim because, according to the patent laws, only the true inventor could be given a patent.[32] Stuart, who feared that "unjust discrimination" and prejudices against slave

---

statement of the ascertained value of the invention, and his loss and profit. The Patent Act of 1836 instructed that the public interest should be considered before granting extensions; but at the same time, "it is just and proper that the term of the patent should be extended, by reason of the patentee, without neglect or fault on his part, having failed to obtain, from the use and sale of his invention, a reasonable remuneration for the time, ingenuity and expense bestowed upon the same, and the introduction thereof into use."

Extended or renewed patents naturally generated a great deal of controversy, pitting assignees and licensees, and other members of the public, against the patentee. Members of the Supreme Court themselves were divided about the division of rights between patentees and owners of machines protected by patents that would normally have expired after fourteen years, but were extended for a total of as much as 42 years (Wilson v. Rousseau, 45 U.S. 646, 1846.)

[31] *Colt v. Mass. Arms Co.*, 6 F. Cas. 161 (1851.)

[32] Jeremiah Black, the Attorney General, noted that "if such a patent were issued to the master, it would not protect him in the courts against persons who might infringe it" (because according to the law a patent was invalid unless the patentee were the first and true inventor) [9 Op. Atty Gen. 171; 1858].

owners "might cloud the understanding of a man from a different latitude," subsequently applied to Congress, "praying that the patent laws may be so amended as that a patent may issue to the master for a useful invention by his slave."[33] The Committee on Patents altered his plea to ensure that the slave would be given the patent, even if the usufruct of the property was to be ceded to the slave owner as required by the state laws.[34] But, in any case, the bill was never passed into law.

Table 3.6 illustrates these points, by presenting the distribution of plaintiffs, defendants, and issues. Approximately one third of all unsuccessful patent cases prior to 1837 were lost because the plaintiff was not the "first and true" inventor.[35] Courts were also concerned about the novelty of the invention. Joseph Story considered the question of novelty in *Earle v. Sawyer* (1825), in the context of an improvement which was acknowledged to be new, but contested on the grounds that it was too obvious an invention to warrant a valid patent. According to the 1793 statutes, merely superficial changes in form, matter, or proportion to obtain the same result were not a discovery – "substantial" novelty was therefore the major prerequisite.[36] Putting an existing invention (such as ether) to a new use (as an anaesthetic) was not itself patentable. From the point of view of the patentee, the substantial novelty requirement undoubtedly led to some ambiguity, especially when dealing with individuals attempting to patent around the existing invention.

[33] See Stuart's letter of August 29, 1857. The Stuart correspondence, which is replete with the ironies that were unperceived by the slave-owning class, is reproduced in Dorothy Cowser Yancy, "The Stuart Double Plow and Double Scraper: The Invention of a Slave," *Journal of Negro History*, vol. 69 (1) 1984: 48–52. See also the *Journal of the Senate of the United States of America, 1789–1873*; December 14, 1858, p. 45.

[34] See *Journal of the Senate of the United States of America, 1789–1873*, January 31, 1859: Bills and Resolutions, Senate, 35th Congress, 2nd Session, Bill S 548: "To authorize the issue of patents, in certain cases, to negro slaves for the use of their owners." Section 3 of the bill proposed "That all applications for a patent under this act shall, in addition to the facts now required to be set forth by other applicants, be required to state that the inventor is a negro slave and the name or names of his owner or owners; and the oath of such inventor shall be verified by the oath of his owner or owners to the best of his or their knowledge and belief." The bill was read twice but never passed into law, although a similar measure was adopted in the Confederate patent statutes in 1861.

[35] The following statistics are based on an examination of all patent cases reported for 1790–1837. Although defense counsel were creative in their attempts to undermine the patent grant (offering twenty grounds for invalidation, in one instance), the summary statement of the court tended to be more focused, so it was possible to isolate one predominant reason for the decision.

[36] A noteworthy clause of the patent act ruled that alien residents lost their rights if they failed to employ the patent usefully, but such restrictions did not apply to citizens of the United States (*Tatham v. Loring*, 23 F. Cas. 720, 1845.) For an excellent discussion of the standards for patentability, see Kenneth J. Burchfiel, "Revising the 'Original' Patent Clause: Pseudohistory in Constitutional Construction," *Harvard Journal of Law and Technology*, vol. 2 (Spring) 1989: 155–218.

Table 3.6. *The Distribution of Court Cases Across Issues, Outcomes, and Identities of Plaintiff and Defendant*

	Lower Courts						Supreme Court					
	1800–39		1840–60		All		1800–39		1840–60		All	
	(#)	(%)	(#)	(%)	(#)	(%)	(#)	(%)	(#)	(%)	(#)	(%)
Plaintiffs												
Patentee	40	81.6	156	59.1	196	62.6	7	70.0	19	38.8	26	44.1
Assignee	8	16.3	101	38.3	109	34.8	1	10.0	16	32.7	17	28.8
Licensee	–	–	3	1.1	3	1.0	–	–	2	4.1	2	3.4
Indep. Manuf.	–	–	1	0.4	1	0.3	2	20.0	12	24.5	14	23.7
Unknown	1	2.0	3	1.1	4	1.3	–	–	–	–	–	–
Defendant												
Patentee	8	16.3	24	9.1	32	10.2	2	20.0	15	30.6	17	28.8
Assignee	3	6.1	31	11.7	34	10.9	–	–	18	36.7	18	30.5
Licensee	3	6.1	26	9.8	29	9.3	1	10.0	3	6.1	4	6.8
Indep. Manuf.	31	63.3	128	48.5	159	50.8	6	60.0	13	26.5	19	32.2
Unknown	4	8.2	55	20.8	59	18.9	1	10.0	–	–	1	1.7
Issue												
Infringement	31	63.3	105	39.8	136	43.5	7	70.0	25	51.0	32	54.2
Injunction	5	10.2	106	40.2	111	35.5	–	–	3	6.1	3	5.1
Damages/Costs	3	6.1	8	3.0	11	3.5	1	10.0	5	10.2	6	10.2
Jurisdiction/ Procedural	3	6.1	16	6.1	19	6.1	–	–	8	16.3	8	13.2
Contract	3	6.1	7	2.7	10	3.2	–	–	7	14.3	7	11.9
Other	4	8.2	22	8.3	26	8.3	2	20.0	1	2.0	3	5.1

(continued)

Table 3.6 (continued)

	Lower Courts						Supreme Court					
	1800–39		1840–60		All		1800–39		1840–60		All	
	(#)	(%)	(#)	(%)	(#)	(%)	(#)	(%)	(#)	(%)	(#)	(%)
Decision												
Plaintiff Won	15	32.6	147	56.8	162	53.1	—	—	14	28.6	14	23.7
Reasons Why Defendant Won												
Did Not Infringe	2	4.4	44	17.0	46	15.1	1	10.0	11	22.5	12	20.3
Patent Void	15	32.6	12	4.6	27	8.9	2	20.0	3	6.1	5	8.5
New Trial	7	15.2	22	8.5	29	9.5	2	20.0	5	10.2	7	11.9
Procedural Issue	2	4.4	7	2.7	9	3.0	—	—	4	8.2	4	6.8
Jurisdiction	1	2.2	5	1.9	6	2.0	2	2.0	4	8.2	6	10.2
Plaintiff Lacks Title	1	2.2	7	2.7	8	2.6	1	10.0	—	–	1	1.7
Insufficient Grounds	2	4.4	9	3.5	11	3.6	2	20.0	7	14.3	9	15.3
Other	1	2.2	6	2.3	7	2.3	—	–	1	2.0	1	1.7
Validity of Patent												
Held Void	15	31.9	13	4.9	28	9.0	2	20.0	4	8.2	6	10.2
Upheld	12	25.5	100	38.0	112	36.1	1	10.0	12	24.5	13	22.0
Validity Not at Issue	10	21.3	119	45.2	129	41.6	5	50.0	30	61.2	35	59.3
No Decision	10	21.3	31	11.8	41	13.2	2	20.0	3	6.1	5	8.5

(continued)

Table 3.6 (continued)

Grounds for Challenging Validity of Patent

	1800–39		1840–60		All	
	(#)	(%)	(#)	(%)	(#)	(%)
Novelty/Original Inventor	15	36.6	100	65.8	115	59.6
Specification	17	41.5	21	13.8	38	19.7
Prior Use/Abandonment	3	7.3	14	9.2	17	8.8
Extension/Reissue	3	7.3	13	8.6	16	8.3
Other	3	7.3	4	2.6	7	3.6

Notes: The data comprise a sample of reported lawsuits included in LexisNexis®. The total number of cases in the table is 372 (59 Supreme Court cases; and 313 lower court cases, which include three district court cases and 310 circuit court cases.) I have omitted interference cases (which involve appeals made from the decision of the Commissioner of Patents in disputes between two inventors who both claim priority in making an invention) because the sole issue was whether a particular applicant should be given a patent – the grant of a patent in an interference process could not revoke the rights of an already granted patent – and *ex parte* appeals from decisions by the Commissioner of Patents. Plaintiffs and defendants comprised claimants to the patent property (patentee, assignee, and licensee), independent manufacturers (businesses without any legal claims to use the patent), and unknown (others who were not specified in the lawsuit.) The issue of the lawsuit is categorized as infringement, injunctions in courts of equity, conflicts about damages, costs, and expenses, questions about the appropriate jurisdiction of the court, procedural issues (such as whether a witness is to be allowed to give evidence), contractual disputes (for example, whether a licensee is violating a contract by selling in a particular location) and other (mainly motions for a new trial based on questions such as an alleged error in the judge's summary instructions.) Decisions about the validity of the patent include: the patent was held void (indicating that the plaintiff could not bring a charge *in the specific case being decided* based on the patent); the patent right was upheld; cases in which the validity of the patent was not contested (such as a lawsuit where the alleged infringer acknowledged that the plaintiff's patent was valid, but claimed that his machine did not infringe because it was substantially different); and cases in which no decision was reached about the validity of the patent (an example being an outcome where the case was remanded to another court, or if the decision was not mentioned in the report.) In addition to the above, decisions went in favor of defendants if they were found to use inventions that did not infringe the plaintiff's, if the plaintiff lacked the legal title to the patent (such as an assignee who had violated the contract of assignment with the patentee), or if there was an inadequate basis for the claims of the plaintiff. See the text for a discussion of the grounds on which the validity of the patent was challenged. Totals vary because of missing variables in some categories of the table.

For instance, a patent could be held not to be infringed if substitution resulted in greater efficiency, for this amounted to a new patentable invention in itself.[37] However, Justice Story rejected the argument that patents required inventive inputs or efforts that went beyond those that could be produced by an artisan who was skilled in the arts. Story was not persuaded by the "metaphysical" notion of patentability, for the standard "proceeds upon the language of common sense and common life, and has nothing mysterious or equivocal in it.... It is of no consequence, whether the thing be simple or complicated; whether it be by accident, or by long, laborious thought, or by an instantaneous flash of mind, that it is first done. The law looks to the fact, and not to the process by which it is accomplished."[38] This commonsense standard was entirely appropriate for an era in which ordinary nontechnical craftsmen and women could make valuable innovations based on simple know-how, and provided a striking contrast to the British approach.

A departure from this policy occurred when *Hotchkiss v. Greenwood* (1851) proposed that "unless more ingenuity and skill in applying the old method ... were required in the application of it ... than were possessed by an ordinary mechanic acquainted with the business, there was an absence of that degree of skill and ingenuity which constitute essential elements of every invention." The frequency of citations in Figure 3.2 indicates that the Hotchkiss ruling long remained an isolated decision, but after the 1870s it became the precedent for decisions that invalidated patent grants on the grounds of nonobviousness and later for the absence of a "flash of genius." Although the purist might view the move toward the more stringent nonobviousness criterion as not strictly in keeping with a democratic orientation, the heightened standards likely functioned as a more effective filter in view of the great increase in technical qualifications and patenting rates among the population during the postbellum period. The efficacy of Hotchkiss citations as a gauge of standards of nonobviousness is reinforced by the sharp decline in their frequency in the 1980s, when modern courts allowed a flood of patents for obvious inventions such as business methods.

The question of novelty and obviousness was related to the wider question of what constituted a true invention, as described by the patent statutes. To nineteenth-century courts, patentable technology incorporated ideas and discoveries that were vested in tangible form, and "a mere abstract idea" or processes independent of a means of fixation could not be treated as the

---

[37] See *Kneass v. Schuylkill Bank*, 1 Robb 303 (1820); *Poppenhusen v. New York Gutta Percha Co.*, 4 Blatch. 253 (1859); however, in *Gray v. James*, 1 Robb 120 (1817), the defendants argued unsuccessfully that they had improved on, rather than infringed, Jacob Perkin's "useless" nail-cutting machine. In accordance with the federal statutes, James Stimpson's 1831 patent was held not to be infringed by an invention using some of the elements of his railroad invention in conjunction with an element that was substantially different in form (*Stimpson v. Balt. & Susq. Railroad Co.*, 51 U.S. 329 1850.)

[38] *Earle v. Sawyer*, 8 F. Cas. 254 (1825.)

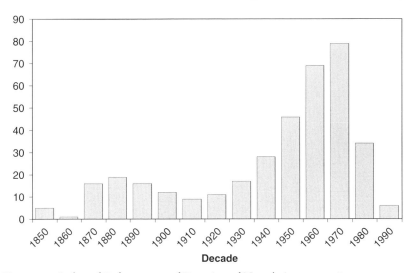

Figure 3.2. Index of Enforcement of Doctrine of Nonobviousness, 1850–2000 (Citations to *Hotchkiss v. Greenwood* decision)

Notes and Sources: Citations were obtained from *LexisNexis*® and comprise a decadal count of cases that refer to *Hotchkiss v. Greenwood*, 52 U.S. 248 (1851).

exclusive property of any one person, for this would limit diffusion and learning without any measurable social return. Consequently, several of the early "business method" cases dealt with copyright laws. In 1859 Charles Selden obtained copyright in a book entitled *Selden's Condensed Ledger, or Bookkeeping Simplified*. His method of bookkeeping was infringed by Baker, who denied the validity of copyright protection for a business method. The Supreme Court agreed that copyright protection extended only to expression, and did not provide an exclusive right in the idea itself: "The copyright of a work on mathematical science cannot give to the author an exclusive right to the methods of operation which he propounds, or to the diagrams which he employs to explain them, so as to prevent an engineer from using them whenever occasion requires."[39] Exclusive rights could only be granted through examination for novelty according to patent laws, although the court refused to decide whether the bookkeeping method was indeed patentable.

When patents were granted for inventions that seemed to be for contracts or business methods, they were uniformly overturned by the courts, unless the idea or principle could be construed as vested in a tangible medium. The Patent Office granted an 1891 patent to Levy Maybaum of Newark for inventing a "means for securing against excessive losses by bad debts," which he assigned to the U.S. Credit System Company. The patent covered a method

[39] Baker v. Selden, 101 U.S. 99 (1879.)

of computing the industry norm for operating losses and constructing tables that allowed comparisons relative to the industry average. The owners of the patent brought an infringement claim before the courts, but the patent was dismissed as "a method of transacting common business, which does not seem to be patentable as an art."[40] In litigation regarding the validity of an invention for "time limit" transfer tickets for use by street railways, the defendants also sought to decry the patent as a business method. The Circuit Court admitted that if the defense claim was true, then the patent had to be invalidated. As another judge expressed it, "advice is not patentable." However, it was decided that, though the case was borderline, the ticket was patentable as an item to be manufactured.

The courts distinguished between conflicts involving patentees on both sides – where enterprises operated under competing patents – and conflicts in which the patent right was deliberately infringed. In the latter instance "vindictive damages" were granted, but awards for damages tended to be mitigated or dismissed if the infringement was unknowing, or if the defendant was operating with the sanction of another patent. Such decisions undoubtedly led to frustration of the part of the patentee whose rights were infringed. For instance, Livingston appealed the verdict of the lower court assessing damages at almost $4,000 for his infringement of Woodworth's patent.[41] The Supreme Court upheld his appeal against damages because the infringer was the assignee of a patent filed by Hutchinson; this, the court felt, indicated that the act was not one of "wanton infringement." Nevertheless, when infringement was clearly in violation of patentee rights, judicial interpretation followed the spirit rather than the technicalities of the law. For instance, infringers were not allowed "colorable evasions" by switching their operation to another location: "Where a party has been enjoined from the use of a machine in one district, its use in another district . . . will not be allowed while the injunction against him remains in full force."[42]

Another reason for rejecting a patentee's case depended on whether the patentee had used the invention before obtaining a patent. It might be argued that a system that defends original inventors (unlike England) also encourages secrecy, and diminishes the value of obtaining a patent.[43] However, both the law and the judiciary followed the Constitution in the belief that patented inventions were pivotal in gaining industrial and economic

---

[40] United States Credit System Co. v. American Credit Indemnity Co., 53 F. 818 (1893.)

[41] *Livingston v. Woodworth*, 56 U.S. 546 (1853.)

[42] Woodworth v. Edwards, 30 F. Cas. 567 (1847.)

[43] The English Statute of Monopolies granted the right to patent for the "sole working and making of new manufactures." Hence, the emphasis was on the nature of the invention rather than the inventor, and patents could be granted to noninventors such as the importers of inventions new to England.

leverage; thus, they strongly favored promptness in obtaining a patent.[44] To do otherwise would be to jeopardize the inventor's position against infringers, or even to risk voiding the patent as a "lache" (undue procrastination in asserting a legal claim), or on the grounds of prior use, or abandonment. Inventors who did not take out a patent, the Supreme Court ruled in *Pennock v. Dialogue* (1829), were not entitled to legal protection. Inventors were well advised not only to obtain patent protection but also to demonstrate "reasonable diligence" in so doing. After 1839 the law permitted two years for experimenting to reduce the invention to practical use. Still, the sewing machine industry proved to be particularly unfortunate in this regard. Walter Hunt, one of the first to be involved in developing a viable machine, sold his unpatented improvement of 1834. He was later refused a patent for the invention, because he was held to have abandoned it to the public by virtue of the sale. Wickersham lost priority to Singer in 1859 for the same reason. The rulings against delays, abandoned rights and prior use ensured that inventive rights were vested in patents rather than in secret processes, and expanded the information set available to potential inventors. At the same time, it increased the risks inherent in inventive activity that was directed toward longer-term payoffs.

A further factor in reaching a verdict was the extent and accuracy of the patent specification. Over a third (34.1 percent) of the early patent cases were lost because of overly broad specifications that failed to distinguish between the inventor's contribution and the original device. Patentees in the early period were unfamiliar with the newly established laws, especially with regard to the role and format of the patent specification.[45] Patents were held

---

[44] *Pennock v. Dialogue*, 27 U.S. 1 (1829): "If an inventor should be permitted to hold back from the knowledge of the public the secrets of his invention; if he should, for a long period of years, retain the monopoly, and make and sell his invention publicly; and thus gather the whole profits of it, relying upon his superior skill and knowledge of the structure; and then, and then only, when the danger of competition should force him to procure the exclusive right, he should be allowed to take out a patent, and thus exclude the public from any further use, than what should be derived under it, during his fourteen years; it would materially retard the progress of science and the useful arts; and give a premium to those who should be least prompt to communicate their discoveries." McLean delivered the majority opinion in *Shaw v. Cooper*, 32 U.S. 292 (1833), urging "Vigilance is necessary to entitle an individual to the privileges secured under the Patent Law. It is not enough that he should shout his right by invention, but he must secure it by law." This confirmed the prior ruling on this issue: "No man is permitted to lie by for years and then take out a patent. If he has been practising the invention with a view of improving it, ... that ought not to prejudice him. The intent of the delay is a question for the jury, also whether allowing use before patent does not amount to abandonment to the public," *Morris v. Huntington*, 17 F. Cas. 818 (1824.) Many inventors filed caveats in the Patent Office to record their progress on the invention long before formally filing for a patent.

[45] The English legal system also experienced early difficulties with questions such as incorrect specifications, which were resolved as patentees became more familiar with the requirements,

to be void on these grounds because faulty specifications fostered monopolies by claiming what had already belonged to others, and because they projected inaccurate information to potential inventors: "Specification should be complete so as not to injure other inventors" (*Whitney v. Emmett*, 1831.) This was entirely consistent with policies that fostered the rights of *all* patentees, including those out of court. Although a patent was limited to the extent of its published claims, judges united in interpreting those claims in favor of the patentee. According to Chief Justice Marshall, "In the construction of a patent, where the words are ambiguous, the intention of the parties is entitled to great consideration," and other courts concurred that "Patents should be construed liberally to support the claims of meritorious inventors."[46] Even if the patent were declared void on the grounds of a technicality, the inventor had the option of obtaining a reissued patent.[47] In *Moody v. Fiske* (1820), the machinist brought a suit against infringers of his 1819 patented improvements on the double speeder for roping cotton.[48] Moody lost the case because, contrary to the patent laws, he had bundled improvements on two machines in the one patent; however, he was able to withdraw the voided patent and obtain another with an accurate specification. The case was tried the following year under the new patent, and Moody was awarded punitive damages.

The patent laws had specified that patents were to be granted for "new and useful" inventions, but antebellum judges adopted a laissez-faire approach to the issue of the utility of the invention. In the 1817 case, *Lowell v. Lewis*,

---

and as patent agents grew in number. The implication is that certain features of the legal institution were independent of the attitudes of the courts, and could only be resolved with time and improved familiarity of the structure of law. Thomas Blanchard was nonsuited through two mistakes in his renewed patent, rectified the defects, and was awarded a decision on retrial of the case. "His honor Judge Story, on making his remarks, paid the following high compliment to Mr. Blanchard, viz.: 'That after much trouble, care, and anxiety, he will be enabled to enjoy the fruits, unmolested, of his inventive genius, of which he had a high opinion; and it afforded him much pleasure in thus being able publicly to express it'," Henry Howe, *Memoirs of the Most Eminent American Mechanics*, New York: A. V. Blake (1841.)

[46] Evans v. Eaton, 16 U.S. 454 (1818.) For decisions specifically instructing the jury in the liberal interpretation of patentee rights, see *Blanchard v. Sprague*, 3 Sumn. 535; *Brooks v. Bicknell*, 3 McL. 350; *Parker v. Stiles*, 5 McL. 44; *Corning v. Burden*, 15 How. 252; *Davoll v. Brown*, 1 W. & M. 53; *Parker v. Sears*, 1 Fish. 93; *Whitney v. Mowry*, 2 Bond 45; *Wintermute v. Redington*, 1 Fish. 239; *Goodyear v. Providence Rubber Co.*, 2 Cliff. 351; *Latta v. Shawk*, 1 Bond 259; *Cornell v. Downer & Bemis*, 7 Biss. 346; *Washburn v. Gould*, 2 Robb 206; *Emerson v. Hogg*, 2 Blatchf. 1.

[47] The statutes explicitly incorporated an option to obtain a reissue in Section 3 of the 1832 Act, but this merely legitimized rules that the courts had already established in cases such as *Morris v. Huntington*, 17 F. Cas. 818 (1824) and *Grant v. Raymond*, 31 U.S. 218 (1832.) The reissue was intended merely to correct the original patent, and could not be used to enlarge the claims of the first, although it could be amended to omit previous improvements. Assignees' rights were protected in the event of a reissue.

[48] *Moody v. Fiske*, 1 Robb 312 (1820.)

Joseph Story charged the jury that the utility of the invention "is a circumstance very material to the interest of the patentee, but of no importance to the public. If it is not extensively useful, it will silently sink into contempt and disregard."[49] It was thus the role of the market, rather than the courts, to determine the ultimate success of the patent. Consequently, judges followed the non-negative ruling that "all the law requires is that the invention should not be frivolous or injurious to the well-being, good policy, or sound morals of society." This policy was continued by the Patent Office, which similarly did not attempt to gauge the social or technical value of the invention, deciding conflicting claims predominantly on the basis of novelty.

However, as Justice Story stressed, the utility of the invention was of great interest to transactors in the market. According to the law, therefore, sale of property rights in the patent was held to be dependent on the value of the patent. Proof that the patent was of no utility was sufficient to void any contract to transfer those rights: "A lack of utility in a patented improvement may avoid a promissory note given for a conveyance of an interest in the patent;" and "If a patented thing be wholly useless, that will avoid a promise to pay money for an interest in the patent."[50] Moreover, the Supreme Court of Massachusetts ruled that "although the plaintiff might have purchased and sold the supposed patent right thinking it to be valuable property, still he could not recover in this action, for the defence did not rest on the ground of fraud, but on the ground that the defendant had received no value."[51] As such, the secondary market in patent rights was based on the legally valid assumption that the patent embodied some intrinsic technical value. This implies that the relatively deep market in patent assignments did not merely reflect speculative bubbles, but investments in productive capital.

The judiciary was united in its ranking of "mere licensees" who had no claim to property rights, but still were somewhat ambivalent about the rights of assignees, as opposed to inventors. It was argued that treble damages were due to inventors, but assignees could only recover actual damages. Some judges contended that "sacred rights of property and genius" did not apply

---

[49] The question of utility arises from the constitutional phrase, "the useful arts;" the statutes likewise permit patents for "new and useful" discoveries. Justice Story's ruling in Lowell v. Lewis was maintained by later courts as the legal standard: "The law, however, does not look to the degree of utility; it simply requires that it shall be capable of use," *Bedford v. Hunt*, 1 Robb 148 (1817); and "The popular demand for an article is, in the long run, the best test of utility," *Turrel v. Spalth*, 14 O.G. 377 (1878.) In any event, non-utility was hardly the most cogent defense against the patent right, for "Where the defendant has used the patented device, it does not lie in his mouth to dispute its utility," *Lowell v. Lewis* (1817.) I have read only one case, *Langdon v. de Groot*, 14 F. Cas. 1099 (1822), which was dismissed on the grounds of lack of utility. See however, the discussion in Page v. Ferry, 1 Fish. 298 (1857.)

[50] See Burnham v. Brewester, 1 Verm. 87 (1828) and Fallis v. Griffith, Wright 303 (Ohio, 1833.)

[51] *Dickinson v. Hall*, 21 Mass. 217 (1833.)

entirely to assignees, secular maximizers who dealt with the patent-right as a matter of business and speculation. Patents could be extended by patentees, never by assignees or licensees. In the eastern circuit, Justice Story had argued that the benefit of patent renewals should not be extended to assignees unless so specified in the assignment contract, whereas in the middle circuit Chief Justice Taney held otherwise. As insurance, assignees often brought their suit to term under the aegis of the inventor's name (occasionally with a token appearance in court for dramatic effect, as counsel for one of Samuel Morse's assignees suggested.) This distinction between assignees and inventors does underline that inventors and inventive property were regarded as a separate and "meritorious class."

The intricate patterns of assignment revealed in litigation records also shed some light on attitudes toward patentees and the degree to which patent rights were enforced. The development of trade is predicated on recognized rights of property: the market for patent rights, therefore, also signals the existence of enforceable property claims in antebellum inventions. Patent Office assignment records and law reports both reveal that an extensive and deep market in patent assignments and licenses functioned during the antebellum period. The Patent Office recorded over twenty-one hundred assignments in 1845 alone, almost 15 percent of cumulative patent awards up to that year.[52] Secondary (and tertiary) markets in patent rights flourished, sanctioned by law from the inception of the patent system, involving complex networks of subdivided rights that were bought, resold, and even bequeathed.[53] Amos Whittemore, for example, sold the rights to his textile card machinery in 1812 for $150,000.[54] If antebellum patents were indeed unenforceable, as some have argued, it is unlikely that patent rights would have been assigned for large sums, or resold at even higher prices.

ANALYSIS OF LEGAL SYSTEM

Although quantitative evidence of litigated cases provides an inadequate gauge of judicial attitudes toward the wider universe of patent disputes, the within-litigation sample of cases may nevertheless prove informative about

---

[52] See, for instance, *Potter v. Holland*, 19 F. Cas 1160 (1858), in which Allen Wilson's improvement for sewing machines was involved in some eight transactions within five years. The patent right itself could not be subdivided, but undivided parts could be limited by territories. Although assignees of partial interest in the patent were treated as licensees, they could bring equitable suits for injunctions against infringers (*Ogle v. Ege*, 4 Wash. 584, 1826.) The figure for assignments is from the *Report of the Commissioner of Patents*, 1846.

[53] The Act of 1793 had allowed "the assignee, having recorded the said assignment in the office of the Secretary of State, shall thereafter stand in the place of the original inventor, both as to right and responsibility . . . " as did assignees of assignees. Early assignees of overlapping patents mutually re-assigned rights in some instances to avoid potential litigation – for a description of such an 1829 patent pool, see Wilson v. Rousseau, 45 U.S. 646 (1846.)

[54] Henry Howe, *Memoirs of the Most Eminent American Mechanics*, New York: A. V. Blake (1841.)

Table 3.7. *Logit Regressions (Dependent Variable – Decide)[a]*

Independent Variables	(1)	(2)
Intercept	−0.279	−0.199
	(0.79)	(0.32)
Circuit		
First	−0.249	−0.252
	(0.69)	(0.60)
Second	−0.141	−0.139
	(0.24)	(0.21)
Third	−0.364	−0.461
	(1.13)	(1.64)
Courts		
Equity	−0.202	−0.319
	(0.82)	(1.89)
Supreme		−1.162
		(9.43)***
Supt2	−0.939	
	(2.36)	
Supt3	−1.160	
	(7.71)***	
LAW36	0.898	0.739
	(9.78)***	(6.72)***
LOGFREQ		0.164
		(4.74)**
	−2 Log L = 591.1***	−2 Log L = 582.8***

*   Significant at the 5 percent level
** Significant at the 1 percent level
*** Significant below the 1 percent level
[a] Probability of verdict for patentee

Notes: Numbers in parentheses are Wald χ-square statistics. The coefficients represent the log-odds of a verdict for the patentee. The sample totals 444 cases, with a value of 0 if the case was decided against the patentee, and a value of 1 if the decision was in favor of the patentee's claims before the court. The data exclude *ex parte* and interference appeals from decisions by the Commissioner of Patents.

Sources: See the text, footnote 4 and notes to Table 3.6 for definition of variables and sources.

relative enforceability. This section presents results from a multivariate logit model (Table 3.7) in which the dichotomous dependent variable, DECIDE, represents the likelihood of a verdict for the patentee in the courts.[55] The

[55] Maximum likelihood methods yield the reported coefficients, β_i, which represent the log-odds of a favorable decision conditional on the vector of independent variables, X. Since

$$d(\log P/(1 - P) = \beta_i dX_i,$$
$$dP/dX_i = \beta_i P(1 - P)$$

where P is evaluated at the mean probability. The dummy variables are dichotomous variables that take on a value of zero or 1.

discussion assesses four hypotheses about litigated patent cases. First, legal decisions were consistent across regions, tending to increase the degree of certainty in the system; second, an estimate of the degree to which the change in the patent system affected the probability of a decision favoring the patentee; and third, patent rights were more effective for minor inventions, but not for important discoveries. Finally, I consider whether common-law courts, in which patent rights were debated, generated more favorable decisions relative to courts of equity and the Supreme Court.

The dummy variables representing different Circuit Courts in the regressions reveal an overall consistency between regional decisions, especially in New England and New York, where 58.8 percent of all reported litigation occurred. Patentees were less likely to recover in Pennsylvania, but the difference is not statistically significant. Dummy variables representing individual judges also were statistically insignificant, which is likewise compatible with uniformity in judicial decisions. The implication is that the delegation of patent issues to the federal level did indeed foster a national market in patent rights, as the framers of the commerce clause intended. As early as 1807, Oliver Evans of Philadelphia prosecuted his patent claim before courts in Maryland and Virginia, as well as in his native Pennsylvania, whereas the Blanchard lathe was contested in five states.[56] The tendency toward ultimately consistent policies across regions mitigated the uncertainty of patentees whose inventions were used throughout the country.

To reiterate, the results confirm that decisions at law tended toward a level of national consistency ensured by constitutional edict. Observed outcomes were related less to the attitudes of the judiciary, and more to features of the patent claim specified by congressional statutes, which had defined true invention in terms of novelty and utility. The registration system of 1793 undoubtedly created difficulties because of a mismatch between patent grants and "new and useful" inventions. Judges attempted to maintain incentives under such a system by sending strong signals that only patent rights in inventions that conformed with statutory requirements would be maintained and enforced. The 1836 institutional change to an examination system, according to the results for the LAW36 variable, did increase the probability of a favorable outcome for plaintiffs.[57] However, rather than differences in

---

[56] Thomas Jefferson, an unknowing infringer, was called to account by Evans for royalties due, according to "Jefferson the Inventor and his Relation to the Patent System," Levi N Fouts, *Journal of the Patent Office Society* vol. 4, 1921–1922: 316–331.

[57] It may be argued that the change in the legal features of the patent system was initially merely redistributive in effect. In the pre-1836 regime, the burden of prosecuting a claim against infringement and/or interference was on the patentee. After 1836, infringement was still the domain of the courts, but interferences were decided by employees of the Patent Office, that was funded (and sometimes made a profit) from fees paid by patentees. The benefit to the patentees was equal to this sum, plus the cost of litigating interferences times

enforceability, the figures signal that the underlying system had changed to conform with the greater specialization and complexity of patent claims. The Patent Office would now filter out those claims that failed to meet the standards for novelty. As a result, a different portfolio of cases would be brought for trial, and for different reasons.

Some researchers have argued that the patent system merely induces the creation of "marginal inventions," whereas the legal system "discriminated and still discriminates against those who open and pioneer major fields, as distinguished from the inventors of gadgets."[58] The LOGFREQ variable – representing the frequency of litigation – is included as a proxy for important inventions. In an environment that was unambiguously supportive toward true inventors, a decision to uphold the right of any specific patentee conveyed a caveat to actual and potential infringers; it also provided an inducement for litigants to settle out of court in future disputes about the patent.[59] The 210 inventions litigated in a single lawsuit were presumably settled out of court in any subsequent disputes. Yet, as noted earlier, the phenomenon of multiple litigation was a notable feature of the American legal system. A list of frequently litigated inventions clearly represents inventions of exceptional commercial value. McCormick's harvester, the Blanchard lathe, Evans' hopperboy, sewing machine patents filed by Singer and Wilson, cotton gins, and Morse's telegraph all proved to be as well known in the annals of litigation as of invention. Three antebellum outliers were even more controversial: the Parker waterwheel (21); Goodyear's vulcanization of india-rubber (21); and William Woodworth's planing machine, the subject of an astonishing seventy-eight reported lawsuits.

The greater the value of any discovery, the greater the effort to circumvent the inventor's monopoly by contriving a new method to achieve the same effect. Whether or not infringers could successfully attack the patent depended on the courts, which followed a consistent policy of attempting to secure just rewards for individuals whom they recognized as public

the reduction in the probability of litigation due to screening by examiners (interferences.) Given that this probability was very small in the first place, it is not entirely clear that this shift in the liability was of much social significance at the time. This is borne out quantitatively by the relatively minor fall in the time series of patent applications. However, as inventions became more complex technically, and as the Patent Office expanded its operations to include the dissemination of information, it may be expected that the relative efficiency of the post-1836 system increased. The point is that the efficiency of the later system does not imply the inefficiency of the former regime.

[58] Frank D. Prager, "Trends and Developments in American Patent Law from Jefferson to Clifford (1790–1870)," *American Journal of Legal History*, vol. 6, January (1962.)

[59] Judges at times attempted to encourage litigants to settle out of court, especially when an injunction might result in the closure of large numbers of enterprises, for example, *Parker v. Brant*, 18 F. Cas. 1117 (1850): "we feel a reluctance to stop two hundred mills ... without giving the defendants a chance of making a settlement or compromise," and *Woodworth v. Edwards* (1847.)

Table 3.8. *Jurisdiction in Patent Cases, 1790–1860*

Decade	Common Law	Equity	Supreme Court	Appeals CP*
1790–1799	1	–	1	–
1800–1809	4	1	–	–
1810–1819	22	2	3	–
1820–1829	21	5	5	–
1830–1839	10	1	2	–
1840–1849	70	73	13	16
1850–1859	77	111	39	101
1860	13	7	1	33
Total	218	200	64	150

Notes: Appeals CP* indicates appeals from the Commissioner of Patents, including interferences.

benefactors. Justice Grier, in deciding for the plaintiff declared with open indignation that "It is only when some person, by labor and perseverance, has been successful in perfecting some valuable manufacture, by ingenious improvements, and labor-saving devices, that their patents are sought to be annulled by digging up some useless, musty, forgotten contrivances of unsuccessful experimenters."[60] As the regression results imply, these unsuccessful experimenters tended to be equally unsuccessful at law. For instance, more than 70 percent of the Woodworth patent cases – litigated in twelve states – were decided in favor of his assignees. "Great inventors" such as Eli Whitney, Oliver Evans, and Charles Goodyear were equally well known in the annals of litigation as of invention. The records for the great inventors underline the conflict between monopoly grants and social welfare that later courts would recognize and attempt to resolve, especially in equitable disputes.

PATENTS, MONOPOLY, AND SOCIAL WELFARE

Patent rights were protected under the "supreme law of the land," but the intent to promote the progress of science and useful arts required a more subtle approach than the monolithic protection of all patentee-plaintiffs. The 1840s saw an increase in the number of patentees resorting to courts of equity, to obtain temporary or permanent injunctions against unauthorized users of their inventions (Table 3.8.) Joseph Story, recognized as the greatest proponent of equity jurisprudence of the period, identified the most marked contrast between the United States and Britain to be policies in Chancery, or before courts of equity. In England, Story pointed out, suits brought before Chancery were uncertain, and dependent on the whims of

---

[60] *Adams v. Jones*, 1 F. Cas. 126 (1859.)

individual judges. Unlike this "judge-made law," federal circuit courts in the United States were granted original jurisdiction at law and at equity since 1819, and both were conducted on "scientific" principles of precedent.[61] The negative coefficient on the EQUITY dummy variable suggests a lower likelihood of decisions in favor of plaintiffs at equity relative to common law (the excluded variable.) Common-law cases dealt with infringement and disputes about the validity of the inventor's property rights. Even though the same judge frequently presided in both courts, rulings in equity were a more complex matter because equitable disputes forced the courts to weigh the rights of the individual inventor against the rights of fellow inventors and the community at large.[62]

Preliminary injunctions could be obtained pending common-law litigation, if patentees stood to suffer severe losses. But judges were alert to the possibility of "irreparable harm to the defendant, in breaking up his trade or business."[63] Oliver Parker's request for a wholesale injunction against one hundred mill owners was disallowed. The patent was within weeks of expiring; the judge was thus reluctant to issue an injunction that would adversely affect so many enterprises, when the patentee received no benefit from closure of the mills and would later be compensated by the payment of damages if it were indeed proven that the patent was infringed (*Parker v. Sears*, 1850.) In *Woodworth v. Hall* (1846), Justice Woodbury ruled that where some doubt existed regarding the merits of the injunction, the courts should be inclined against, rather than in favor of the plea. The logit results

---

[61] Joseph Story, *American Jurist*, vol. 1 (January) 1829: 1–34. Lubar disagrees with Story: "The growth of equity – it's triumph over the common law – allowed judges to take patent law into their own hands. A hearing before a judge, followed by an injunction, became the general rule in patent cases," "The Transformation of Antebellum Law," *Technology and Culture*, October 1991. Compare: "Whether the complainant's patent is good and valid so as ultimately to secure to him the right he claims, is not a question for decision upon the equity side of this court. That is a question which belongs to a court of law, in which the parties have a right of trial by jury" (Sullivan v. Redfield, 1 Paine 444); "Where there is reasonable doubt as to the novelty of the patent or its infringement, a preliminary injunction will not be granted" (*Winans v. Eaton*, 1 Fish. 181; see also *Woodworth v. Hall*, 1 W. & M. 248; *Potter v. Fuller*, 2 Fish. 251; *Brooks v. Bicknell*, 3 McL. 250; *Hovey v. Stevens*, 1 W. & M. 290.) These decisions all indicate that equity *followed* the common law. Moreover, Clifford pointed out, Chancery jurisdiction and practice in the U.S. circuit courts were the same in all states (*Blanchard v. Sprague*, 1 Cliff. 288.) For patentees, "The prevention of a multiplicity of suits is one of the most salutary powers of a court of equity," *Thomas v. Weeks*, 7 F.C. 154 (1827.)

[62] The Supreme Court underlined this in *Kendall v. Winsor*, 21 How., 322, Sup. Ct. (1858): 'Whilst the remuneration of genius and useful ingenuity is a duty incumbent upon the public, the rights and welfare of the community must be fairly dealt with and effectually guarded. Considerations of individual emolument can never be permitted to operate to the injury of these.'

[63] *Goodyear v. Dunbar*, 1 Fish P.C. 472 (1859.) The citations for the other cases in this paragraph are Parker v. Sears, 1 Fish. 93 (1850); and Woodworth v. Hall, 2 Robb 495 (1846).

confirm Woodbury's ruling, and in general judges preferred to rule that either party post a bond as surety pending a decision at law.

In the absence of antitrust statutes, equity provided a more flexible channel for mediating between the inventor's exclusive rights and a general monopoly. The plaintiff in *Smith v. Downing* (1850), an assignee of Morse, sought a permanent injunction against the defendants, who operated a telegraph under assignment from Royal E. House. After a detailed exposition of the incremental nature of the development of the telegraph, the court refused the injunction. Exclusive patent rights allowed the inventor to benefit from the acknowledged property in his improvement; at the same time such property did not extend to the entire field, because this would grant the marginal improver a monopoly that would halt general progress in the area. House's telegraph was not only different from Morse's, but technically superior; hence to mandate an estoppel against his ingenuity and the defendants' enterprise would be an "extraordinary" measure.

Decisions before the Supreme Court reveal a similar attempt to resolve the paradox of promoting inventive rights without suppressing economic progress.[64] The variables SUPT2 and SUPT3 represent the interaction between a dummy variable for the Supreme Court, and the 1837 to 1849 and 1850 to 1860 periods, respectively. The statistically significant negative coefficient for the 1850s relative to the earlier period (SUPT3) suggests that the early insouciant judicial optimism about the coincidence between private and public welfare had waned by the second half of the century.[65] By then, the courts had experienced the network of litigation launched by patentees

[64] *Reported Decisions by Judges in the* U.S. *Federal Courts, 1790–1860*

Judge	No. For	(%)	No. Against	(%)	No Decision
Grier*	18	58	13	42	–
Ingersoll	12	55	9	41	1
Leavitt	12	80	3	20	–
McLean*	27	59	17	37	2
Nelson*	50	53	37	39	7
Sprague	9	69	3	23	1
Story*	22	46	24	50	2
Taney*	7	44	7	44	2
Washington*	8	35	15	65	–
Woodbury*	20	69	9	31	–
Others**	55	50	50	46	7
Total	240	53	187	41	22

Notes: * Indicates Justices on the Supreme Court of the United States.
** Excludes appellate cases from the Commissioner of Patents, D.C.

[65] Baldwin declared: "Congress have declared the intention of the law to be to promote the progress of the useful arts by the benefits granted to inventors; not by those accruing to the public, after the patent has expired, as in England."

and their assignees, such as William Woodworth and the Parker brothers, to protect national monopolies. Justice Woodbury was prompted to dictate: "The rights of inventive genius, and the valuable property produced by it, all persons in the exercise of this spirit will be willing to vindicate and uphold, without colorable evasions and wanton piracies; but those rights on the other hand, should be maintained in a manner not harsh towards other inventors, nor unaccommodating to the growing wants of the community."[66] Horwitz presents evidence for similar developments in judicial decisions regarding riparian property rights and corporate charters. Although initially monopoly grants had been considered essential to promote economic development, "the restrictive consequences of these grants were becoming apparent by the second quarter of the nineteenth century."[67] However, it was not until much later that patent decisions addressed the concerns of the judiciary in riparian and corporate disputes, because antebellum courts initially argued that patents belonged to a different class of rights: rather than monopolists, patentees were "public benefactors," whose property it was the duty of the court to defend in order to promote technological change. Ultimately, the judiciary came to openly recognize that the enforcement and protection of *all* property rights involved trade-offs between individual monopoly benefits and social welfare.

Early courts had to grapple with a number of difficult issues, such as the appropriate measure of damages, disputes between owners of conflicting patents, disagreements between employers and employees, and how to protect the integrity of contracts when the law was altered. Changes inevitably occurred when litigants and judiciary both adapted to a more complex inventive and economic environment. We saw one example of this in the evolution of the doctrine of nonobviousness. Another accommodation related to the treatment of employees who invented new and useful improvements. According to Morton Horwitz, in the antebellum period the judiciary reinterpreted existing legal rules in property, torts, and contracts in an instrumentalist fashion to place the burdens of expansion on workers and farmers. This "ruthless" transformation meant that the economically progressive classes were able to "dramatically . . . throw the burden of economic development on the weakest and least active elements of the population."[68] This claim is questionable in general, but especially inaccurate in relation to patent contracts, where the rights of all inventors were protected by the Constitution.

For most of the nineteenth century, courts followed the ruling that employees were the rightful owners of any inventions they created on the job, and

---

[66] *Woodworth v. Edwards*, 30 F. Cas. 567 (1847.)

[67] Morton J. Horwitz, *The Transformation of American Law, 1780–1860*, Cambridge, Mass.: Harvard University Press (1977), p. 130.

[68] Horwitz, *Transformation*, p. 101.

at best employers were granted a license to use inventions they had commissioned.[69] The 1880 Dahlgren Gun case presents an interesting episode that illustrates the limitations of the work for hire doctrine in the context of patent property.[70] John A. Dahlgren, a resident of Philadelphia, was employed as a U.S. naval officer with the express task of improving on heavy artillery ordnance. The U.S. Government covered all expenses for experiments, tests and construction of the guns. The Navy manufactured and used 3,948 Dahlgren guns on its war vessels, successfully transforming its military strategy at significantly lower cost. In recognition of his efforts, the inventor was paid a premium of $1,000 in annual salary above the regular pay for his rank, and also was promoted several times. Dahlgren allowed the Navy to use the inventions but obtained several patents for his improvements on his own account. The Attorney General argued that the inventions and the patents rightfully belonged to the government because they were created in the line of duty, under the supervision of the Navy, and at the public expense. Despite the strong evidence in favor of the government's position, the work for hire argument was rejected, and the court focused on an assessment of the appropriate level of compensation. The U.S. government was the only consumer of the inventions, so it was difficult to compute the opportunity cost of its infringement. The U.S. Court of Claims settled on an estimate of the incremental value of the improvements, and awarded Madeleine Dahlgren, the widow of the inventor, the sum of $65,000 to compensate for previous use of the patented inventions and to transfer the rights to the government.

Toward the early part of the twentieth century, as corporate laboratories and team invention were becoming more standard, courts allowed employers greater rights. These included free access to patented inventions that had been produced within the scope of employment, and arguments that some employment contracts implied assignments of patent rights.[71] However,

[69] As the law reporter for *McClurg v. Kingsland*, 42 U.S. 202 (1843), pointed out: "If a person employed in the manufactory of another, while receiving wages, makes experiments at the expense and in the manufactory of his employer; has his wages increased in consequence of the useful result of the experiments; makes the article invented and permits his employer to use it, no compensation for its use being paid or demanded; and then obtains a patent, these facts will justify the presumption of a license to use the invention."

[70] *Mrs. Madeleine V. Dahlgren, Administratrix, v. The United States*, 16 Ct. Cl. 30 (1880.) "The gun invented by said Dahlgren proved of great advantage to the United States, by reason of its larger shell to the same amount of metal, of its greater safety, of its greater power, of its economy in powder, and of its requiring a smaller number of men to the same amount of metal."

[71] *Solomons v. United States*, 137 U.S. 342 (1890): "But this general rule is subject to these limitations: If one is employed to devise or perfect an instrument or a means for accomplishing a prescribed result, he cannot, after successfully accomplishing the work for which he was employed, plead title there too as against his employer. That which he has been employed and paid to accomplish becomes, when accomplished, the property of his employer. Whatever rights as an individual he may have had in and to his inventive powers and that which

regardless of such changes that accommodated the evolution of technology and its organization, the system remained true to the Constitution in the belief that the defense of rights in patented invention was important in fostering industrial and economic development. Those who allege that patent rights comprised little more than advertisement value therefore seriously misrepresent the efficacy of the early patent system. The federal courts acknowledged that inventive efforts varied with the extent to which inventors could appropriate the returns on their discoveries, and attempted to ensure that patentees were not unjustly deprived of the benefits from their inventions. Increases in public welfare, it was proposed, were a derivative of private initiative and enterprise: to encourage and benefit individual inventiveness thus promoted economic growth and democratic development. The early focus on securing the rights and benefits of patentees, rather than on the social costs of monopoly grants, ensured the effectiveness of markets in inventive rights, and enhanced the private return on patent protection. If inventive activity was indeed responsive to material incentives during early American industrialization, then the legal system played an important part in stimulating greater technical change by reinforcing the effectiveness of the patent system.

they are able to accomplish, he has sold in advance to his employer." See also Catherine L. Fisk's excellent and comprehensive paper, "Removing the 'Fuel of Interest' from the 'Fire of Genius': Law and the Employee-Inventor, 1830–1930," 65 *U. Chi. L. Rev.* 4 (1998): 1127–98.

4

Democratization and Patented Inventions

"We have democratized the means and appliances of a higher life."
– Horace Greely (1853)

Early American history was remarkable for the unprecedented surge in technological creativity that affected every facet of life, from the production of paper collars to the speed and comfort of transatlantic voyages.[1] Political commentators of the time were confident that technology and republican values of liberty and equality of opportunity were linked in a mutually reinforcing fashion: not only would a democratic system result in the greatest incentives for the creation and diffusion of useful ideas, but inventive activity also would increase the welfare of all citizens regardless of social class. The greatest good for the greatest number would be best attained if efforts were concentrated on pragmatic objectives like increasing national output, and minimizing the cost of production through improvements in ploughs, shoemaking implements, and brewing. Tocqueville had thought it necessary to explain why "the example of the Americans does not prove that a democratic people can have no aptitude and no taste for science, literature, or art." By way of contrast, Joseph Holt, the U.S. Commissioner of Patents in 1857, was gratified that American inventors had not paid much attention to "articles of mere luxury." He was further pleased to note that "the inventive genius of the country . . . has been confined to no class or pursuit . . . [that it] is one of the boons of our republican institutions, may be affirmed without the hazard of contradiction."[2]

Who were the individuals responsible for the dramatic increase in the rate of technological change in the United States during this period, and what motivated their contributions? Many believe that this rapid increase in invention was propelled by fundamental breakthroughs, which yielded a

[1] A detailed and pathbreaking study of the sources of inventive activity in the antebellum period is provided in Kenneth L Sokoloff, "Inventive Activity in Early Industrial America: Evidence from Patent Records, 1790–1846," *JEH* v. 48 (December) 1988: 813–50.
[2] *Report of the Commissioner of Patents for the Year 1857*, p. 2 and p. 7.

profusion of secondary innovations.[3] They argue that these major improvements in the stock of knowledge available to prospective inventors, or an
outward shift of the supply curve, was the primary source of the increase
in inventive activity during the early nineteenth century. Others, however,
view the era as one in which knowledge required for invention was basic
and widely diffused across the population, and in which the supply of ideas
was relatively elastic. Unparalleled increases in market demand and producer competition helped identify new problems that could be solved with
an existing and readily available stock of knowledge.[4] The "fuel of interest"
or the prospect of expanding profit opportunities induced growing numbers
of people to invest more in inventive activity and innovation. The virtuous
circle of democracy and technology induced ordinary individuals to reorient
their efforts to improve both inputs and final goods, and ensured that their
cumulative efforts increased the standard of living for all classes of society.

The U.S. patent system was instrumental in directing the efforts of a diverse
array of individuals toward extracting returns from their improvements.
Samuel L. Clemens, who is more noted as the author of classic works in
American literature, earned $50,000 from his 1873 patent on a self-pasting
scrapbook, more than the income from any one of his books.[5] Clemens,
a resident of Hartford, Connecticut, was involved in a patent interference
suit with Henry Lockwood of Baltimore regarding mutual patent applications for "adjustable and detachable straps for garments." The author
expended much of his fortune in fruitless investments in patented mechanisms, including the purchase of a partial assignment in the patent rights to
New York inventor Charles Sneider's 1881 improvement in relief-line printing and embossed plates. The most spectacular of these failed investments
was in the Paige Compositor, a typesetting machine with eighteen thousand
parts and such a voluminous specification that the Patent Office employees referred to it familiarly as "the Whale." A less-renowned inventor, Frank
Fuller of New York, had equal hopes for his 1868 "Universal Gardener," and
his patent specification assures that he "intended to manufacture these beautiful instruments by the million, and to furnish them to every household in
the United States; and it is fair to conclude that by thus supplying the means
for the cultivation of useful and beautiful plants, trees, and shrubs, a new

---

[3] For a closely argued and readable exposition of this approach, see Joel Mokyr, *The Lever of Riches: Technological Creativity and Economic Growth*, Oxford: Oxford University Press (1990.)

[4] Historians of American technology generally emphasize that many people were familiar with the basic technology of the period and capable of producing useful inventions. See David Hounshell, *From the American System to Mass Production, 1800–1932*, Baltimore: Johns Hopkins University Press (1984.)

[5] Specification forming part of Letters Patent No. 140,245, dated June 24, 1873; application filed May 7, 1873.

love for those Floral apostles, that, in dewy splendor, "Weep without woe, and blush without a crime," will be developed, encouraged, stimulated, and that the sum of human happiness will thereby be increased."[6] Other patentees may have been less lyrical, only making stolid references to "utility," but their common objective was to benefit themselves through benefits to society.

Kenneth Sokoloff and I used a random sample of U.S. patent records from 1790 to 1846 to explore the nature of technological change in the early nineteenth century, and the results of our joint study are discussed here.[7] The data set included the number of career patents that each inventor obtained, as well as a measure of sectoral specialization. We matched a subset of the patentees with city directories to obtain information on their occupations. Although it is difficult to distinguish empirically between the different kinds of knowledge at issue, these data help to investigate the role of advances in knowledge in stimulating inventive activity. The focus is on technical knowledge or skills that, like specific human capital, were costly to obtain, largely restricted to members of particular occupations or industries, and scarce among the general population.[8] This approach excludes consideration of a broad range of other types of knowledge or information, such as those concerning market conditions and general technology, but this is not an arbitrary choice. Rather, the interest here is in whether most advances were made by a small group of elite individuals who were exceptionally well-qualified for invention because of their technical expertise or background. The intent is to address the hypothesis that fundamental advances in knowledge caused the surge of invention at the onset of economic growth, relative to the alternative hypothesis of a democratization of invention that incorporated a wide cross-section of the population.

---

[6] Patent No. 81619 September 1, 1868, Improvement in Garden Implements.

[7] Kenneth L. Sokoloff and B. Zorina Khan, "The Democratization of Invention during Early Industrialization: Evidence from the United States," *Journal of Economic History*, vol. 50 (2) 1990: 363–78; B. Zorina Khan and Kenneth L. Sokoloff, "Institutions and Democratic Invention in 19th Century America," *American Economic Review*, vol. 94 (May, Pap. and Proc.) 2004; B. Zorina Khan and Kenneth L. Sokoloff, "The Early Development of Intellectual Property Institutions in the United States," *Journal of Economic Perspectives*," vol. 15 (3) 2001: 233–246. For discussions of the data set, see Sokoloff (1988.)

[8] The logic is that individuals with large investments in invention-generating capital will tend to focus their attention on ideas that involve capital equipment – both because they probably have a comparative advantage in such inventions, and because such inventions are more likely to yield returns to the inventor. For an interesting treatment of the bias toward invention involving large scale or capital equipment, see Stephen A Marglin, "What Do Bosses Do?: The Origins and Functions of Hierarchy in Capitalist Production," *Review of Radical Political Economics*, vol. 6 (Summer) 1974: 60–112. For a discussion of the importance of investment in capital equipment, see Kenneth L Sokoloff, "Investment in Fixed and Working Capital During Early Industrialization: Evidence from U.S. Manufacturing Firms," *Journal of Economic History*, vol. 44 (June) 1984: 545–56.

The United States.

To all to whom these Presents shall come, Greeting.

Whereas Samuel Hopkins of the City of Philadelphia and State of Pennsylvania hath discovered an Improvement, not known or used before, such Discovery, in the making of Pot ash and Pearl ash by a new Apparatus and Process; that is to say, in the making of Pearl ash 1st by burning the raw Ashes in a Furnace, 2d by dissolving and boiling them when so burnt in Water, 3d by drawing off and settling the Ley, and 4th by boiling the Ley into Salts which then are the true Pearl ash; and also in the making of Pot ash by fluxing the Pearl ash so made as aforesaid; which Operation of burning the raw Ashes in a Furnace, preparatory to their Dissolution and boiling in Water, is new, saves little Residuum, and produces a much greater Quantity of Salt: These are therefore in pursuance of the Act, intitled "An Act to promote the Progress of useful Arts", to grant to the said Samuel Hopkins, his Heirs, Administrators and Assigns, for the Term of fourteen Years, the sole and exclusive Right and Liberty of using and vending to others the said Discovery, of burning the raw Ashes previous to their being dissolved and boiled in Water, according to the true intent and meaning of the Act aforesaid. In Testimony whereof I have caused these Letters to be made patent, and the Seal of the United States to be hereunto affixed. Given under my Hand at the City of New York this thirty first Day of July in the Year of our Lord one thousand seven hundred and Ninety.

G Washington

City of New York July 31st 1790.—

I do hereby certify that the foregoing Letters patent were delivered to me in pursuance of the Act, intitled "An Act to promote the Progress of useful Arts"; that I have examined the same, and find them conformable to the said Act.

Edm: Randolph   Attorney General for the United States.—

Illustration 7. The first U.S. patent was issued to Samuel Hopkins in 1790. The document is signed by George Washington, Attorney General Edmund Randolph, and (on the reverse side) by Secretary of State Thomas Jefferson. (*Source:* U.S. Patent Office.)

PATENTEES IN THE ANTEBELLUM PERIOD

The first U.S. patent was issued on July 31, 1790, just a few months after Congress enacted the first patent legislation. Samuel Hopkins, a forty-six-year-old Quaker potash-maker from Philadelphia, was granted a patent for a furnace apparatus and improved method of making potash and pearl ash (a more refined form of potash), an ingredient in the manufacture of soap, glass, fertilizer, and saltpeter for gunpowder.[9] The Patent Examination Board approved the patent, and the document was personally signed by George Washington, Thomas Jefferson, and Attorney General Edmund Randolph. Hopkins actively pursued the returns from his invention by manufacturing the product himself, and he also authorized others to employ his ideas. He licensed the rights to use his patent for a down payment of $50, and an additional $150 for the next five years; or for equivalent payments in potash produced.[10] In 1791 he petitioned the Quebec Parliament in Canada for exclusive protection for his invention and was granted a privilege for six years. The United States would become one the leading producers of potash during the antebellum period. Hopkins's patent was only the first of some 1,811 patents relating to potash, out of approximately 650,000 patents that would be granted in the United States over the course of the nineteenth century.

A salient feature of this impressive growth of patenting in the United States during the early stages of industrialization is that the advance was far from continuous. Patenting appears to have been markedly procyclical, and virtually all of the increase between 1790 and 1850 was realized during two concentrated intervals.[11] These spells of rapid growth in patenting do not appear to have been stimulated by specific developments in technology or in the diffusion of information. Instead, they were related to macroeconomic events that led to the expansion of markets and increased expected benefits relative to the cost of securing a patent. Patenting surged after Jefferson's embargo of 1807 raised the effective demand for domestic manufacturers and expanded domestic markets. These patterns suggest that inventors were responsive to economic factors and that demand-induced advances changed the potential return to inventive activity. This does not imply, however, that

---

[9] See David M. Maxey, "Inventing History: The Holder of the First U.S. Patent," *Journal of Patent and Trademarket Office Society*, vol. 80 (March) 1998: 155–70. Thomas Jefferson had an interest in the manufacture of potash, which he abandoned in favor of nail production because the latter required lower financial capital investments.

[10] The licensing information is from Henry M. Paynter, "The First U.S. Patent," *Invention and Technology*, Fall (1990): 18–22.

[11] Other scholars have noted the procyclicality of patenting. See the references provided in Chapter 2. For different perspectives on the procyclicality of invention, see Robert C Allen, "Collective Invention," *Journal of Economic Behavior and Organization*, vol. 4 (1983):. 1–24; and Andrei Shleifer, "Implementation Cycles," *Journal of Political Economy*, vol. 96 (December) 1986: 1163–90.

the states of knowledge during these cycles were irrelevant. On the contrary, if the level of technical knowledge had not been sufficient to yield inventions worthy of being patented during these intervals, expansion in the size of the market economy alone would not have led to significant increases in the number of patents filed. Hence, the procyclicality of patenting during the early phase of American industrialization suggests that much of the invention carried out was based on currently available technical knowledge, and involved the application and marginal extension of existing techniques. In conventional economic terms, the states of knowledge and technology allowed for a relatively elastic supply of patentable ideas. For instance, Cyrus Hamlin, a junior at Bowdoin College (Class of 1834) who intended to be a missionary, was able to build the first working model of a steam engine in Maine, as a class project.[12]

Table 4.1 presents the total number of patents that a patentee received (through 1846) as a measure of long-term commitment to patenting and invention-generating capital.[13] The most important finding here is the growing prominence of individuals with only one or two patents over their career. For example, whereas patentees with only one career patent accounted for 46.1 percent of those awarded between 1790 and 1804, their share rose to 53.3 percent during the embargo years when patenting sharply increased, and remained in the high fifties and low sixties for the remainder of the period. With the rest of the distribution generally stable (especially the lack of trend in the proportion going to individuals with over ten patents), this change in composition implies that patentees with relatively modest long term

---

[12] Hamlin graduated in 1834 from Bowdoin College, where he was a student library assistant for Prof. Henry Wadsworth Longfellow. During his junior year he was inspired by a single lecture and an article in the *Journal of the Franklin Institute*, Philadelphia ("Lardner on the Steam Engine".) The Hamlin steam engine was produced at a cost of $72. He noted that "I would not like to have any mechanic look at it without remembering that it is the first steam engine ever made in the state of Maine and that I made it without competent tools or competent knowledge." His autobiography is entitled *My Life and Times, by Cyrus Hamlin, missionary in Turkey*, Boston; Chicago: Congregational Sunday-school and publishing society, c. 1893.

[13] The term "invention-generating capital" encompasses any human or physical capital that raises individual productivity in inventive activity. Such effects can be either sector-specific or general. The patents of patentees who were active at the beginning or the end of the period are not fully recorded in the data, and accordingly their career patent (or future patent) totals are undercounted. Although careful examination indicated that patentees active after 1842 suffered from this problem (leading to a downward bias in the estimate of their career patents), not much of an effect was detected for those who filed before. Hence, it is only the observations from 1843 onward that are suspect, and these were deleted from the regressions below. Because any bias during the 1790s would work against the argument presented here, no adjustments to the sample were made. The 1836 change in the patent system also could have accounted for some decline in the average number of career patents. However, because most of the shift toward patentees with relatively few patents was realized during the embargo years, it is unlikely that either the institutional change or the cut-off at 1846 could provide an adequate explanation.

Table 4.1. *Distribution of Patents by Patentee Career Total, Britain and America, 1750–1850*

Period	1 Patent		2 Patents		3 Patents		4–5 Patents		6–9 Patents		More than 10 Patents	
	n	row %	n	row %	n	row %	n	row %	n	row %	n	row %
1750–1769												
England	181	71.0%	51	20.0%	10	3.9%	9	3.5%	4	1.6%	0	0.0%
United States	–	–	–	–	–	–	–	–	–	–	–	–
1790–1811												
England	144	52.2%	65	23.6%	27	9.8%	23	8.3%	13	4.7%	4	1.4%
United States	263	51.0%	98	19.0%	62	12.0%	39	7.6%	36	7.0%	18	3.5%
1812–1829												
England	75	42.9%	33	18.9%	23	13.1%	19	10.9%	14	8.0%	11	6.3%
United States	823	57.5%	249	17.4%	102	7.1%	109	7.6%	78	5.5%	70	4.9%
1830–1842												
England	83	46.1%	37	20.6%	18	10.0%	20	11.1%	7	3.9%	15	8.3%
United States	1102	57.4%	317	16.5%	156	8.1%	153	8.0%	108	5.6%	85	4.4%
1843–1850												
England	100	51.8%	28	14.5%	21	10.9%	21	10.9%	10	5.2%	13	6.7%
United States	329	60.5%	96	17.7%	48	8.8%	39	7.2%	13	2.4%	19	3.5%
All Years												
England	583	49.1%	214	18.2%	99	10.9%	82	10.4%	48	6.5%	43	6.1%
United States	2,517	57.1%	760	17.2%	368	8.3%	340	7.7%	235	5.3%	192	4.4%

Sources: For U.S. sample, see Kenneth L. Sokoloff and B. Zorina Khan, "The Democratization of Invention During Early Industrialization: Evidence from the United States, 1790–1846," *Journal of Economic History*, vol. 50 (no. 2), 1990: 363–378. The English sample was drawn from Bennett Woodcroft, *Titles of patents of invention*, chronologically arranged from March 2, 1617 (14 James I.) to October 1, 1852 (16 Victoriae); London: Printed by G. E. Eyre and W. Spottiswoode. Published at the Queen's Printing Office (1854.)

investments in inventive activity responded most strongly to whatever conditions were stimulating the increased economy-wide commitment to invention and innovation. The contrast between the procyclicality of their activity and the virtual absence of procyclicality on the part of those with large long term commitments provides further evidence that patentees with smaller numbers of career patents were especially sensitive to economic conditions. This feature is reflected in the sharp rises in the share of patents going to patentees with ten or more patents during the two prolonged economic downturns (averaging 64 percent.)

Table 4.1 also provides further perspective on the more restrictive English system. The changes in these distributions over time reveal a sharp contrast between the two economies. In the United States, the share of all patents accounted for by patentees with only one career patent grows during the initial phase of industrialization (from 46.1 percent in 1790–1804 to 57.5 percent in 1812–1829), whereas this share shrinks dramatically in England during the analogous stage (71.0 percent in 1750–1769 to 42.9 percent in 1812–1829.) The divergence in experience may reflect differences across countries both in how broadly the commercial opportunities created by economic growth extended across social classes, as well in the capital requirements for securing a patent. Given the severe difficulties ordinary citizens in England faced, relative to their wealthy countrymen, in obtaining property rights to their inventions, it is perhaps not surprising that such property rights were more concentrated. The figures in Table 4.1 suggest that the holding of patents was more concentrated in England than in the United States. If one considers the much lower patenting rates in England, the disparity as regards concentration of patents seems all the greater.

The increase in the proportion of patents awarded to individuals with few career patents leads one to question the significance of technical knowledge and skills in accounting for the growth of inventive activity experienced during the period. After all, if capital investments in the acquisition of technical skills and knowledge were associated with an advantage in generating patentable inventions, or in exploiting them commercially, then one might expect that patentees with greater demonstrated long-term commitments to patenting would account for an increasing share of patents over time. Moreover, the same logic should apply to improvements in the diffusion of knowledge: if individuals with investments in technical knowledge were better equipped to understand the significance of new information and act on it, then their patenting should have been disproportionately affected by a geographically wider diffusion of information. However, given that the share of these "high-commitment" patentees actually declined, it would appear that the most salient developments behind the rise in the number of patents during the era were those that stimulated individuals with relatively limited investments in such invention-generating capital.[14]

It is true that patentable ideas vary enormously in terms of both private and social returns, and that evidence of a growing proportion of patents does not necessarily imply an increase in the relative value of inventions contributed by "low-commitment" patentees. Indeed, some might argue that this segment of patentees yielded few important inventions, and registered a disproportionate share of trivial or frivolous patents. This possibility is difficult to evaluate but indirect evidence seems to bear against it. As Chapter 3 indicated, the 1836 change in the patent law involved a tightening of the requirements for an award. If the patents previously filed by "low-commitment" individuals were of generally inferior quality, one would expect the new standards to screen out a larger fraction of their applications and to produce a decline in their share of total patents awarded. Because there was no such drop, however, this method of gauging the significance of patents supports the inference drawn above about the growing prevalence and importance of inventions from individuals whose lifelong commitments to patenting were relatively modest.

OCCUPATIONS, SPECIALIZATION, AND CAREER PATENTING

The occupations of patentees provides another means of assessing the changes in the structure and sources of patentable invention. These data were retrieved from city directories for a subset of urban patentees who appeared in the sample. Table 4.2 presents the distributions of these matched urban patentees across six occupational groups for four periods between 1790 and 1846. The occupational structure clearly changed substantially over the years of major growth in inventive activity.[15] Men in commerce and the professions (primarily merchants) enjoyed a dominant (50 percent) share of patents through the 1790s and up to the embargo, but their position began to erode steadily as various classes of artisans and other occupations increased their patenting much more rapidly. By 1836–1846, the commerce share had declined to 18.6 percent, and on the eve of the Civil War merchants

[14] If information on the occupational distribution of the underlying population was available, estimates of occupational patenting rates could be computed. The lack of detail about those individuals who did not receive patents, as well as the absence of information on occupation for most of the sample, raise questions for the interpretation of the regressions presented later.

[15] Part of the trend in the occupational distribution may be because of the development of a market for inventions, which reduced the requirement for capital to commercially exploit a patent. Licensing and sales of patent rights became more common over the period. See Dirk J Struik, *Yankee Science in the Making*, Boston: Little, Brown (1948), for further discussion.

Table 4.2. Distribution of Patents by Patentee Occupation: All English Patentees and U.S. Urban Patentees, 1750–1850

	1750–1769		1790–1804		1805–1822		1823–1836		1836–1850	
	n	row %	n	row %	n	row %	n	row %	n	row %
Commerce and Professional										
England	131	54.8%	110	41.8%	74	40.9%	89	47.7%	70	39.1%
U.S.	–	–	13	50.0%	60	38.7%	59	24.6%	43	18.6%
Engineers/Machinists										
England	7	2.9%	28	10.6%	26	14.4%	37	20.7%	47	26.3%
U.S.	–	–	1	3.9%	17	11.0%	34	14.2%	40	17.3%
Artisans										
England	76	31.8%	95	35.1%	63	34.8%	33	18.4%	54	30.2%
U.S.	–	–	9	34.6%	48	31.0%	80	33.4%	67	29.1%
Manufacturers/Metal Dealers										
England	17	7.1%	26	9.9%	17	9.4%	17	9.5%	5	2.8%
U.S.	–	–	2	7.7%	17	11.0%	40	16.7%	49	21.2%
Other Occupations/None listed										
England	8	3.4%	4	1.5%	1	0.6%	3	1.7%	3	1.7%
U.S.	–	–	1	3.9%	13	8.4%	27	11.3%	32	13.9%

Notes: The U.S. figures show the occupations of patentees that were traced from city directories, and therefore refers to urban patentees. The English figures in the 1823–1836 column pertain to patents filed in 1840, whereas the U.S. figures are based on patents filed through July 3, 1836, when the new U.S. patent law took effect. The occupational category for "artisans" includes manufacturers of nonmetal products, and "commerce and professional" includes merchants and gentlemen.

Sources: See Sokoloff and Khan (1990) for U.S. sample; The English sample was drawn from Bennett Woodcroft, *Titles of patents of invention*, chronologically arranged from March 2, 1617 (14 James I.) to October 1, 1852 (16 Victoriae), London: Printed by G. E. Eyre and W. Spottiswoode. Published at the Queen's Printing Office (1854.)

accounted for no more than 3.2 percent of all patentees.[16] Given that the concentration of patenting in urban centers also diminished over the period, the movement away from the predominance of the merchant class would likely be even more pronounced if rural patentees were also considered.

This shift in the occupational composition of urban patentees reflected the emergence of greater opportunities for deriving returns from invention. Before the embargo, when domestic markets were least developed, patenting was carried out primarily by merchants and men in similar occupations, who were best placed to commercially exploit useful ideas. Not only did they have the most direct access to wider markets, but they also were more likely to control sufficient financial resources to support inventive activity and development. Francis Lowell, the founder of the Boston Manufacturing Company, epitomizes this class of merchant. Having decided to expand from a profitable base in foreign trade into textile manufacturing, he and a fellow merchant proceeded, without previous mechanical experience, to study the existing technology and construct their own power loom. For the most part, enterprises such as Lowell's were inspired by, and prospered because of, superior business acumen – not advanced technical knowledge. The expansion of domestic markets accelerated during the interruptions in foreign trade, and more and more individuals began to compete to satisfy the extensive demand for their products.[17] Such competition included attempts to exploit expanded opportunities for extracting returns from invention. These changes, which included improvements in markets for goods, capital, and patents (i.e., licensing), should have boosted invention and patenting throughout the population, but such gains were likely to be proportionally higher for trades other than commerce. From this perspective, it is hardly surprising that the shift in the occupational mix of patentees was accompanied by a major rise in patents per capita.

Typical histories of technology point to the achievements of skilled machinists and engineers in generating new discoveries. Yet, the evidence suggests that the technical skills and knowledge associated with occupational classes such as machinists and metal workers/dealers were not at all necessary for patentable invention – especially early in the period, but even as late as 1870. Before the embargo, these more technical groups, whose skills and knowledge were relatively scarce and costly to acquire, received only 3.9 and 7.7 percent respectively of patents. Their relative significance among urban patentees did increase substantially over time, perhaps signaling the enhanced importance of a technical background for effective invention, but even by the 1840s when mechanization was spreading quickly, their shares

[16] Merchants comprised 3.2 percent of patentees in 1860, and 2.4 percent in 1870. See B. Zorina Khan, "Creative Destruction: Technological Change and Resource Reallocation During the American Civil War" (unpublished manuscript).
[17] Sokoloff (1988) addresses this issue.

had risen to only 17.3 and 21.2 percent. The totals are large, but their impact is reduced somewhat when one remembers that machinists and metalworkers/dealers were certainly overrepresented among urban patentees relative to the general population of patentees, and that urban patentees were a much smaller proportion of the population of patentees in the 1840s than they were earlier.[18] Moreover, relatively few patents were awarded to these trades before the 1830s. Thus, it is doubtful whether advances in knowledge of machinery and metalworking alone could go far in explaining the increase in patenting during the first several decades of the nineteenth century. Rather, this progress depended more on the pragmatic problem-solving abilities of a kind pervasive among artisans of the era. As some scholars have emphasized, "varied and dextrous mechanical abilities were all but universal . . . the Industrial Revolution in its infancy produced surprisingly few basic technical skills not already familiar to American mechanics."[19] Indeed, the most significant inventions of the age, such as Whitney's cotton gin or Blanchard's lathe, were typically based on commonly available information applied to a specific problem.

American exceptionalism is apparent in this dimension as well. Table 4.2 reveals that English patentees were markedly more likely to be from the relatively elite commercial and professional classes than their U.S. counterparts, whereas the latter were more likely to be artisans, manufacturers, or from a miscellaneous category. The cross-country difference in occupational composition may be partially attributable to differences in industrial composition, but differences in effective access to the use of the patent system and to commercial opportunities generally seem responsible for most of it. The substantial share of English patents going to "gentlemen" (generally between 20 and 30 percent over the period) is particularly relevant on this point. Information on occupations was available for all English patentees, whereas the U.S. data refer to urban patentees alone; thus the comparison is all the more striking because the table underestimates the differences in the occupational composition of patentees.

Another way to explore the significance of technical skills and knowledge is to examine the extent of specialization by patentees.[20] One would expect

---

[18] Metropolitan patentees received 31.3 percent of all patents during 1805–1811, but only 22.1 and 28.0 percent in 1830–1836 and 1836–1842. In 1860, blacksmiths, engineers and machinists accounted for 2.8 percent, 2.4 percent, and 9.4 percent respectively of a national sample of 720 general inventors, drawn from both urban and rural areas. Artisans and ordinary laborers (30.4 percent) comprised the single largest occupational category.

[19] George S. Gibb, *The Saco-Lowell Shops: textile machinery building in New England, 1813–1949*, Cambridge, Mass.: Harvard University Press 1950), p. 10.

[20] Although there were patentable inventions that could be used in more than one sector, the classification here mitigates against the potential mismeasurement of specialization. Closely related inventions were consistently allocated to the same sector. Hence, even though a patentee might have registered several patents for steam engines to be used in different sectors,

that most knowledge at the technological frontier would be sector-specific, or limited in direct applicability to other inventions in the same sector.[21] Hence, if such human capital did give one an advantage in producing an invention in a particular field, and was costly to acquire, a patentee who had invested in this type of specific human capital would tend to concentrate his efforts, as well as his patents, in a single sector. The more important these sector-specific skills and knowledge were, the more specialized by sector the patents of inventors were likely to be. For example, individuals who had acquired an expertise in power looms or sewing machines would tend to file manufacturing patents, as opposed to patents in agriculture, construction, transportation, or miscellaneous categories (the other four sectors in our classification scheme.) There are of course exceptions to this generalization, such as inventions pertaining to the steam engine, but such cases were numerically insignificant relative to the patents awarded for designs of final products or capital equipment, such as cultivators, guns, clocks, stoves, and looms.

Overall, the small proportions of patentees who specialized imply that much of the human capital tapped by inventors could be applied to problems in a wide range of economic activities. Whatever knowledge or skills were needed for patentable invention appear to have been either general in nature or easily acquired. Although one might perhaps expect an increase in the specialization of patenting over time, the multivariate regressions presented in Table 4.3 reveal that after controlling for region, urbanization, access to transportation, the number of patents over which the degree of specialization is measured, and other relevant variables, there was no trend (judging from the insignificant time period dummies.) The finding may seem puzzling, because it conflicts with the usual conceptions of what the advance of knowledge and the development of markets leads to, but it is robust to variations in specification of the equation. Moreover, the degree of specialization was greater among occupations associated with technical knowledge (i.e., machinists), great inventors, and for individuals with a high career commitment to patenting – which all conform well with expectations about the relationship between large investments in invention-generating capital and specialization.

all of his patents would be in the same category, and he would be considered specialized. Inclusion of dummy variables for sectors in the specialization regressions did not alter the qualitative results.

[21] Dutton argued that patentees during early British industrialization were relatively unspecialized. Although his classification scheme and estimates of the degree of specialization are not directly comparable, it appears that American patentees were even less specialized. His data also imply that British patents were more concentrated among patentees with more career patents than was the case in the United States. Although many factors may contribute to these differences, the much higher cost of securing a patent in Britain presumably played some role. (See the discussion in Chapter 2.)

Table 4.3. *Regressions: Sectoral Specialization of Patentees, 1790–1846*

	(1)	(2)	(3)	(4)
Constant	106.86	79.83	79.06	81.62
	(16.53)	(9.66)	(9.09)	(9.83)
Proportion of Labor	−1.13	−0.18	−0.14	−0.17
Force in Manufacturing	(−1.61)	(−2.12)	(−1.69)	(−2.07)
Log (Annual Patents Per		7.83	7.73	7.84
Million in County)		(5.15)	(5.16)	(5.16)
Number of Patents (Squared)	0.98	0.94	1.03	1.02
	(3.84)	(3.72)	(4.12)	(4.00)
Number of Patents	−12.00	−11.87	−12.95	−12.77
	(−4.98)	(−4.97)	(−5.46)	(−5.26)
Time Dummies:				
1790–1804	−2.14	4.82	5.05	4.32
	−(0.44)	(0.97)	(1.02)	(0.87)
1805–1822	−3.54	−3.49	−2.16	−3.34
	(−1.46)	(−1.45)	(−0.9)	(−1.4)
1823–1836	3.41	1.1	2.07	1.32
	(1.63)	(0.52)	(0.99)	(0.62)
Regional Dummies:				
N. New England	1.43	2.77	−1.76	−3.12
	(−0.32)	(−0.61)	(−0.40)	(−0.69)
S. New England	−2.06	−8.00	−8.97	−9.11
	(−0.72)	(−2.61)	(−2.96)	(−2.93)
New York	1.49	−2.09	−2.33	−2.45
	(−0.55)	(−0.75)	(−0.85)	(−0.88)
S. Mid-Atlantic	4.40	4.15	5.16	3.49
	(1.31)	(1.25)	(1.56)	(1.05)
Urbanization Dummies				
Urban	5.37	4.69	3.64	4.04
	(2.59)	(2.29)	(1.79)	(1.95)
Metropolitan	1.39	−4.66	−7.53	−4.38
	(−0.66)	(−1.94)	(−3.07)	(−1.82)
Transportation Dummies				
Located on Navigable	−7.67	−8.28	−7.01	−8.45
River or Canal	(−2.36)	(−2.57)	(−2.20)	(−2.63)
Located on Ocean	−14.53	−14.09	−13.42	−13.03
	(−3.40)	(−3.33)	(−3.21)	(−3.06)
Occupational Dummies				
Commerce/White Collar			2.48	
			(0.81)	
Traditional Artisan			7.90	
			(2.83)	

(*continued*)

Table 4.3 (*continued*)

	(1)	(2)	(3)	(4)
Machinist			13.03	
			(3.63)	
Metalworker/Dealer			20.23	
			(4.64)	
Great Inventor Dummy				6.09
				(2.00)
R^2	0.09	0.11	0.14	0.12
N	1016	1016	1016	1016

Notes and sources: The regressions were estimated over all of the individual patent records for patentees in northeastern states who had more than one patent appearing in the sample. The constant refers to a patent filed during 1836–1846 in rural Pennsylvania counties without access to a navigable waterway, by an individual for whom either no occupation was retrieved or his occupation was different from those represented by the occupational dummy variable. The traditional artisan category refers to artisans who worked with non-metallic raw materials or did fine crafts (e.g., jewelry and watches.) Regression coefficients are reported with the t-statistics in the second row. For a discussion of the sample, see Kenneth L. Sokoloff, "Inventive Activity in Early Industrial America," *Journal of Economic History*, vol. 48 (Dec.) 1988: 813–50.

This may not resolve the problem of why an expected movement toward greater specialization is not apparent, but it does point to an interesting possibility. As patenting grew over time, there may have been two streams of development at work which obscured each other in the general time trend. On the one hand, the expansion of markets, which both enhanced the commercial value of inventions and enlarged the population who could benefit from patents, attracted new entrants to patenting whose long term investments in invention-generating capital were relatively modest and whose efforts were not as specialized as their better-equipped or more-committed peers. On the other, a growing competitive advantage to technical knowledge and skills may have served over time to promote greater investments in such sector-specific human capital, and greater specialization, particularly by patentees in very competitive districts. As long as the former mechanism continued to operate with sufficient strength, it might on average net out the effect of the latter in the aggregate pattern.

This perspective appears consistent with the results that Kenneth Sokoloff reported in his influential research on the sources of inventive activity.[22] Sokoloff used access to waterways as an index of market expansion. He demonstrated that patents per capita rose substantially and rapidly in rural counties when these areas gained access to waterways. Moreover, cities invariably emerged in these counties along the water routes, and this further boosted patenting rates. The analysis in this chapter strongly suggests that

[22] Sokoloff, "Inventive Activity," 1988.

patentees who filed in rural counties located on navigable waterways were less specialized than their counterparts in counties without such low-cost access to major markets. This finding is of special interest, because the transitional nature of such counties captures an important stage in the process through which levels of inventive activity increased during early industrialization. The pattern of coefficients in these new regressions implies that the degree of patentee specialization did not move solely in one direction. Note that the coefficient on the waterway dummy is large and negative, while those on the urban and patents per capita variables are positive. This implies that the average patentee became less specialized when water transportation first became available, but that this change was reversed over time as urbanization and further investment in inventive activity progressed. In the initial phases, an increase in the proportion of the population awarded patents brought about a rapid increase in patenting. The influx of new inventors produced a temporary shift in the composition of patentees toward individuals who had not made major investments in sector-specific capital and accordingly were less specialized in their efforts. During the later phases, as local populations adjusted to the greater opportunity and competition of an urban and extensive market, residents made larger investments in invention-generating capital, and focused more on their particular niches.

Table 4.4 includes regressions with the log of the number of career patents as the dependent variable. The assumption here is that career patents provide a reasonable gauge of the long term commitment of an individual to patenting. Admittedly, there is considerable noise in this measure, and it can be influenced by many factors including both the long-term investments in invention-enhancing capital made by that individual as well as the size of the market; nevertheless, it does provide valuable information about the people who were involved in patentable invention. The regression results support the argument that part of the increase in inventive activity was associated with a broadening of the segments of the population that previously had not been oriented toward invention. This increase in patenting was especially marked in districts during their years of transition to high levels of patents per capita. Perhaps one of the most telling signs of a growing proportion of the population who filed patents is the finding that the number of career patents per patentee peaked during 1790–1804. In all of the specifications estimated, the coefficient on the dummy variable for these years was positive and statistically significant, whereas those for the later periods were close to zero and insignificant. The number of career patents for the average patentee evidently decreased during the embargo years and this decrease was sustained afterward. The most obvious inference is that the first major surge of patenting was disproportionately accounted for by patentees with relatively modest investments in invention-generating capital and this accordingly pushed the average number of career patents down.

Although the time trend is a decline to a lower plateau, most of the other coefficients suggest that career patents increased with conditions normally

Table 4.4. *Regressions: Analysis of the Number of Lifetime Patents per Patentee, 1790–1842*

	Dependent Variable: Log of Lifetime Patents to Patentee			
	(1)	(2)	(3)	(4)
Constant	0.275	−0.048	−0.023	−0.026
	(3.27)	(−0.41)	(−0.20)	(−0.24)
Proportion of Labor	0.001	0.001	0.002	0.000
Force in Mfg	(0.83)	(0.52)	(1.35)	(0.41)
Log (Annual Patents		0.100	0.080	0.098
Per Million in County)		(3.98)	(3.25)	(4.20)
Time Dummies:				
1790–1804	0.192	0.287	0.273	0.328
	(2.16)	(3.12)	(3.04)	(4.03)
1805–1822	0.030	0.028	0.039	0.049
	(0.66)	(0.61)	(0.88)	(1.15)
1823–1836	0.027	−0.011	0.025	0.002
	(0.65)	(−0.27)	(0.60)	(0.05)
Regional Dummies:				
N. New England	0.074	0.042	0.059	0.075
	(1.08)	(0.61)	(0.89)	(1.20)
S. New England	0.213	0.138	0.119	0.072
	(4.31)	(2.62)	(2.31)	(1.47)
New York	0.139	0.095	0.102	0.067
	(3.03)	(2.02)	(2.21)	(1.57)
S. Middle Atlantic	0.107	0.111	0.158	0.166
	(1.71)	(1.77)	(2.58)	(2.79)
Urbanization Dummies:				
Urban	0.186	0.185	0.142	0.142
	(5.32)	(5.31)	(4.13)	(4.34)
Metropolitan	0.116	0.025	−0.088	−0.055
	(2.85)	(0.54)	(−1.88)	(−1.28)
Transportation Dummies:				
Located on Navigable	−0.001	−0.021	−0.026	−0.020
River or Canal	(−0.01)	(−0.41)	(−0.52)	(−0.44)
Located on Ocean	−0.010	−0.022	−0.052	−0.060
	(−0.13)	(−0.32)	(−0.78)	(−0.97)
Occupational Dummies:				
Commerce/White Collar			0.605	0.492
			(8.60)	(7.37)
Artisan			0.259	0.224
			(0.23)	(3.81)
Machinist			0.830	0.607
			(9.16)	(7.72)

(*continued*)

Table 4.4 (*continued*)

	Dependent Variable: Log of Lifetime Patents to Patentee			
	(1)	(2)	(3)	(4)
Metalworker/Dealer			0.420	0.330
			(3.98)	(3.19)
Great Inventor				1.009
				(18.13)
R^2	0.035	0.040	0.090	0.211
N	2957	2957	2957	3140

Notes and sources: The first three regressions were estimated over all of the patents in the general sample filed in northeastern states between 1790 and 1842. The data for the fourth regression also includes the additional observations from the "great inventors" sample. The constant refers to a patent filed during 1836–1842 in a rural Pennsylvania county without access to a navigable waterway by an individual for whom either no occupation is available or whose occupation is different from those represented by the occupational dummies. See the note to Table 4.3.

associated with development. For instance, career patents were higher in urban and metropolitan areas, in counties in which patents per capita were higher, and in Southern New England and New York, which were the most economically developed regions. A variety of explanations could be offered for the greater commitments to patenting in such districts, including differences in occupational composition, or the responses of individuals to the greater opportunities derived from larger markets and richer supplies of capital and information. Whatever the particular source, the small coefficient on local patents per capita implies that most of the variation in patents per capita across place and time was because of differences in the proportion of the population who patented – not because of differences in the number of patents per patentee.[23]

This aspect of the spread of high levels of inventive activity during early industrialization is reflected in the small and statistically insignificant coefficient on the dummy variable for proximity to a navigable waterway. Such access to transportation raised county-level patents per capita between 55 and 90 percent, after controlling for other relevant variables, but the regression in column (1) indicates that it did not lead to any increase in career patents per patentee. Again, the obvious implication is that when rural counties were opened up to low-cost transportation, the rapid growth in patenting in those areas was realized virtually exclusively through individuals who had limited experience with patenting, if not invention as well. These additional

---

[23] Accounting decompositions of the differences between metropolitan, urban, and rural counties, or between regions, yield the same conclusion.

patentees were similar to their peers in rural counties without proximity to waterways in terms of their long term commitment to invention, but there were many more of them in proportional terms. Those with low-cost transportation were, however, less specialized; they may, at first, have been more inclined to change their livelihoods after the opening to the wider market. Accordingly, theories that the early rise in inventive activity was largely confined to a technical elite whose knowledge and skills made them best positioned to recognize and act on new information seem particularly inappropriate for the initial phase of the transition in rural counties. On the contrary, a broad cross-section of the general population appears to have shifted resources toward inventive activity and patenting when opportunities improved.

Although most of the growth in patents per capita appears to have been because of increased activity by individuals with a relatively small number of career patents, the regressions demonstrate that individual investments in capital conducive to invention, such as technical skills, did yield more patents. This is a crucial point, because the inferences we can draw about long-term commitments to invention from information on career patents rest on an assumption that such returns were present. The independent variables explained a good deal of the variation in the number of career patents, and the large positive coefficients for machinists, metalworkers/dealers, and "great inventors" suggested that patentees who obtained technical or specialized knowledge produced significantly more patents than those who did not. The similar result for patentees who were in commerce or the professions may indicate that those inventors who had invested in capital allowing for better commercial exploitation of patentable ideas were also likely to register more patents.

THE DEMOCRATIZATION OF INVENTION

The picture that emerges from this closer study of patenting and patentees is one that highlights the more modest achievements of ordinary citizens, the cumulative effect of which resulted in a transformation of antebellum society. Despite the lofty rhetoric of its founders, American democracy was far from perfect, and history records its numerous injustices. Nevertheless, in a country whose laws and customs were rife with discrimination against numerous classes, the patent system stands out as one of the most democratic and objective institutions in the United States. As the chapter on patent laws emphasized, the patent system did not exclude or differentiate among applicants on the basis of personal characteristics such as race or gender. As early as 1821, Thomas L. Jennings may have been the first black inventor to receive a patent, for a method of "dry scouring" or cleaning clothes. Jennings

was a free black who had been born and lived all his life in New York City.[24]
Another black inventor, Henry Blair of Glen Ross, Maryland, patented an
improvement in seed planters in 1834. Blair also obtained the fifteenth patent
granted under the new examination system of 1836, for a cotton planter.[25]
Because the reforms were enacted in July and Blair's patent was granted in
August, the implication is that the patentee was well informed about the new
system despite his illiteracy (he signed the document with an "X".) It also
implies that the Patent Office, which was aware of his race, was prompt in
examining and granting his application within a two-month interval.[26]

Henry E. Baker, a black college graduate, was appointed as an assistant
examiner in 1876.[27] According to Baker, only eight hundred to twelve hundred patents were credited to black inventors in the period from 1863 to
1913. Baker himself pointed out that the Patent Office itself was "color-
blind," so other factors must be cited for the relatively low numbers. Some
of these causes were specific to the majority of blacks, including their lack of
education and technical training, legal disabilities, and the deleterious effect
of social and economic prejudices against their advancement. However, some
of these disadvantages were because of broader issues that characterized
much of the South, including the degree of inequality and the lack of effective markets. Such factors are reflected in the much lower rates of patenting
per capita throughout the entire South during the nineteenth century: in
the 1850s, for example, New England accounted for over 175 patents per
million residents, relative to approximately 15 patents per million Southern residents. The South, the least democratic region in the United States,
might paradoxically provide the strongest evidence for the favorable effects
of individual liberty and opportunity on incentives for contributions at the
technological frontier in the nineteenth century.

In summary, the beginning of rapid growth in inventive activity in
the United States coincided with a substantial broadening of the pool of

---

[24] In the 1850 census, Thomas L. Jennings is recorded as fifty years old, and married. He is likely
to have been quite prosperous, as he ran a boarding house in a respectable neighborhood
in the fifth Ward of New York City. Other residents included his wife, Elizabeth, and two
family members, one of whom (eighteen-year-old James E. Jennings) was a teacher.

[25] Letters Patent No. 15, dated August 31, 1836. Blair was the only person identified in the
Patent Office records as a "colored man."

[26] However, even if the examiners wanted to discriminate, it would have been impossible to do
so: there was no means of distinguishing black patentees, who could mail in their materials
or employ intermediaries to prosecute their applications. Indeed, it is difficult for scholars
today to study black patentees for this very reason.

[27] It is not clear that this made a difference to black patentees, because many of them preferred
not to be identified in terms of their race, as Baker found. In 1900 and 1913, the USPTO
conducted a survey of black inventors that Henry Baker was involved in collating and analyzing. See Henry E. Baker, "The Negro in the Field of Invention," *The Journal of Negro
History*, vol. 2, no. 1 (January 1917): 21–36.

individuals in the population who engaged in inventive activity. Indeed, nearly all of the initial rise in patenting per capita was produced by the increased proportion of the population who patented, as awards became less concentrated among commercial/professional occupations, urban residents, and individuals with multiple career patents. This shift in the composition of patentees was especially pronounced in the Northeast, in transitional counties that had gained access to water transportation, but lacked significant urban centers. At a later stage, there may have been a trend back toward patentees with more career patents, especially in the more developed areas, because of change in the occupational composition of the population or the evolution of higher returns to investment in invention-generating capital in such districts. But even at the beginning of the twentieth century, quite modest improvements could lead to value-enhancing productivity gains and product variety. As contemporaries noted, one of the virtues of a mass market was that a small gain when multiplied by large numbers could result in significant returns for a supposedly minor invention.

The analysis in this chapter implies that the knowledge and skills necessary for patentable invention at the beginning of industrialization were widely dispersed among the general population. Conventional histories and biographies in technology for this period typically highlight the achievements of outstanding individuals who were technically adept, such as machinists, metalworkers, and engineers. Although training in technical occupations did yield many more patents for the particular individual, such backgrounds were not at all required. The state of technology combined with the endowment of the population evidently permitted a rather elastic supply of patentable ideas. In short, the evidence points to a less heroic but correspondingly more democratic vision of technological progress in nineteenth-century America. These special circumstances undoubtedly made it easier for the economy to realize major and broadly based increases in invention and productivity growth, once conditions raised the private return to inventive activity and encouraged a reallocation of resources in that direction.[28]

All of this enhances our understanding of the record of inventive activity, but the fundamental issue still remains. What do these patterns in the characteristics of patentees indicate about the source of the sharp rise in patenting during the early nineteenth century? The answer will certainly not be as simple as the question, but it is clear that the explanation will have to focus on what pulled men and women with relatively common sets of skills and knowledge into invention. For this reason, the evidence makes it more difficult to maintain that exogenous advances in technical knowledge,

---

[28] See Kenneth L Sokoloff, "Productivity Growth in Manufacturing During Early Industrialization: Evidence from the American Northeast, 1820–1860," in Stanley Engerman and Robert Gallman (eds.), *Long-Term Factors in American Economic Growth*, Chicago: University of Chicago Press (1986.)

or in its diffusion, caused the beginning of the process. Instead, we should highlight the disproportionate responses of "low-commitment" individuals to market demand in these transitional counties and during business cycles. The greater sensitivity of this key group of patentees provides further support for the claim that the expansion of markets and the "fuel of interest" were powerful inducements to the increase of inventive activity during early American industrialization.

The patent system was a key institution in the progress of technology, but it also facilitated the creativity and achievement of otherwise disadvantaged groups such as blacks and women. The federal laws toward patentees placed all applicants on an equal footing, so if equality of access is considered to be an important feature of a democratic society, the patent system was indeed a prominent symbol of American democracy. Nevertheless, patent institutions did not exist in a social vacuum, and the norms, laws, and practices that operated at the state level as well as in society at large, could hinder or even unravel the strong protections that bolstered patent property at the federal level. A more subtle appreciation of the strengths and weaknesses of the patent system and the sources of technological change would require a closer examination of the experiences of the numerous inventors who faced widespread discrimination and obstacles that may have limited their opportunities to invent or to benefit from their efforts. The next two chapters therefore focus on one such group, the thousands of women who filed for patents during the nineteenth century.

5

Women Inventors in America

"It is our intention to . . . clear away existing misconceptions as to the originality and inventiveness of women."
                    – Woman's Building, World's Columbian Exposition (1893)

Previous chapters proposed that the degree of democratization comprised a
key difference in the European and American intellectual property systems.
American democracy in the nineteenth century, both in political and economic terms, excluded large groups of residents, ranging from black slaves
to the impoverished to women. It is therefore all the more significant that
federal patent protection was available to all citizens without regard to race,
class or gender. Indeed, the Patent Act of 1790 explicitly stated that "upon
the petition of any person or persons that *he, she, or they*, hath invented
or discovered any useful art, . . . it shall be lawful . . . to cause letters patent
to be made out in the name of the United States" (my emphasis.) Thus,
the nation's intellectual property laws were more inclusive than the norms
and laws of society in general, and provided equal opportunities and incentives for women inventors to participate in the "new economy" of the nineteenth century. The patent system helped women who were technologically
creative to realize returns from their efforts and at the same time gives us
insights into the contributions and travails of a relatively neglected class of
society.

Feminists, then and now, seized on the example of women inventors as
a symbol of gender equality, but they were more impressed by heroines of
invention. In 1893 when the Columbian Exposition was held in Chicago, the
World's Congress of Representative Women met there as part of the celebra-
tion of the advances of women in culture, industry and technology. Leaders
of the suffrage movement, including Jane Addams, Elizabeth Cady Stanton,
Lucy Stone, and Susan B. Anthony, addressed the Congress. Nineteenth-
century advocates of women's rights wanted to publicize women who
patented industrial machines that were complex and technically demanding.
They denigrated inventions that were stereotypically feminine, and preferred
"to make no note of the inventions of women unless it is something quite distinguished and brilliant. We must not call attention to anything that would

cause us to lose ground."[1] However, a focus only on the "distinguished and brilliant" inventions makes us lose sight of an even more compelling facet of the democratization of invention: the large numbers of ordinary women who attempted to gain "fair compensation" from minor improvements in familiar household items.

A number of social historians have investigated "feminine ingenuity," or the role of women as producers and consumers of technology, but this topic still warrants further research.[2] First, little systematic attention has been paid to contributions that women themselves have made to technological progress. An early play reflects the common perception that inventive ability is rare among women: "The End of the Tether, or, A Legend of the Patent Office: *an original drama in two acts for male characters only.*"[3] More recently, Donald Cardwell devoted a mere two paragraphs to women inventors out of a total of over five hundred pages, on the grounds that "female technologists of any distinction are hard to find."[4] Inventions and innovations feature prominently in accounts of the industrial revolution, but economic historians who addressed the relationship between women and technology have limited their attention to the impact of technical changes on the labor market participation of single women during industrialization.[5] Thus, despite extensive research on the role of women in the antebellum labor force, "one is left with a strong sense that the industrial revolution is primarily a men's story."[6] Second, some historians have argued that inventions made only a nominal impact on the lives of the vast majority of women who married and exited the labor market to become full-time housewives.

[1] Jeanne Weiman, *The Fair Women*. Chicago: Academy, 1981, p. 429.
[2] See Martha Moore Trescott (ed) *Dynamos and Virgins Revisited: Women and Technological Change in History*, Metuchen, N.J.: Scarecrow, 1979; Carroll W. Pursell, Jr. "Women Inventors in America." *Technology and Culture*, vol. 22, no. 3 (July 1981): 545–49; Autumn Stanley, *Mothers and Daughters of Invention: Notes for a Revised History of Technology*, Metuchen, N.J.: Scarecrow, 1993; Denise E. Pilato, *The Retrieval of a Legacy: Nineteenth Century American Women Inventors*, Westport, Conn.: Praeger, 2000; Judith McGaw, "Inventors and Other Great Women: Toward a Feminist History of Technological Luminaries," *Technology and Culture*, vol. 38 (1) 1997: 214–31; Judith McGaw, "Women and the History of American Technology," *Signs*, Summer 1982: 799–828. Shorter treatments include Otis Mason,"Woman as an Inventor and Manufacturer," *Popular Science Monthly*, May 1895: 92–103; Joseph Rossman, "Women Inventors," *Journal of the Patent Office Society*, vol. 10, 1927: 18–30; "Household Inventions," *Scientific American* vol. 74, February 15, 1896: 99; "Female Inventive Talent," *Scientific American*, September 17, 1870: 184.
[3] The play is by G. C. Baddeley, London, T. H. Lacy, c. 18? 19 pages (my italics.)
[4] Donald Cardwell, *History of Technology*, London: Fontana, 1994, p. 506.
[5] The literature on technological change and the labor market participation of women is enormous. See, for example, Elizabeth Faulkner Baker, *Technology and Woman's Work*, New York: Columbia University Press, 1964.
[6] Thomas Dublin, *Transforming Women's Work*, Ithaca, N.Y.: Cornell University Press, 1994, p. 2.

Rather, "woman's work" was insulated from the widespread technological progress that increased productivity in the market economy. Diaries, catalogues, and letters are marshaled to support the claim that the diffusion of household inventions was slow or nonexistent, especially in rural areas.[7] Other scholars argue that women did not benefit from technological change because even when innovations were adopted the amount of time spent on housework remained largely unchanged.[8] More evidence on the market for household inventions would be useful, as these issues have important implications for our assessment of the welfare of women who worked at household tasks.

Descriptions of the technology of household production and consumption patterns can provide valuable insights.[9] However, such sources may themselves suffer from unexpected gender biases that may have contributed to the notion that the "industrial revolution is primarily a men's story." For instance, in the antebellum period, supporters of women's rights argued that household work was as valuable as market work, but after the Civil War

[7] A much-cited work in this area contends that few household aids to ease the burden of housewives were manufactured or available until the start of the twentieth century: "Mechanical cooking utensils existed in the second half of the nineteenth century, but few houses had them. Eggbeaters, cherry stoners, apple parers and corers, butter churns, meat choppers – all these and more were patented in large numbers. But mechanical devices rarely appear in lists of necessary equipment for nineteenth-century kitchens." See Susan Strasser, *Never Done: a History of American Housework*, New York: Pantheon Books, 1982, p. 44. It is argued that manufacturers refused to produce household items because it was unprofitable and that their prices were beyond the budget of the average household. In the twentieth century demand conditions changed, according to this view, because homemakers were enticed to enter the mass market through persuasive advertising that preyed on their feelings of guilt. See also Susan Strasser, "An Enlarged Existence? Technology and Household Work in Nineteenth-Century America," in Sarah F. Berk (ed.), *Women and Household Labor*, Beverly Hills, Calif.: Sage Publications, 1980: 29–52.

[8] Ruth Schwartz Cowan, *More Work for Mother: The Ironies of Household Technology from the Open Hearth to the Microwave*, New York: Basic Books, 1983. For a survey of the plethora of studies addressing this question, see Ronald R. Kline, "Ideology and Social Surveys: Reinterpreting the effects of 'Laborsaving' Technologies on American Farm Women," *Technology and Culture*, vol. 38, 1997: 355–85. He points out (p. 379) that earlier researchers felt that "household technology made work easier and enabled a higher standard of living," but the majority of modern studies have chosen to emphasize the supposed "paradox" that time spent on housework did not change. However, the ultimate objective of individuals and households is not simply to save time, but also to consume more higher-quality goods and services. Household innovations lead to two separate effects: one is to reduce time and the other to lower the price of household goods at the margin. This lower marginal price serves to induce a substitution effect toward the innovation that may outweigh labor-saving effects. Even if the time spent on housework was indeed unchanged, housewives undoubtedly would reveal a preference for the scenario after adoption of the innovation, implying higher welfare.

[9] In addition to the studies mentioned earlier, see also Jeanne Boydston, *Home and Work: Housework, Wages and the Ideology of Labor in the Early Republic*, New York: Oxford University Press, 1994, who points out that new household technologies were not labor saving but did increase labor productivity.

leaders of the women's movement deprecated household labor.[10] Instead, the new ideology considered "traditionally 'male' activities as socially privileged – and encouraged women to repudiate traditionally 'female' activities as socially subordinate."[11] The same schism was apparent in the objectives of the organizers of the Women's Pavilion at the World's Columbian Exposition in 1893. The vast majority of women submitted their predominantly domestic inventions in the hopes of profiting from the publicity and sales at industrial exhibitions. The organizers, on the other hand, were disappointed to find that female invention and innovation was primarily directed toward household products such as kitchen tools and apparel. Spokeswomen of the suffragist movement, predominantly from a middle class background, had significantly different interests, experiences, and affinities relative to women from other backgrounds, especially those from rural or frontier areas.[12] By denigrating household work and the invention of household articles, the women's movement contributed to the notion that women were not technologically adept.[13] Even a *Women's Bureau Bulletin* on women's inventions from 1905 to 1921 opined: "If the steady increase in the numbers of patents granted women is accounted for merely by the increase in the number of patented hairpins, hair curlers, and such trifles in feminine equipment, it is without large significance either to civilization or as an indication of women's inventive abilities."[14]

Ironically, the views of suffragists of the past century are reflected in current economic models of productivity and technological change. Economic

[10] For an excellent discussion of related issues, see Reva Siegel, "Home as Work: The First Woman's Rights Claims Concerning Wives' Household Labor, 1850–1880," vol. 103 Yale Law Journal (1994): 1073–1217.

[11] Reva Siegel, "Home as Work," p. 1061.

[12] According to Reva Siegel, p. 1080: "the strategies the postwar movement employed to reduce the division of labor in the family reflected disparaging judgments about "women's work" of the sort the antebellum movement originally contested. Paradoxically, the movement's new understanding of autonomy and dependence was as entangled in the gender discourse of its culture as the older vision it repudiated, and tacitly class-biased as well: to achieve this new form of autonomy, women were to delegate the work of household maintenance to other women."

[13] Judith McGaw, "Inventors and other Great Women," p. 219: "Similarly, emphasizing woman's capacity to invent outside the domestic sphere, an approach characteristic of earlier feminist efforts and one that shapes Macdonald's and also Stanley's work, evidently left popular conceptions of inventors and invention virtually unaltered."

[14] U.S. Department of Labor. *Women's Contributions in the Field of Invention*. Women's Bureau Bulletin, No. 28. Washington, D.C.: GPO, 1923, p. 13. The Women's Bureau authors took heart from the finding that "there is not an important sphere of industry, commerce, or the sciences unrepresented in these classifications." For, "the invention of a new hook and eye, a new garment appurtenance, a new kitchen appliance or other household device, finds no place among these grants. Excluding all such articles, although they unquestionably stimulate productive activity, and confining the list strictly to the operating methods and materials of manufacturing industries, gives peculiar significance to this group of inventions patented by women," p. 21.

theory emphasizes the role of manufacturing processes and intermediate capital goods in promoting growth. Empirical work mirrors this bias, due to well-known problems that lead to the understatement of improvements because of changes in quality and the introduction of new goods. The chapter on "great inventors" showed that inventors tend to be judged in similar terms, for biographies of the heroes of technology focus on supposedly discrete "macroinventions" and on large-scale capital inputs in major industries such as iron and steel.[15] At the same time, the experience of both important and ordinary inventors suggests that a good deal of technological change is incremental and consists of the accumulated effects of numerous improvements, and that a significant part of economic welfare is related to small changes in the quality or nature of final goods. Thus, a full understanding of the remarkable transformation in the daily experience of both men and women in the past two hundred years cannot be obtained without understanding incremental improvements in their dress, the shelf-life of processed foods, household utensils, and other aids to housework, and a host of other supposedly minor "microinventions."

In short, although scholars have made valuable contributions to the study of domestic technology, we still need further research on empirical patterns. The lack of quantitative analysis is partly due to the paucity of relevant data in an era when women were rendered "invisible" by legal and social conventions. Patent records, despite their acknowledged limitations, are inherently useful in this regard because they provide a continuous source of information over time about market-related activities of women. Time series and cross-sectional analyses of women's patenting allow us to trace variation in female market participation across regions and sectors. Comparison of the record for female patentees to male patentees affords insights into the sources of inventive activity. Patents also provide information about creativity, entrepreneurial activity, and the pursuit of profit. The very attempt to obtain a patent signaled a commercial orientation, but some women also obtained returns from their discoveries by selling the patent rights, licensing others to use the invention, or by developing enterprises to promote the inventions themselves. These market activities at the same time yield information about the users of inventions, as well as indirect evidence about the transformation within the home that was likely to accompany the diffusion of domestic innovations.

This chapter addresses the relationship between women and technology through their patterns of patenting and commercialization. The main data set comprises a sample of 4,196 patents filed in the United States by some 3,300 women inventors between 1790 and 1895, supplemented with information

---

[15] For a discussion of "macroinventions" and "microinventions" see Joel Mokyr, *Lever of Riches: Technological Creativity and Economic Progress*, New York: Oxford University Press, 1992.

from city directories and assignment records.[16] We are given insights into inventors' occupations and marital status, and patterns of patenting over

[16] Toward the end of the nineteenth century, the U.S. Patent Office published a list of women patentees: *Women Inventors to whom Patents Have Been Granted by the United States Government, 1790 to July 1, 1888*; and two appendices that extended the coverage through March 1, 1895 [WIP]. The publication included 3975 patents filed either by women inventors alone, or with co-inventors (both male and female), omitting initials and androgynous names. WIP catalogues the patent number (available after 1836), the names of inventors, coinventors, and assignees (if the patent right was transferred at time of issue), along with their state and city of residence, a brief description of the invention, and the date the patent was issued.

This list is, however, incomplete, and should be regarded as a sample, rather than a complete census, of the population of female patents. The preconception that women invent few technically complex devices probably influenced the exclusion of some androgynous names. However, in most cases, the omitted names are quite common, evidently female names, the result of careless tabulation on the part of the patent office clerks drawing up the list. I found WIP omitted roughly 56 percent of patents issued to women in 1870. The omission of patents granted to women subsequently decreases, at least in the years I checked: 21.0 percent in 1876; 14.5 percent in 1888; 9.9 percent in 1889; 14.3 percent in 1890; and 9.8 percent in 1891. My sample, which totals 4,198 patents, includes missing data drawn largely from 1888 to 1891. The basic patterns are not altered by the inclusion or exclusion of the missing data, suggesting that the omissions were not systematic. I categorized inventions according to sector of final use, and also obtained information on the numbers of patents per person, and the length of inventive career (defined at the period between the first and last patent.)

The patent office records normally also include information on assignments that are made when the patents is issued. The data regarding assignments of women's patents are unreliable for the earlier years, and are entirely missing from WIP between July 1, 1888 to October 1, 1892; however, I retrieved these assignments from the Patent Office Gazette. Additional information on assignments transacted after the date the patent was issued was obtained from the records held at the National Archives. Since these records are voluminous, I constructed a random sample of assignments by selecting inventors whose surname started with the letter "B". This is a standard procedure, as there is no reason to suspect that individuals vary systematically depending on their initials.

City directories from 1875 to 1890 provided additional information on marital status and occupations for the inventors of some 900 patents. Cities include: Washington, D.C.; Indianapolis; Oakland; San Francisco; Los Angeles; Brooklyn; New York; Buffalo; Syracuse; Toledo; Philadelphia; Pittsburg; Providence; New Orleans; Kansas City; Topeka; Jersey City; Newark; Cambridge; Boston; Worcester; Springfield; Framington; Milwaukee; Detroit; Grand Rapids; Minneapolis; St. Louis; Kansas, MO; Cincinnati; Cleveland; Toledo; Denver; New Haven; Chicago; Baltimore; and Portland, Ore.

In the analysis, I attribute the entire invention to one woman. Patent records undoubtedly undercount the numbers of inventions by women, in part because some might have allowed male relatives to file the invention. Some researchers have speculated about women who might have been responsible for discoveries formally attributed to men, such as Caroline Greene/Eli Whitney, Mrs Elias Howe/Elias Howe. However, the patent law explicitly voids a patent that is not filed by the true and original inventor, so it is far more likely that an undeserving male was listed as a co-inventor on the patent, rather than as the sole inventor. According to Charlotte Smith, a nineteenth-century lobbyist for women's rights, women inventors tended to record their inventions using only initials, in which case the invention could not be traced according to gender. In any event, patent records, although imperfect, still provide the best *objective* and unspeculative source of knowledge about women in invention.

time, region, and industry. Such inventive activity by female patentees was based on the relative advantage of women in perceiving and satisfying household wants. Thus, women responded to market demand by devising inventions specific to their role in the home and family, including domestic contrivances that substituted for scarce household help in rural areas.[17] Women inventors also engaged in entrepreneurial behavior such as multiple patenting and the launching of manufacturing enterprises. The commercialization of their inventions allows us to investigate claims that the market for household innovations was thin with limited diffusion. Current research tends to depict domestic workers as passive toilers who were excluded from the benefits of the industrial revolution until the start of the twentieth century. The scenario that emerges from an examination of women inventors is more optimistic, for it implies that, both as producers and consumers, women applied considerable creativity to improve productivity and products within their households. This conclusion is all the more impressive when one realizes that, as not all inventions are patented, the roster of patents and assignments analyzed merely provide an index of inventive activity and innovation in the nineteenth century.

WHO WERE WOMEN PATENTEES?

The first woman inventor in America to be officially recognized was Sybilla Masters, a resident of colonial Pennsylvania, whose husband obtained two English patents on her behalf in 1717.[18] In 1809 Mary Kies of Connecticut obtained a U.S. patent for weaving straw to make hats and is usually credited with being the first female patentee in the United States. Because the patent records do not classify inventors in terms of race or gender, it is especially difficult to identify patentees on the basis of both those characteristics. However, Judy W. Reed, of Washington, D.C., was likely the first black woman to receive a patent, granted in September 1884 for a mechanical dough kneader. Both Judy Reed, a fifty-eight-year-old seamstress, and Allen Reed, her husband, were unable to read or write, and she signed her patent application with a mark.[19]

[17] An alternative hypothesis is that specialization by women inventors was due to "supply factors" or to their lack of more technical skills. If so, we would expect the focus on household innovations to diminish in relative importance over time as women gained better access to education and labor market experience.

[18] See Bruce Bugbee, *The Genesis of American Patent and Copyright Law*, Washington, D.C.: Public Affairs Press, 1967, p. 72. One of the Masters patents dealt with a method for curing corn while the other was an invention for weaving straw into bonnets. The English patents were granted in 1715 and 1716 but published in 1717.

[19] Patent No. 305,474, dated September 23, 1884. The biographical information is from the 1880 Census, which classifies Judy W. Reed's race as Mulatto, and specifies her husband's occupation as a gardener. Shortly after, another black woman, Sarah E. Goode, of Chicago, Illinois, obtained a patent for a cabinet bed (Letters Patent No. 322,177, dated July 14, 1885.) Neither of these women received any other patents.

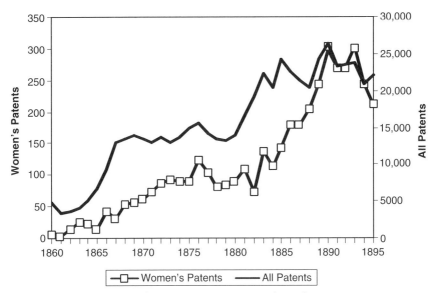

Figure 5.1. Patents Granted to Women Inventors and Total Patents, 1860–1895
Notes and Sources: See text and Footnote 16. U.S. Patent Office, *Annual Report of the Commissioner of Patents*, various years.

Only 77 patents were credited to women inventors from 1790 through 1860, even though 4,773 patents were issued to male patentees in 1860 alone.[20] The Civil War and Reconstruction era proved to be a watershed for patenting by women inventors. More patents (86) were filed between 1860 and 1865 than over the entire seventy years since the inception of the patent system. Many of these inventions were clearly related to the war effort. Mary Jane Montgomery of New York, who would later consider herself a professional inventor, obtained patent protection for her 1864 invention of a "war vessel." Clarissa Britain of St. Joseph, Michigan, received nine patents, including an 1863 patent for an ambulance. Sarah J. A. Hussey, a Quaker from Cornwall, New York, noted in her patent specification that her invention was motivated by her "long experience as a nurse in the United States hospitals."[21]

The lower level of female inventiveness relative to men, at least as demonstrated by patent grants, persisted throughout the period. Figure 5.1 shows that the post–Civil War contribution of women inventors amounted to less than 1 percent of all patents granted by the United States Patent Office. However, the growth rate of women's patents exhibited a strong upward

[20] However, even this scant number exceeds the roster for England, where sixty-two women are listed as patent grantees between 1617 and 1852, and several of these obtained patents for inventions that they did not devise themselves, as permitted by English law.
[21] Patent No. 47831, May 23, 1865 for an "improved table for hospitals." Sarah Hussey was buried with full military honors in her hometown in 1898.

trend that exceeded that for male patentees. In the centennial year of 1876, the cumulative total patents for women amounted to just over one thousand, whereas by the 1890s the number of patents issued to women was double that of the preceding decade. The decadal rate of increase for patents by women at this point was more that three times the corresponding rate for men.

Several factors were likely responsible for the rapid growth in the numbers of women participating in patenting over the period, including their higher labor market participation, greater access to education over time, and legal reforms that improved women's economic rights.[22] During this period, more information about prospects for patenting and marketing inventions was being disseminated to a wider audience. Journals such as *Scientific American, Inventive Age*, and *The Patent Record* were dedicated to descriptions and analyses of current patent activities, both in the United States and abroad. Their articles occasionally gave advice to women who wished to obtain patents and exploit their inventions.[23] *Scientific American* in particular issued editorials that highlighted the commercial profitability of "small inventions" that might seem technically undemanding. Exhibitions such as the Philadelphia Centennial Exposition and the World's Columbian Exposition reserved special pavilions for women inventors, and alerted other women to the opportunities available in this sphere of activity.[24] In the early 1870s, the Patent Office hired its first female patent examiner, possibly encouraging women to submit inventions that they might have feared would be viewed with less sympathy by male examiners.[25] Women may have been prompted to invent and patent by the example of other successful female patentees. For instance, in 1887 Caroline Pusey of Philadelphia cited Helen

[22] For a discussion of the impact of the Civil war and the importance of commercial networks and other forms of business acumen, see Lisa Marovitch, "'Let her have brains too': Commercial Networks, Public Relations and the Business of Invention," *Business and Economic History*, vol. 27, 1998: 140–61. The issue of married women's property rights laws is explored in Chapter 6.

[23] For instance, according to the *Patent Record and Monthly Review*, "Few inventions are more profitable than little matters of feminine utility" (cited in Anne L. Macdonald, *Feminine Ingenuity: Women and Invention in America*. New York: Ballantine, 1992, p. 248.)

[24] See Deborah Warner, "Women Inventors at the Centennial," in *Dynamos and Virgins Revisited: Women and Technological Change in History: an anthology*, edited by Martha Moore Trescott, pp. 102–19. Metuchen, N.J.: Scarecrow Press, 1979; and Jeanne Weimann, *The Fair Women*, Chicago: Academy Chicago, 1981.

[25] Sarah J. Noyes, a specialist in chronological devices, entered the Patent Office in 1873, and was appointed First Assistant Examiner of the Electrical Division (*The Woman Inventor*, vol. 1(1) 1890.) Anna R. Nichols was also employed as a patent examiner in the 1870s. In 1869, a time when only 2.5 percent of clerical workers were female, the Patent Office employed fifty-three women copyists at an annual salary of $700. Clara Barton, the founder of the Red Cross, held a regular civil service appointment as a patent clerk as early as 1854. Female clerks who wished to advance in the Patent Office could study to pass the rigorous written test that was required of all patent examiners. Victoria C. Neagle, who was employed as a clerk in the Patent Office in 1870, had become an assistant examiner by 1890.

Blanchard's earlier patents in her own application for a patent to protect an invention that improved the manufacture of socks and gloves.[26]

As Figure 5.1 shows, the rate of change was not constant over time. Patenting for both men and women was responsive to major economic cycles, falling in the business downturns of 1873, 1883, and 1893. Like other patentees, therefore, women inventors appear to have varied their efforts in relation to changes in market demand and to expected profitability. Indeed, the very attempt to obtain a patent signaled an interest in commercial activity, and introduced female patentees to market-related activity. Conversely, it is possible that women who were already involved in commerce or a profession might have had a comparative advantage in perceiving existing demand, in gaining the required skills, and in pursuing the patent application process. The latter hypothesis is true for women inventors such as Margaret Knight, a machine operator in a paper bag factory who invented a machine that manufactured "satchel-bottomed" paper bags. Knight's patenting career spanned forty-five years, and the last of her twenty-two patents was granted in 1915 for improvements in the internal combustion engine, and assigned at issue to the K-D Motor Company of New York.[27] However, most female patentees appear to have been introduced to commercial activity after patenting their inventions.

City directories provided information about the occupations and marital status of a number of women patentees (Table 5.1). One should interpret these data with caution, because of sampling differences across cities covered. They are also likely to overrepresent heads of household, widows, single women, or wives whose jobs were separate from husband or family, as well as urban residents. The individuals traced through the city directories accounted for some nine hundred patents, or a little over a fifth of the entire sample. A third of the patents linked to city directory entries were granted to women with no listed occupations. Another 10.7 percent of these patents belonged to professional women such as doctors, school principals, and painters. Almost one quarter of all patents (21.3 percent) were issued to working-class women such as dressmakers, milliners, and factory workers. The predominance of middle-class women may be an artifact of city directory listings, especially as rural residents are excluded. But it is possible that such women were indeed overrepresented among patentees because they were able to obtain funds for the patent application and likely had greater opportunities for patent management.

---

[26] Caroline S. Pusey, Letters Patent No. 361,080, dated April 12, 1887: "I am aware of . . . Helen A. Blanchard's Letters Patent, No. 167,492, dated September 7, 1875, and No. 174,764, dated March 14, 1876, wherein the margins of the fabric are united together by an overseaming or zigzag stitch."

[27] Patent No. 1,132,558 was granted in 1915, a year after her death at age 76; original application filed July 8, 1911.

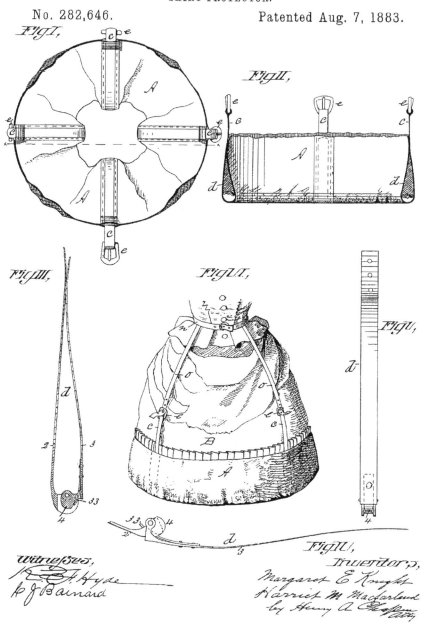

M. E. KNIGHT & H. M. MACFARLAND.

SKIRT PROTECTOR.

No. 282,646.   Patented Aug. 7, 1883.

Illustration 8. Margaret Knight's paper bag machine is frequently held up as an example of women's inventions, but her apparel patent is more typical of the improvements that nineteenth-century women inventors patented. (*Source*: U.S. Patent Office.)

Table 5.1. *Characteristics of Patenting by Women Inventors, 1790–1895*

	Patents	Percent
Patents filed, by Patentee Occupation		
Professional	81	10.7
Education	57	7.5
Owner of firm	184	24.2
Apparel maker	74	9.7
Factory employee	12	1.6
Clerical	24	3.2
Boarding, petty trader	52	6.8
None	275	36.3
Multiple Patenting, patents filed		
One patent filed	2683	66.6
Two or Three	874	21.7
Four to Nine	324	8.0
Ten or More	149	3.7
Patents filed by Marital Status		
Single	127	14.3
Married	207	23.4
Widowed	233	26.3
Married or widowed	52	5.9
Unknown	267	30.1
Coinventors, patents filed		
Female, Related	19	6.0
Male, Related	115	36.3
Female, Unrelated	35	11.0
Male, Unrelated	148	46.7
Assignments made when patent issued		
Clothing	96	29.6
Household machines	19	5.9
Household (nonmachine)	77	23.8
Industrial machines	59	18.2
Tools	10	3.1
Transportation	20	6.2
Health	19	5.9
Miscellaneous	24	7.4

Notes: Patentee occupations and marital status were obtained from city directories. Coinventors: the categories of related or not are based on whether the individuals have the same surname, which will tend to be an underestimate. Patent assignments made after the patent was granted are not included here.

Sources: See text and Footnote 16 for details about data set.

I also retrieved information from city directories on the marital status of these patentees, a variable that might contribute to our understanding varia-tion in patenting by women. Of the linked patents, 127 patents (14.3 percent) were filed by single women, 207 (23.4 percent) by married women, 233

(26.3 percent) by widows, and 52 (5.9 percent) by women whose marital status was either married or previously married. The remaining 267 patents were issued to women whose marital status was unknown. The sample size is too limited to sustain any strong conclusions, but some interesting patterns do emerge. First, half of all patents to single women were filed in New England, and almost two thirds of all widows filed in the Mid-Atlantic region. Unmarried women accounted for almost one half (47.4 percent) of patenting in the New England states, compared to only 3.9 percent of patents in the Mid-Atlantic, 7.6 percent in the South, 14.6 percent in the West, and 15.5 in the Midwest. By contrast, married women filed 35.1 percent of Midwestern patents, and 31.7 percent of Western patents, compared to 19.7 percent in the South, 16.8 percent in the Mid-Atlantic, and 22.6 percent in New England. New England's population might have featured a higher fraction of single women, and women might have been more likely to marry in the West and Midwest, but the figures seem sufficiently distinct to warrant speculation that location-specific factors – such as the laws relating to the property rights of married women – might account for some of the differences.

If inventive activity by women was systematic and related to market demand, it is likely that the regional pattern of female patenting would tend to vary with that of male patentees, and there would be systematic variation in industrial patterns of patenting. Moreover, patentees would attempt to commercialize and profit from their investments in inventive activity. At the same time, as Claudia Goldin pointed out, women's work "cannot be understood as an isolated market responding to economic factors" because of life cycle factors and persistent links with the family.[28] One might expect that those factors would influence both the demand and supply of inventive activity by women. Female inventors were confronted by problems specific to women, such as coping with voluminous, impractical clothing, and managing onerous household tasks in rural areas without access to hired help. Such women would attempt to benefit from their comparative advantage in household-related skills to satisfy the demand for innovations to ease the burden of domestic chores. Sarah M. Clark provides an extreme example of this, for she was a servant who obtained four patents for a cake-stirrer, dough maker, an invention for adjusting mirrors, and an improved cooking stove.[29] Despite their formal occupations, all women tended to have a measure of domestic expertise, so one observes artists, physicians, and milliners adding to the roster of household inventions.[30]

---

[28] Claudia Goldin, *Understanding the Gender Gap: An Economic History of American Women.* New York and Oxford: Oxford University Press, 1990.

[29] Annie F. Craig, another servant, obtained patent No. 215,725 for an "improvement in pictures," in May 1879.

[30] For instance, Mary Evard, a milliner, invented a broiler and toaster; Lucinda Warren, a dressmaker, patented a dishwasher; Adelia Covell, an artist, a dough-making machine; Mary Ross, a physician, obtained patents both for a salve and for an invention related to dressmaking.

The regional distribution of patterns of inventive activity by all patentees appears to have varied with the extent of the market, and Table 5.2 indicates that women patentees were located in the same regions as other patentees.[31] Almost one quarter of all women patentees over the entire period lived in New York state, followed by Illinois (8 percent) and Massachusetts (7.5 percent.) These states were dominant from the Civil War period in terms of patenting by both women and men, but lost ground in subsequent years, when patentees from areas such as Kansas, Michigan, Minnesota, Texas, and Wisconsin increased in relative and absolute numbers. New York, for instance, accounted for 35 percent of all women *patentees* before the Civil War, but only 18 percent by the 1890s. Nevertheless, in the latter period women inventors in the Mid-Atlantic still produced 32 percent of all *patents*, implying that these women were more prolific patentees than the new entrants from the Western and Midwestern states.

A striking congruence between the general and female patterns of patenting is revealed in the dramatic decline in the percent of patents from the New England area, and the less marked fall in the Mid-Atlantic. Conversely, Midwestern states increased their share of patenting to 32 percent (women) and 35 percent (all patentees) by the 1890s. Naomi Lamoreaux and Kenneth Sokoloff first noted this phenomenon in their study of a sample of all patentees in the late nineteenth and early twentieth centuries.[32] They found that the regional change in patenting corresponded to variation in relative per capita income, and suggested that the patenting rates were associated with economic development. Patenting activity possibly responded to a fall in per capita incomes of New England states relative to the increases of the East North Central region. Whatever the underlying reasons, the parallels between the results for females and all patentees suggest that many of the factors that motivated other patentees also influenced women's patenting.

Adjusting the numbers of patents filed in a particular region for population size highlights an intriguing exception to the result that patenting by women followed the general trends. The patenting rate, reported per million women, reveals that a rapid increase in patents relative to population occurred in the Western states after the 1860s. Table 5.3 shows that patents per capita increased in all regions throughout the century, even in the lagging South, but the rate of increase as well as the absolute rates were highest in the West. This was not true of the per capita rates for the general population of patentees, because in their case the absolute number of patents per capita

---

[31] Evidence relating to patterns for all patentees is provided in Kenneth L. Sokoloff, "Inventive Activity in Early Industrial America: Evidence from Patent Records, 1790–1846." *JEH* 48, no. 4 (1988): 813–50; and N. Lamoreaux and K. Sokoloff, "The Location of Invention and Technical Change in Late Nineteenth and Early Twentieth Century America," unpublished paper, presented NBER, July, 1993.

[32] See Lamoreaux and Sokoloff, "Location of Invention."

Table 5.2. *Female Patenting, Total Patenting, and Population, 1800–1895*
(percentage distribution)

	Pre-1860	1860s	1870s	1880s	1890s
New England					
Female patents	26.0	19.2	19.0	11.7	9.9
Female population	10.2	9.8	9.1	8.5	8.2
Total patents	28.0	22.8	19.8	17.8	15.8
Total population	10.0	9.5	8.8	8.2	7.9
Middle Atlantic					
Female patents	62.0	43.6	39.3	34.4	32.2
Female population	24.0	23.8	22.8	22.1	21.7
Total patents	45.9	40.1	38.9	35.7	33.3
Total population	23.7	23.4	22.3	21.6	21.3
East North Central					
Female patents	4.0	20.3	20.8	24.2	23.0
Female population	21.4	22.4	22.5	22.2	21.9
Total patents	12.0	24.1	22.5	24.7	25.8
Total population	22.0	22.9	22.8	22.5	22.1
West North Central					
Female patents	0.0	4.5	6.5	11.1	14.0
Female population	6.6	7.9	9.4	11.0	12.0
Total patents	1.2	4.2	6.2	8.4	9.5
Total population	6.9	8.3	9.8	11.4	12.3
West					
Female patents	0.0	3.0	5.2	5.8	9.0
Female population	1.3	1.5	2.0	2.7	3.3
Total patents	0.3	1.9	3.0	4.1	5.6
Total population	2.0	2.2	2.6	3.3	3.9
South					
Female patents	8.0	9.4	9.1	12.8	12.0
Female population	36.6	34.6	34.2	33.5	33.0
Total patents	12.7	7.0	9.5	9.3	10.0
Total population	35.4	33.7	33.6	32.9	32.5

Notes: The female patent figures exclude 196 patents granted to foreigners, and 40 patents for which no information was available. The decadal figures for 1890 are obtained by inflating the patents granted up to March 1895 by 1.9355.

Sources: Data for total patents at the regional level are computed from the *Annual Report* of the Commissioner of Patents for 1891. For source of women's patents, see text and Footnote 16. Population data are from the Census of the United States, computed at the decadal midpoint by exponential interpolation.

filed in the West still lagged behind that of the more technologically experienced Mid-Atlantic and New England regions. In short, Western women (in per capita terms) held a greater absolute and comparative advantage over non-Western women than men in that region held over non-Western men. These results are consistent with the "Turner thesis" that the frontier

Table 5.3. *Female and Total Patenting per Capita, 1800–1895*
(per million residents)

	Pre-1860	1860s	1870s	1880s	1890s
New England					
Female	0.8	30.3	96.0	79.6	102.9
Total	102.1	484.4	725.3	820.2	698.4
Ratio (%)	0.8	0.2	13.2	9.7	14.7
Middle Atlantic					
Female	0.8	28.5	79.3	90.6	126.3
Total	70.4	346.7	561.3	626.4	547.0
Ratio (%)	1.1	8.2	14.1	14.5	23.1
East North Central					
Female	0.0	14.1	42.7	63.1	89.3
Total	19.9	212.7	317.0	417.3	409.4
Ratio (%)	0.0	6.6	13.5	15.1	21.8
West North Central					
Female	0.0	8.9	31.9	58.6	99.4
Total	6.1	102.6	204.0	277.9	269.2
Ratio (%)	0.0	8.7	15.6	21.1	36.9
West					
Female	0.0	30.1	120.5	126.7	231.5
Total	5.0	175.8	367.7	464.2	504.6
Ratio (%)	0.0	17.1	32.8	27.3	45.9
South					
Female	0.0	4.2	12.2	22.1	31.0
Total	13.0	41.9	91.5	107.0	107.6
Ratio (%)	0.0	10.0	13.3	20.7	28.8
United States					
Female	0.3	15.5	46.0	58.0	85.1
Total	36.4	202.4	322.1	379.1	349.8
Ratio (%)	0.8	7.6	14.3	15.3	24.3

Notes: The total number of patents for women in the 1890–1894 period includes patents issued during January and February, 1895. A small number of patents attributed to women included male coinventors. The ratio comprises female patents per capita as a percentage of total patents per capita within each region. The data include patent grants alone, because separate information is not available for patent applications by women.

Sources: Figures for male patentees were obtained by subtracting the total for women from the annual data in the *Report of the Commissioner of Patents* for various years.

was associated with "a special American character of fierce individualism, pragmatism, and egalitarianism" especially in the case of women, who were relatively more disadvantaged in terms of the laws, norms, and practices of the settled regions.[33] The frontier states were foremost in protecting the

[33] Judith K. Cole, "A Wide Field for Usefulness," *American Journal of Legal History*, vol. 34, 1990:262–294. See also Mari Matsuda, "The West and the Legal Status of Women: Explanations of Frontier Feminism," *Journal of the West*, vol. 24, 1985: 47–56.

rights of women, and this might have stimulated greater efforts to secure their intellectual property in the form of patents. It is possible that the frontier "feminist ethos" and liberal laws served to attract from other parts of the country women migrants who were more innovative and independent than average, thus boosting the rates of inventive activity in the West and Midwest.

This assessment of patenting activity by women inventors also casts some light on the experiences of frontier and rural women. The literature on such women demonstrates ambivalence about the role of technological change in their lives. Historians of technology claim that the market for innovations was largely stratified and related only to wealthy middle-class women and to urban households.[34] Some have argued that men who were prompt to purchase innovations for the farm were resistant to improvements in the home, and that women did not have sufficient power to counter their wishes. The "pessimists" also assert that technological change led to a "deskilling" of women's work and the devaluation of women's contributions to the household. By contrast, some studies point to the importance of women's innovations in key industries such as dairying and food preservation, and to changes in the nature (if not the time) of housework. More broadly ranging sources provide useful supplements to these conclusions. For instance, technological change encompasses improvements in diet and techniques of food preparation, more efficient processes of childrearing, and new designs in furnishing.[35] Clearly, these are complex questions that the evidence at hand cannot directly address, but the results reported here do induce a more optimistic view about the role of rural women as creators and users of innovations. The distribution of women's patenting was far more concentrated in rural areas than was the case for men, and this was especially true in frontier states. Moreover, there appears to have been a ready market for such inventions.

Table 5.4 indicates that less than a quarter of women's patented inventions were in untraditional fields such as tools, industrial and agricultural

[34] For instance, Strasser ("Enlarged Human Existence" p. 30) contends that "the technological potential of the nineteenth-century house was fairly high; it could only be achieved, however, by wealthy people in urban areas." She concludes (p. 37) that "technology had little impact on most women's working life." See also Pamela Riney-Kehrberg, "Women, Technology and Rural Life: Some Recent Literature," *Technology and Culture*, vol. 38, 1997: 942–53 for a survey of studies on the role of technological change in rural women's lives; and John Mack Faragher, "History from the Inside Out: Writing the History of Women in Rural America," *American Quarterly*, vol. 33, 1981: 537–57.

[35] A good example of this literature is Sarah McMahon's study of "Laying Foods By" in early New England households (in Judith McGaw (ed.), *Early American Technology: Making and Doing Things from the Colonial Era to 1850*, Chapel Hill: UNC Press, 1994.) McMahon proposes (p. 177) that "the search for better methods may well have followed the growing impulse, in rural New England households, to adopt new methods and technologies that were being developed in agriculture and cooking in the first half of the nineteenth century."

Table 5.4. *Types of Inventions Patented by Women,*
1790–1895

Type	No.	Percent
Teaching, music or games	226	5.4
Food preparation	29	0.7
Textiles and Sewing	244	5.8
Clothes, general	295	7.0
Hats	46	1.1
Shoes	27	0.6
Dresses	115	2.7
Corsets	241	5.8
Furniture and Household	817	19.5
Kitchen Utensils	410	9.8
Tools and Instruments	140	3.3
Irons	30	0.7
General household machines	68	1.6
Stoves	182	4.3
Sewing machines	118	2.8
Laundry machines	123	2.9
Churns	29	0.7
Industrial machines	261	6.2
Agricultural inventions	61	1.5
Transportation	142	3.4
Medical and Chemical	350	8.4
Construction/building	120	2.9
Miscellaneous	116	2.8

Notes: The inventions are categorized according to industry
of final use.
Source: See text and Footnote 16 for discussion of data set.

machinery, transportation, construction, and chemicals. A noticeable feature
of the patterns of invention by industrial category is that they are predominantly related to the role of women in the home and family. According to a
housewife in 1873, "It is a woman's right to be supplied with labor-saving
appliances to assist her in the labor of the house,"[36] and the patent records
indicate that some women were ready to devise any number of solutions
to the problem of seemingly endless household tasks. For instance, Mrs.
Sophronia Dodge of DeSoto, Iowa, contrived an appliance in 1872 for raising dough, noting that "This apparatus effects a great saving in time. It does
the work thoroughly and perfectly in the coldest weather. The vessel...is
provided with a tightly-fitting cover,...so that a kerosene lamp may be used

[36] "Woman's Rights," p. 1, *New Northwest*, October 31, 1873.

without any ill effects upon the bread."[37] Ella Haller patented a fruit jar in 1873 that contained a gasket to expel air and so increase the shelf life of preserved fruit, and later also patented a self-lighting lamp in 1878. Women also created improvements in household appliances such as stoves, dishwashers, and devices to launder clothes.

If this focus were predominantly because of a lack of education or market experience, one would expect a decline in household patents over time. Instead, one finds that clothing and related items, including hats, shoes, sewing, and textiles, absorbed a fairly constant share of patent efforts. The majority of women patentees tended to produce inventions related to household articles and furniture, the share of which increased over time from 22.8 percent over 1800–1865 to 36.1 percent of all patents between 1890 and 1895. Some may propose supply-side explanations of these patterns, such as the hypothesis that technological change required greater skill inputs and forced women who lacked those technical skills to focus on household gadgets; or else that technology may have developed to the point where it was easier to make marginal contributions to household items. However, these arguments are not entirely convincing, given that the share of household patents was increasing at the same time as patenting by women was rising in absolute terms as well as relative to men. The evidence is consistent with the idea that female inventive activity was responding more to market demand than to trends in technical knowledge, education, or occupations.[38]

Table 5.5 reports the regional decomposition of these patterns. New England states dominated in the production of clothing and related items, and we find by far the largest share of inventions in this industry in the same region. Those who view "female inventions" such as corsets as a niche area without much commercial significance should note that male inventors obtained 91 percent of the 1661 corset patents filed during the nineteenth century. Massachusetts inventors Susan Taylor Converse, Clara Clark, and Emmeline Philbrook all patented corsets that were manufactured by George Frost and Company of Boston.[39] Foy, Harmon, and Chadwick of New Haven employed several hundred female workers to make patented articles by inventors such as Lavinia Foy of Massachusetts, whose seventeen

[37] Patent No. 125,445, dated April 9, 1872.
[38] The Women's Bureau, *Women's Contributions*, and the Office of Technology Assessment and Forecasting, *Buttons to Biotech: U.S. Patenting by Women, 1977 to 1988.* Washington, D.C.: GPO, 1990, reveal that similar patterns (in terms of both the level and composition of women's patents relative to men) exist in the twentieth century, a period during which women's labor force participation has increased dramatically.
[39] Emmeline Philbrook's patents included No. 608005, July 26, 1898 for an eye shade; No. 544071, August 6, 1895 for a friction hinge; No. 503436, August 15, 1893 for an underwaist; No. 335237, February 2, 1886 for a clothes hook; No. 305471, September 23, 1884, for a weather strip; No. 286555, October 9, 1883 for a window screen; and No. 259422, June 13, 1882 for a corset.

Table 5.5. *Regional and Industrial Distribution of Patents, 1790–1895*
(within region percentages)

	1800–59	1860–69	1870–79	1880–89	1890–94	1800–1894
Middle Atlantic	(31)	(116)	(368)	(492)	(424)	(1431)
Clothing	12.9	20.7	17.9	30.0	21.7	23.3
Health	22.6	15.5	7.3	8.5	6.6	8.5
Household	35.5	31.9	40.5	33.1	40.8	37.3
Industrial Machines	22.6	23.4	19.6	15.2	14.6	17.0
Transport	0.0	1.7	8.2	2.9	3.8	4.3
Miscellaneous	6.5	6.9	6.5	10.4	12.5	9.6
New England	(13)	(51)	(179)	(167)	(131)	(541)
Clothing	30.8	47.1	38.6	37.7	34.5	37.9
Health	7.7	3.9	8.4	5.4	3.1	5.7
Household	23.1	37.3	33.0	33.5	34.4	33.6
Industrial Machines	23.1	5.9	10.1	13.2	17.6	12.8
Transport	0.0	2.0	1.7	1.2	3.1	1.9
Miscellaneous	15.4	3.9	8.4	9.0	7.6	8.1
Midwest	(2)	(66)	(256)	(507)	(488)	(1319)
Clothing	50.0	25.8	18.8	23.3	20.0	21.3
Health	0.0	12.1	10.2	7.1	5.1	7.2
Household	50.0	36.4	49.7	46.0	52.9	48.8
Industrial Machines	0.0	12.1	14.1	12.0	11.7	12.3
Transport	0.0	1.5	2.7	3.2	1.6	2.4
Miscellaneous	0.0	12.1	4.7	8.5	8.8	8.0
West	(0)	(8)	(49)	(83)	(119)	(259)
Clothing	0.0	37.5	12.2	15.7	13.5	14.7
Health	0.0	12.5	24.5	12.1	7.6	12.4
Household	0.0	25.0	38.8	43.4	50.4	45.2
Industrial Machines	0.0	0.0	18.4	10.8	10.1	11.6
Transport	0.0	0.0	2.0	2.4	8.4	5.0
Miscellaneous	0.0	25.0	4.1	15.7	10.1	11.2
South	(5)	(25)	(85)	(183)	(162)	(460)
Clothing	0.0	8.0	15.3	21.3	13.6	16.5
Health	0.0	12.0	16.5	8.2	4.3	8.5
Household	80.0	56.0	45.9	47.0	47.5	47.8
Industrial Machines	20.0	16.0	12.9	14.2	16.7	15.0
Transport	0.0	0.0	2.3	3.8	5.6	3.9
Miscellaneous	0.0	8.0	7.1	5.5	12.4	8.3
United States	(70)	(266)	(937)	(1433)	(1324)	(4030)
Clothing	20.0	26.3	21.6	26.5	20.5	23.3
Health	15.7	12.0	10.0	7.8	5.5	8.0
Household	35.7	36.1	41.9	40.1	46.3	42.2
Industrial Machines	22.9	15.8	15.6	13.5	13.7	14.4
Transport	0.0	1.5	4.6	2.9	3.6	3.4
Miscellaneous	5.7	8.3	6.3	9.2	10.4	8.8

Notes and Sources: Total number of patents are in parentheses. The percentages represent the proportion for which the industry accounts within each region. The South includes the District of Columbia. Percentages may not total to one hundred because of rounding. This table excludes patents filed by foreign inventors. See text and Footnote 16 for discussion of the sample.

corset inventions brought her a reputed annual income of $25,000.[40] Catherine Griswold, a New York resident, produced some twenty clothing-related inventions for which she obtained patents, including garment supporters and corset innovations that were manufactured by the Worcester Corset Company. Lena Sittig, yet another New York inventor, was granted several patents for garments, at least one of which was commercialized. The advent of activities such as bicycling and automobile outings meant that the market for new forms of apparel was expanding and profitable.[41] This provided an incentive in 1898 for Susan Emily Francis, of Wellington, New Zealand, to obtain U.S. patent No. 614,097 for a cycling skirt, as (according to her patent specification) "It is well known to lady cyclists that the ordinary skirt is uncomfortable when used for riding and by the movement of the legs is unavoidably raised to an undesirable degree."

The direction of women's inventive activities also may be related partially to the nature of the market for female inventions in the West relative to areas such as the South. Southern white households may have been less inclined to use household innovations because a surplus of low-wage black labor was available as domestic servants.[42] Midwestern and western women, without the benefit of such a supply of readily available help, might have had a greater incentive to substitute devices to help in their household tasks. This proposition may be tested by examining the industrial distribution of patents by region and by level of urbanization. Western women patented only thirty industrial (machine and tool) inventions in the entire century. Their efforts, and to a lesser extent those of women in the Midwest, focused on household appliances, furniture, and utensils. From 1790 to 1895 household inventions amounted to 47.8 percent of patents granted to women in the South, 45.2 percent in the West, and 48.8 percent in the Midwest. Moreover, the share of household patents increased in the 1890s to 50.4 percent in the West, and 52.9 percent in the Midwest, compared to 34.4 percent in New England during the same period.

Among these, Margaret Colvin of Michigan invented a successful washing machine, while others like Hattie Adler of Colorado, Nella Balch of Wisconsin, Margaret Brass of Minnesota, and Ellen Dillon of Iowa, patented clothes driers, washboards and boilers, dishwashing machines, iron-heaters, and

---

[40] Anne Macdonald, *Feminine Ingenuity*, p. 247.

[41] Sarah L. Naly and Mary Scott Jones of Philadelphia Pennsylvania, 1893, Patent No. 499,244 (from specification): The conventional bicycle saddle "is excellently adapted to male riders, it is obviously not suited to female riders. Aside from being uncomfortable to a female rider, it has the important objection that the point or projection at the forward part catches upon the skirts of the rider when getting on or off the bicycle, often causing considerable mishap, to say nothing of mortification." Mary Frances Harris of Carthage, Illinois obtained Patent No. 1073345 in September 16, 1913, for an automobile bonnet that would not muss the hair and acted as a vizor and sunshade.

[42] David M Katzman, *Seven Days a Week: Women and Domestic Service in Industrializing America*. New York and Oxford: Oxford University Press, 1978.

other household improvements. The lack of access to household help possibly accounts for a significant fraction of Western and Midwestern patented inventions in this category, although at first sight this thesis appears to be somewhat undermined by the fact that Southern women also exhibited the same focus. However, the size of the female population in the South was approximately the same as the western and midwestern states combined. This implies that in the latter states the per capita rate of patenting in the household category was four times higher than the equivalent rate for southern patentees. Thus, in per capita terms, the data are consistent with the idea that the industrial composition of female patenting activity in these regions was at least partly because of labor market conditions for domestics and household help. The urbanization patterns also support this hypothesis.

Table 5.6 indicates that the influence of urbanization differed significantly in terms of male and female per capita patenting. Patenting by male inventors occurred disproportionately in cities. Except for the Northeast, where Boston, Philadelphia, and New York City featured prolific inventive activity by both men and women, female patentees were largely located in rural areas, within counties of fewer than twenty-five thousand residents. This tendency increased, and over the entire period only in the Mid-Atlantic was per capita patenting higher in metropolitan areas. One might speculate that access to information flows, capital, and externalities from clusters of innovations, in addition to the presence of active markets, all contributed to the strong relationship between urban centers and male invention. By contrast, women were not able to benefit from those advantages to the same extent, whereas the potential market for inventions that reduced housework was likely greatest in rural areas. However, the data for rural residents also are consistent with the influence of supply factors such as a relatively lower level of specialized education or technical knowledge outside of cities.

Table 5.7, which examines the distribution of women's patents by industry and urbanization, indicates that most of the clothing-related patents were filed in metropolitan areas, as were industrial machines and transportation patents. However, household articles and machines were prevalent in rural areas, with 60.4 percent of household machines and 45.2 percent of household nonmachine inventions patented by rural residents between 1790 and 1879. Both categories of household patents accounted for over half of rural patenting between 1880 and 1895, compared to 36 percent for metropolitan patentees. Rural women exercised their ingenuity to reduce arduous household tasks, as did Sarah Sewell, of Defiance, Ohio, whose invention of a "combined washing-machine and teeter or see-saw" enabled her to offer "amusement and recreation for children and young persons, while at the same time I utilize their exertions, when desired, in washing the family or other clothes."[43]

[43] From the specification of her November 1885 patent, No. 330,626.

	1870s		1890s		1860–1895
	Women	All	Women	All	Women
East North Central					
No City	25.9	237.8	36.2	240.2	26.9
25,000	5.0	889.8	12.3	703.8	7.0
100,000	1.9	724.2	4.8	763.0	3.8
250,000	7.9	–	32.1	1139.4	17.7
Total	10.2	312.2	21.4	429.9	13.8
West North Central					
No City	19.2	129.4	50.7	168.4	33.8
25,000	2.7	239.9	7.3	300.6	4.5
100,000	–	–	10.5	588.9	5.0
250,000	3.8	293.3	13.5	938.4	11.5
Total	6.4	146.5	20.5	248.7	13.7
New England					
No City	13.4	438.5	12.6	382.4	11.8
25,000	40.0	1039.2	57.1	989.9	39.8
100,000	–	–	1.7	870.2	2.6
250,000	69.7	1875.9	43.9	1250.1	37.1
Total	30.8	775.8	23.8	772.0	22.8
Middle Atlantic					
No City	17.8	295.6	27.4	280.6	19.0
25,000	4.5	603.9	13.1	681.9	7.1
100,000	4.6	1009.0	8.6	795.2	4.8
250,000	53.7	1137.4	73.2	943.5	52.8
Total	20.2	563.4	30.6	607.0	21.0
South					
No City	6.1	53.2	15.4	63.5	10.4
25,000	0.5	266.4	4.4	452.5	1.9
100,000	0.5	563.8	1.2	434.2	0.7
250,000	0.6	492.8	3.5	421.8	2.1
Total	1.9	85.8	6.1	103.1	3.8
West					
No City	50.0	236.3	54.7	265.3	54.2
25,000	–	–	52.8	452.5	35.9
100,000	81.6	876.4	10.1	–	24.3
250,000	–	–	68.1	1056.9	39.4
Total	32.9	366.7	46.7	381.6	38.4

Notes: The data for women refer to the entire decade of the 1870s and 1890s. The columns for females are computed by dividing the number of patents within that urbanization category by total state population. The columns for all patentees refer to the 1870–1871 and 1890–1891 periods respectively.

Sources: See text and Footnote 16 for data on women. The data for all patentees are from Naomi R. Lamoreaux and Kenneth L. Sokoloff, "Location and Technological Change in the American Glass Industry During the Late Nineteenth and Early Twentieth Centuries," *Journal of Economic History*, Vol. 60 (Sept. 2000): 700–729.

Table 5.7. *Distribution of Women's Patents by Industry and Urbanization, 1790–1895*

	1790–1879				1880–1895			
	Rural	Urban	Metro	Total	Rural	Urban	Metro	Total
Clothes								
Number	82	59	145	286	200	101	351	652
Row Percent	28.7	20.6	50.7		30.7	15.5	53.8	
Col. Percent	15.1	31.7	26.7	22.5	18.2	24.5	28.2	23.6
Household Machines								
Number	90	16	43	149	135	35	76	246
Row Percent	60.4	10.7	28.9		54.9	14.2	30.9	
Col. Percent	16.5	8.6	7.9	11.7	12.3	8.5	6.1	8.9
Household (Nonmachine)								
Number	165	59	141	365	437	130	374	941
Row Percent	45.2	16.2	38.6		46.4	13.8	39.7	
Col. Percent	30.3	31.7	26.0	28.7	39.8	31.5	30.0	34.1
Industrial Machines								
Number	56	12	82	150	107	46	114	267
Row Percent	37.3	8.0	54.7		40.1	17.2	42.7	
Col. Percent	10.3	6.5	15.1	11.8	9.7	11.1	9.1	9.7
Tools								
Number	26	7	21	54	39	20	49	108
Row Percent	48.2	13.0	38.9		36.1	18.5	45.4	
Col. Percent	4.8	3.8	3.9	4.2	3.6	4.8	3.9	3.9
Transportation								
Number	15	4	28	47	34	12	42	88
Row Percent	31.9	8.5	59.6		38.6	13.6	47.7	
Col. Percent	2.8	2.2	5.2	3.7	3.1	2.9	3.4	3.2
Health								
Number	79	13	45	137	60	31	94	185
Row Percent	57.7	9.5	32.8		32.4	16.8	50.8	
Col. Percent	14.5	7.0	8.3	10.8	5.5	7.5	7.5	6.7
Miscellaneous								
Number	31	16	38	85	86	38	147	271
Row Percent	36.5	18.8	44.7		31.7	14.0	54.2	
Col. Percent	5.7	8.6	7.0	6.7	7.8	9.2	11.8	9.8
Total								
Number	544	186	543	1273	1098	413	1247	2758
Percent	42.7	14.6	42.7	100	39.8	15.0	45.2	100

Sources: See text and Footnote 16.

Several studies have likewise attested to the prevalence of innovations in the household and in specific tasks that would have affected the welfare of women, such as cheesemaking and other forms of household production.[44] Technological change also influenced women's lives by increasing the availability of goods and services in the market place. It has been proposed that it was not until the 1890s that women tended to be the major purchasers of products for the home, and thus likely to buy items that would relieve household burdens.[45] However, as early as 1861, Mrs. Beeton pointed out that professional laundry women "apply mechanical means" to the treatment of clothes.[46] The decadal census suggests that mechanical means also may have extended to homes in areas in which housewives did not have access to professional laundry women. By 1860, establishments to produce washing machines could be found in small counties in frontier regions such as Iowa, Kansas, and Indiana. For instance, four manufactories were located in Iowa in 1860 producing washing machines valued at $11,890, at a time when there were no more than 132,000 dwellings in the state. This can be contrasted with Pennsylvania, which included four establishments producing goods valued at $14,000 for a population of 515,319 dwellings. Similarly, the 1880 census shows four firms in Wisconsin that produced refrigerators. The production of household goods of this nature accounted for a disproportionate fraction of manufacturing in the more rural states relative to their share in the urbanized East. Indeed, a recent study of business gazetteers contends that "most manufactures of washing machines took place in small rural communities" by transient firms that were not sufficiently long-lived to appear in the decadal census.[47] Thus, rural women patentees were addressing their efforts to arenas that seemed most likely to provide rewards for

---

[44] Joan Jensen, *Loosening the Bonds: Mid-Atlantic Farm Women, 1750–1850*, New Haven, Conn.: Yale University Press, 1986, examines changes in the technology of the dairy industry and Sally McMurry, *Transforming Rural Life: Dairying Families and Agricultural Change, 1820–1885*, Baltimore: Johns Hopkins University Press, 1995, traces improvements in cheese making technology that affected rural women.

[45] Carolyn M. Goldstein, p. 130, "From Service to Sales: Home Economics in Light and Power, 1920–1940," *Technology and* Culture, Vol. 38 (January) 1997: 121–152, states that "commonplace wisdom held that women purchased about 85 percent of American household products overall." In Canada, a region that historically lagged behind the United States in technological progress, new household goods had already "profoundly" improved the standard of living of women by the first decade of the twentieth century. Suzanne Marchand, "L'impacte des Innovations Technologiques sur la vie Quotidienne des Québecoises du Début du XXe Siècle [1910–1940]," *Material History Bulletin*, vol. 28, 1988: 1–14, finds that "L'apparition de nouveaux biens d'équipement ménager, au cours de la période 1910–1940, a contribué à altérer profondément les conditions de vie quotidienne des femmes québecoises."

[46] See Arwen P Mohun, *Steam Laundries: Gender, Technology and Work in the U.S. and Britain, 1880–1940*, Baltimore: Johns Hopkins University Press, 1999, p. 29.

[47] Orville Butler, "The Changing Gender of Authority in American Home Appliance Technology: Dishwasher and Washing Machine Patents from 1860–1950," in Santimay Chatterjee et al. (eds.), *Studies in History of Sciences*, Calcutta, 1997, p. 172.

investments in inventive activity, as well as contributing to a ready market for products that would improve frontier life.

A professional approach to invention is often linked with multiple patenting and the attempt to extract profits and income from one's discoveries. Although only 15 percent of women patentees qualify as multiple inventors, these included Eliza Murfey, who filed twenty-three patents dealing with packing materials for journal boxes (many assigned to the Manhattan Packing Manufacturing Company), Catherine Griswold (twenty patents), Anna Dormitzer (sixteen patents for window washing apparatus), Helen Blanchard (twenty-one sewing machine patents), Harriet Tracy (twenty-eight patents on sewing machines and safety elevators), Margaret Knight (twenty-two, mainly for machinery), and Maria Beasley (fourteen, barrel-making machinery.)[48] Patentees of clothing and related items, household articles and machines (such as dishwashers and clothes driers), and industrial machines all tended to have higher numbers of patents. Because it is unlikely that an inventor would persist in patenting inventions that were worthless, multiple patenting supports the idea that the market for household articles and domestic machines was extensive and profitable. Even professional inventors such as Maria Beasley patented bread-making appliances. Women inventors from other countries who filed for American patents also directed their attention to similar markets.

Table 5.8 presents regressions with the log of the total number of patents awarded to each inventor as the dependent variable. The figures indicate that multiple patenting was most prevalent in the Mid-Atlantic and New England areas, and increased in metropolitan areas as well as in regions with greater per capita patenting rates. It is not coincidental that the midwestern and western states recorded the least number of multiple patentees, whereas the time trend shows a fall in such activity: the inventors from these states entered the market in larger numbers over time, but they tended to file only single patents. As an earlier chapter showed, a similar process of "democratization" had transformed the patenting process in the antebellum period, when patenting by all patentees similarly increased because of greater numbers of new entrants, rather than greater patenting by individuals.[49]

Inventors will tend to file for a patent if the expected benefit exceeds the cost of the patent. In efficient capital markets, the creator of a useful invention will be able to borrow to finance the patent and its development. However, women inventors undoubtedly faced greater obstacles in obtaining funding for their inventions, and might not have been able to afford the cost of patent fees, the application process, and payments to a patent attorney, which could amount to as much as $100 (about a quarter of average annual nonfarm wages in the late nineteenth century.) Some patentees

---

[48] These figures refer to total patents filed over the inventors' lifetimes.
[49] See Chapter 4.

Table 5.8. *Regressions of Total Patents by Individual Inventors*

Dependent Variable: Log of Number of Patents per Person		
	(1)	(2)
Constant	0.27	0.30
	(5.34)	(4.69)
Region		
New England	0.29	0.25
	(6.47)	(5.67)
Mid Atlantic	0.25	0.16
	(6.65)	(4.25)
West North Central	0.05	0.04
	(1.04)	(0.96)
East North Central	0.02	0.09
	(0.58)	(1.95)
West	−0.05	−0.02
	(0.90)	(0.34)
Time Trend		
1870s decade	−0.01	0.01
	(0.29)	(0.21)
1880s decade	−0.14	−0.14
	(2.94)	(2.99)
1890s decade	−0.20	−0.19
	(4.11)	(4.05)
Log Per Capita Patents	0.03	0.02
	(5.40)	(3.28)
Industry		
Industrial Machines		0.28
		(8.77)
Household Machines		0.11
		(3.01)
Apparel and Textiles		0.12
Urbanization		(4.57)
Urban		0.09
		(2.57)
Metropolitan		0.21
		(6.31)
	N = 4001	N = 4001
	$R^2 = 0.06$	$R^2 = 0.11$

Notes: Rural refers to a location with fewer than twenty-five thousand inhabitants (the excluded variable), whereas Urban indicates an urban location of between twenty-five thousand and one hundred thousand; Metropolitan represents districts of one hundred thousand or more residents. Population in the per capita patents variable is computed by exponential interpolation between census years.

Sources: See the text and footnotes.

bypassed capital markets and attempted to promote their inventions through establishing their own enterprises. Other women were able to obtain technical and possibly financial assistance from coinventors, or individuals who accepted payment in the form of partial assignments of the patent right. Over three hundred patents were attributed to coinventors, the majority of whom were unrelated males, typically from the same city and state (see Table 5.1.) Coinventors included machinists, engineers, pattern makers, tool makers, manufacturers, and artisans, who, according to patent law, needed to have made a significant contribution to the invention to warrant joint billing.

Assignees purchased or were granted rights to the invention without any claim to inventive inputs. Patent rights could be assigned any time during the patent's life, including at the time of granting, and 323 patents were assigned when issued. The identities of assignees yield insights into the market for invention, so I matched assignees and occupations from the city directories. Patent lawyers and agents comprised a quarter of the assignees at the time the patent was granted, suggesting that women inventors may at times have traded part of their property rights as payment for patent application fees and services. Manufacturers, another third of assignees at time of issue, either purchased a share in the invention because it was of value, or were granted the share in return for funding, or for undertaking to market and commercialize the invention in return. Maria Beasley reached an agreement in 1881 to transfer half of the rights in an uncompleted invention to James Henry of Philadelphia, in return for an advance of funds to complete the machine. On the same day, she filed an agreement for the boot pasting machine with another set of assignees, who paid $500 for 10 percent of the future patent.[50] Leonia Mabee of Paris, Texas, similarly ceded half the rights in her bedstead patent in exchange for a third ownership in a $100,000 company that was to make and promote the invention.[51]

Table 5.9 shows that the likelihood that a patent would be assigned at time of issue was higher if individuals had larger numbers of patents, and if they obtained patents for industrial or household machines. Women who patented clothing improvements were especially successful in marketing their inventions. The assignments data suggest a reputation effect, where assignees

---

[50] Maria Beasley in the assignment of May 5, 1881, promised to finish the invention within three months. In the contract she reserves the rights for foreign patents to herself. Beasley showed a shrewd business sense in her numerous assignments. Despite the existence of married women's property laws, her husband explicitly signed codicils to her contracts stating that he repudiated any claim that he might have in the transactions, thus making absolutely certain that the agreements could not be overturned based on coverture. Maria sold the U.S. rights in her footwarmer patent to Osborn Conrad of Philadelphia on December 14, 1878, a few months after the application was approved; in this case she did not retain all of the foreign rights, selling the Canadian rights for $100.

[51] See Macdonald, *Feminine Ingenuity*. Mabee obtained a total of four patents, between 1891 and 1901, including a self-basting pan, and two waterproof extensions for petticoats.

Table 5.9. *Logistic Regressions of Patent Assignments*

Dependent Variable: Probability of a Patent Being Assigned at Issue			
	(1)	(2)	(3)
Constant	-2.57^{***}	-2.82^{***}	-2.91^{***}
	(89.3)	(107.9)	(111.29)
Region			
New England	0.27	0.12	0.05
	(1.57)	(0.29)	(0.04)
Mid Atlantic	0.09	-0.02	-0.12
	(0.20)	(0.01)	(0.39)
West North Central	-0.26	-0.26	-0.21
	(1.04)	(1.02)	(0.65)
East North Central	-0.33	-0.31	-0.30
	(2.31)	(2.00)	(1.94)
West	-1.07^{***}	-1.02^{***}	-1.05^{***}
	(7.29)	(6.67)	(6.94)
Time Trend			
1870s decade	0.22	0.21	0.15
	(0.78)	(0.72)	(0.38)
1880s decade	0.30	0.31	0.25
	(1.59)	(1.72)	(1.01)
1890s decade	0.12	0.17	0.09
	(0.25)	(0.48)	(0.14)
Log (number of patents)		0.26^{***}	0.22^{***}
		(12.92)	(9.09)
Industry			
Machine Inventions		0.53^{***}	0.54^{***}
		(11.16)	(11.43)
Household Inventions		-0.31	-0.27
(nonmachine)		(1.46)	(1.17)
Apparel and Textiles		0.36^{***}	0.34^{***}
		(6.56)	(5.70)
Urbanization			
Urban			0.27
			(1.99)
Metropolitan			0.35
			$(5.72)^{***}$
	N = 4030	N = 4030	N = 4030
	$\chi^2 = 26.67^{***}$	$\chi^2 = 64.27^{***}$	$\chi^2 = 69.45^{***}$

*** significant at 1 percent level
** significant at 5 percent level
Notes: The response profile includes 323 assignments at time of issue (N = 1) and 3,707 patents that were not assigned (N = 0) at time of issue. The data do not include assignments that took place after the patent was granted. The urbanization variables represent counties that included at least one town with twenty-five to one hundred thousand residents (Urban), and over one hundred thousand residents (Metropolitan.) Wald χ^2 statistics are in parentheses.
Source: see text and footnotes.

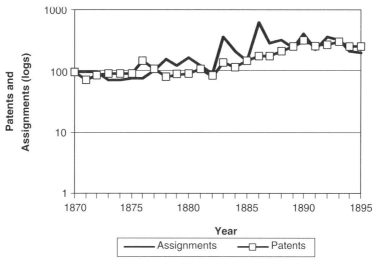

Figure 5.2. Patents and Assignments by Women Inventors, 1870–1895
Notes and Sources: The data for assignments are estimated from a sample of assignments drawn from the listing of inventors whose initial was "B" in the Index to Assignments of Patents held by the National Archives, Record Group 241. Further information was obtained from the Digest of Assignment of Property Rights in Patents, various years. To obtain the totals, I annualized the monthly average and multiplied by twenty-six. Because initials of "B" are probably more common than letter such as "Q" or "Z," these figures likely overstate the actual somewhat; however, the trend would not necessarily be affected by this feature of the calculations.

were more likely to purchase the patent at time of issue if the inventor had proven her ability by obtaining more than one patent. Assignments at time of issue were less likely in the West than in other areas. As may be expected, assignment at time of issue was positively related to urbanization, and was highest in metropolitan areas. Although urban patentees had an advantage in early patent assignments, these results do not imply that rural patentees were unable to benefit commercially from their patents, as assignments were more likely to occur during the life of the patent. Sarah Dake, a resident of Eureka, Wisconsin, obtained an 1872 patent for corsets that was the subject of some thirty-eight contracts in the three years after it was issued. Valuable evidence on trade in inventions also can be gleaned from advertisements, directories, mail order catalogues, and probate records.

Figure 5.2 indicates that the growth rate of assignments drawn up by women inventors after the patent was issued matched the rapid increase in patents granted. For instance, Ellene A. Bailey negotiated several assignments, including the entire rights to a powder puff for "$450 and other considerations" to William Smith and Co. of New York, and sold her patent for a corset protector to John Moore and Addison Tuttle of New York a year later. Amelia Baglin chose to license the rights to her 1879 patent for hats to firms throughout the Northeast. Julia Banfield sold the exclusive rights to her 1878 improvement in corset busks to a Connecticut firm, with the proviso that the company should in turn sell her all the busks she needed for her own manufacturing of corsets, "at a reasonable price." The assignment records do not support the view that patented inventions were unrelated to household production because few were ever used.[52] These contracts highlight the commercialization of numerous household inventions such as washboards, roasting pans, churns, ironing boards, pillowsham holders, improvements in ovens, and various ingenious furniture contrivances. Patent agencies and journals such as *Scientific American* undoubtedly inflated the expected gains from such inventions, but their claim that domestic improvements could prove to be extremely profitable was true for a significant number of women inventors.

The city directory sample allows us to trace the activities of women who exercised entrepreneurial as well as inventive abilities, and attempted to profit from their inventions. Almost one quarter of the linked patents (24.2 percent) were filed by women who established businesses, marketed their inventions or manufactured innovations for sale as final products. Emily Stears obtained six patents over the course of thirty-three years, the last of which in 1911 was for a sanitary rubber seal for sinks. She was listed in the 1885 New York directory as the "patentee and sole manager" of a Brooklyn enterprise that sold odorless, steamless cooking vessels. Sarah Drewry, another New York inventor included in the directory, manufactured and sold surgical instruments. Some of these women demonstrated a degree of independence that was perhaps uncommon, such as Elizabeth van Vleck, patentee of a corset improvement, who listed her Chicago company as "Van Vleck and Daughter." The assignment records show that Carina B. Banning purchased patents from other inventors – such as a dustpan patent for $100 – for her Boston firm, the Banning Stationary Dustpan and Manufacturing Company.

Many of the women inventors who exhibited at the Centennial exposition in Philadelphia had the objective of obtaining profit and economic self-sufficiency. Indeed, approximately 35 percent of the devices at the exhibition were actually produced, some for more than twenty years. Similarly, a study of women inventors at the World's Columbian Exposition in 1893, found that demand for displayed innovations was high, including six hundred customers for Mary Harris' refrigerator and some sixty thousand orders for the Hambell egg and cake beater.[53] The inventions and inventors who participated in expositions undoubtedly were atypical, but evidence of these

---

[52] "Ennoble Domestic Work," p. 116, *Woman's Journal*.

[53] Weimann, *The Fair Women*, p. 432. Mary M. Harris obtained Patent No. 492,286, dated February 21, 1893; there is no record of a patent for the Hambell device.

purchases illustrate the existence of a market that could be tapped through personal transactions, local newspapers, trade journals, and publications devoted to inventors' efforts.

A number of women inventors had recourse to the courts to counter attempts to infringe their rights, and thus incidentally created records of their entrepreneurial activities. Patent lawsuits show that some patentees aggressively defended their inventions against infringers in order to appropriate the returns. For instance, the market for women's corsets, which was competitive and expanding rapidly, provoked several instances of litigation.[54] Lavinia Foy, patentee of several corset improvements that were produced by the Foy, Harmon, and Chadwick corset manufacturing company, was involved in a number of lawsuits to protect the inventions against infringers. Similarly, Mrs. Catharine Judson obtained a patent for corset clasps in 1876, and successfully defended her property against the Bradford enterprise of Massachusetts which was using the invention without authorization. *Mayer v. Hardy*, 127 N.Y. 125 (1891), reveals that in January 1879 Mrs. Judson licensed the patent to Saly Mayer and three others for a royalty fee of 50¢ per gross, promising to protect the patent against infringers. Later in the same year she sold the entire patent right to Garret Hardy, of New York. Helen M. Macdonald's 1874 improvement in dress protectors was the subject of a series of lawsuits in Massachusetts and New York, even in 1886, only two years before the patent was due to expire. Miss Macdonald acted as her own counsel in *Macdonald v. Sidenberg et al.*, 16 F. Cas. 48 (1879), and was successful in obtaining an injunction against the defendants who claimed that their variant of the protector was both different and prior to hers.[55]

Technological change is far broader than the realm of patented inventions, and encompasses improvements in diet and techniques of food preparation, more efficient processes of childrearing, and new designs in furnishing. It is all the more striking that, even within the narrow confines of patents, women were far from being passive bystanders in an environment that was deprived of the innovations that were revolutionizing the workplace. Female inventors were applying their creative insights to the sphere in which they

---

[54] For a discussion of the market for corsets, see Bernard Smith "Market development, Industrial Development: The Case of the American Corset Trade, 1860–1920," Business History Review, vol. 65 (Spring 1991): 91–129.

[55] The invention protected "the dress from being cut or damaged by contact with earth or brick or concrete pavements and sidewalks, said invention being impervious to moisture . . . it also forms a very neat trimming and improves the hanging of the skirt," according to the patent application, May 6, 1873 (cited in *Shepard v. Carrigan*.) The patent application was rejected twice, and the subsequent amended specification was the subject of an interference suit with M. Herbert Chase, over whom she finally prevailed and was granted the patent (*Macdonald v. Blackmer*, 16 F. Cas. 38, 1876); the patent was bequeathed to Carrigan, the appellee in the Supreme Court case, *Shepard v. Carrigan*, 116 U.S. 593, 1886; see also *Eastern Paper Bag Co. v. Standard Paper Bag Co.*, 29 F. 787, 1887.

had the greatest experience. In the process, many were introduced to the market as they attempted to pursue profits. The experience of patentees suggests that the spheres of household and market were closely linked for both men and women. The extensive archives of assignment contracts provide further evidence of an active market in domestic inventions. Taken together, these data indicate that, like the economy at large, the household economy was the locus of both invention and innovation. Although in other regards women may have been excluded from the benefits of a democratic society, the U.S. patent system offered them similar opportunities to excel as their male counterparts.

6

Patentees and Married Women's Property Rights Laws

"So great a favourite is the female sex of the laws"

– Sir William Blackstone (1765)

"This is law, but where is the justice of it?"

– Ernestine Rose (1851)

The previous chapter showed that women patented few inventions in the antebellum period, but the growth rate of female patenting toward the end of the nineteenth century was rapid and exceeded that of male inventors. Suggestions about the factors that might explain the paucity of documented patents credited to women have ranged from the cultural to the economic. Nineteenth-century feminists attributed patterns of female patenting partly to the legal status of married women. According to Matilda Joslyn Gage, a prominent suffragist, "It is scarcely thirty years since the first State protected a married woman in the use of her own brain property. Under these conditions, legally incapable of holding property... that woman has not been an inventor to an equal extent with man is not so much a subject of surprise as that she should have invented at all."[1] Although such declarations were partly motivated by political rhetoric, they do warrant further investigation. In an era when relatively few women remained single, their status and economic welfare were significantly affected by laws regarding married women's rights.[2] Under nineteenth-century common law, a married woman

---

[1] Matilda Joslyn Gage, "Woman as an Inventor," *North American Review*, vol. 136 (318) 1883: 488–89. "Nor is woman by law recognized as possessing full right to the use and control of her own powers. In not a single State of the Union is a married woman held to possess a right to her earnings within the family; and in not one-half of them has she a right to their control in business entered upon outside of the household. Should such a woman be successful in obtaining a patent, what then? Would she be free to do as she pleased with it? Not at all. She would hold no right, title, or power over this work of her own brain. She would possess no legal right to contract, or to license any one to use her invention. Neither, should her right be infringed, could she sue the offender," p. 488. "How does the law recognize women?... It is only a little over a quarter of a century since the first state in this Union protected a married woman in the use of her own brain property. Is it any wonder then, that woman is not equal with man as an inventor," *The Woman Inventor*, vol. 1 (1) 1891.

[2] According to Lee Chambers-Schiller, *Liberty, A Better Husband: Single Women in America*, New Haven, Conn.: Yale University Press, 1984, the numbers of unmarried women were very

was bound by the rules of coverture, which vested her legal rights in her husband. As such, he controlled any property she acquired before or after marriage, as well as her earnings. Married women were prohibited from entering into contracts, or engaging in trade on their own account, as "sole traders."

Reforms in many dimensions were effected at the state level after 1830, and enlarged the ability of married women to own separate property, to trade or engage in businesses on their own account, and to keep the earnings from their labor. Legal historians contend that the Married Women's Acts did not result in significant improvement in the economic status of women. However, if women were motivated by potential profits, it might be expected that their efforts would respond to changes in the legal system that expanded their property rights, and offered greater access to potential income from their participation in the market. This claim was implicitly supported by feminists of the period who attributed the relatively low numbers of inventions created by women to deficiencies in the legal system. Higher inventiveness also might result from laws permitting women to engage in business or professions. Previous chapters showed that many inventions were trade-related, in which case greater job experience would tend to promote inventive activity. The relationship might incorporate both demand and supply factors, where participation in a profession enhanced the ability to perceive demand and further promoted the skills required for invention. Moreover, commercial exploitation of patent property depended on the right to contract and to sue, in order to produce the invented article, to assign the patented invention, or to prosecute infringers. In short, legal reforms were likely to affect women's behavior by altering the economic costs and benefits associated with their involvement in commercial activity.

The results also provide insights into the maximization process that informed household behavior. A finding that married women were responsive to policies that granted them rights to property or earnings separately from their husbands would be consistent with a model in which individual, rather than household, utility was maximized. Conversely, the result that improvements in legal status were unrelated to behavior satisfies at least the necessary condition for a model in which household utility is jointly maximized. More generally, studies of the effect of changes in laws are especially important to our overall understanding of the role of institutions in the period of nascent industrialization. One would like to know whether legal reforms influenced market participation in general, and how different the paths of industrialization and economic growth might have been, had the set of opportunities available to women been expanded earlier. The focus on

---

low in the eighteenth century, and increased slightly thereafter: 7.3 percent of women born between 1845 and 1849 never married; 8.0 percent born between 1845 and 1849; 8.9 percent between 1855 and 1859; and 11 percent over 1865–1875 (p. 3).

a narrower aspect of this problem – the relationship between patenting and married women's property rights laws – sheds some light on the larger issues of the effects of legal reforms on the role of women in the market economy.

LEGAL STATUS OF NINETEENTH-CENTURY WOMEN

For much of the nineteenth century, married women were subject to the "disability of coverture," which vested their rights in their husbands. According to a standard eighteenth-century legal reference, "by marriage, the husband and wife are one person in law; that is, the very being or legal existence of the woman is suspended during the marriage."[3] If women were granted the right to control their own property, other authorities argued, it would lead to an independence that threatened the institutions of marriage and the family. The court opined in *Cole v. Van Riper*, 44 Ill. 58 (1867): "It is simply impossible that a married woman should be able to control and enjoy her property as if she were sole, without practically leaving her at liberty to annul the marriage." Married women were explicitly excluded from many occupations on similar grounds. When the U.S. Supreme Court denied Mrs. Myra Bradwell's 1872 appeal to be admitted to the Illinois bar as a practising attorney, Justice Bradley felt compelled to add a separate concurrence to point out that "the family institution is repugnant to the idea of a woman adopting a distinct and independent career from that of her husband...the paramount destiny and mission of woman are to fulfil the noble and benign offices of wife and mother. This is the law of the Creator."[4]

A market economy is based on the security of contracts, yet, during this critical period when the American system evolved from farm-based

---

[3] Sir William Blackstone, *Commentaries on the Laws of England*, vol. 1, New York: W. E. Dean Publishers, 1836: p. 355. The epigraph refers to p. 366: "even the disabilities which the wife lies under are for the most part intended for her protection and benefit: so great a favourite is the female sex of the laws."

[4] *Bradwell v. Illinois*, 83 U.S. 130 (1872). The Supreme Court of Illinois pointed out that American laws were drawn from the British doctrines, and "female attorneys at law were unknown in England, and a proposition that a woman should enter the courts of Westminster Hall in that capacity, or as a barrister, would have created hardly less astonishment than one that she should ascend the bench of bishops, or be elected to a seat in the House of Commons." Justice Bradley's concurrence (in which he was joined by Justices Swayne and Field) went further by chivalrously declaring that "The natural and proper timidity and delicacy which belongs to the female sex evidently unfits it for many of the occupations of civil life The paramount destiny and mission of woman are to fulfil the noble and benign offices of wife and mother. This is the law of the Creator. And the rules of civil society must be adapted to the general constitution of things, and cannot be based upon exceptional cases." Interestingly, Chief Justice Salmon P. Chase (who had joined Bradley, Field, and Swayne in their dissent to the Slaughterhouse opinion, 83 U.S. 36; 1872) was the only dissenter here. Chase, who was seriously ill at the time and died shortly after, had strongly supported the abolitionist movement, and promoted the passage of married women's property legislation while he was governor of Ohio.

production toward industrial capitalism, the majority of women could not enter into viable commercial contracts once they were subject to a contract of marriage.[5] Single women benefited from the same property rights as men, but a married woman could neither devise nor sell her property, sue nor be sued. She could not file for bankruptcy, and her husband was liable for any debts incurred; conversely, the claims of her husband's creditors extended to her property.[6] For most women, property was earned in the course of marriage, and not simply inherited; yet married women had no right to any wages or income they earned, leading to economic dependence on the husband even if the wife were involved in non-household production. Prior to the changes in the law, the disabilities of married women extended to their rights to benefit from the sale, purchase, or commercialization of their patented inventions. Some exceptions to the doctrine of marital disability were available through equity courts. Colonial common law inherited the feudal bias of English common law, but early equity courts modified feudal laws based on status to take into account fundamental needs of commercial exchange, such as the defense of property rights and contracts. However, solutions at equity were limited to a small class of the population, mainly the daughters of wealthy parents who established separate trusts through the courts to ensure the protection of settlements and bequests.[7]

The reasons for reforms in women's economic rights are important because of their implications for the direction of causality between patenting and changes in the law. Legislation in the 1830s and 1840s did not address the issue of women's participation in market exchange nor women's right to hold income or property on their own account. Rather, the intent of these laws was to secure the property of a married woman from her husband's creditors in order to protect family income during the economic downturn of the late 1830s.[8] Control remained with the husband, and courts interpreted the legislation narrowly to ensure that ownership did not signify independence from the family. Mississippi's 1839 law, one of the first that was passed, typified this class of legislation, for it merely protected slave holdings of

---

[5] According to Joel Prentiss Bishop, *Commentaries on the Law of Married Women*, Boston: Little, Brown & Co., 1873–1875, p. 41, "Being under the power of her husband, she can have no will of her own, and by reason of this lack of freedom of will she cannot contract."

[6] See, for example, John F. Kelly, *A Treatise on the Law of Contracts of Married Women*, Jersey City, N.J.: D. Linn & Co, 1882.

[7] For instance, less than 2 percent of affianced couples in South Carolina employed marriage settlements between 1785 and 1810. Marlynn Salmon, "Women and Property in South Carolina: The Evidence from Marriage Settlements, 1730–1830," *William and Mary Quarterly*, vol. 39 (4) 1982: 655–85; also see Salmon, *Women and the Law of Property in Early America*, Chapel Hill: University of North Carolina Press, 1986.

[8] See Richard Chused, "Married Women's Property Law: 1800–1850," *Georgetown Law Journal*, vol. 71 (2) 1983: 1359–1425; and Norma Basch, *In the Eyes of the Law: Married Women's Property Rights in Nineteenth Century New York*, Ithaca, N.Y.: Cornell University Press, 1982, p. 207.

white married women from seizure by creditors. Southern states especially may have been more concerned with guarding the rights of debtors rather than the rights of women.[9] In short, one might regard the first wave of married women's laws in the 1830s and early 1840s as a species of bankruptcy legislation. These laws created commercial uncertainty, however, because the potential for fraud by debtor households toward creditors increased.[10] This problem was ultimately resolved by granting wives the further right to control their separate estates. As *Small v. Small*, 129 Pa. 336 (1889) noted, the legislature "saw that a married woman's coverture stood in the way of a full, free and expeditious transaction of affairs...that in order to make contracts with her legal and binding it was necessary for every mechanic and every tradesman to have knowledge of the most intricate questions of law; and that to recover even the smallest account against her required the services of a skillful lawyer. These were the mischiefs they undertook to remedy."

Laws that subsequently granted women access to their earnings and promoted their participation in commercial activity evolved from expansions in the scope of the earlier and specific legislation. Some researchers contend that married women's property laws comprised a minor part of codification or efforts to revise and simplify the law of property in general, in order to make access more democratic. The statutory reforms also have been related to an emerging view in the mid-nineteenth century of a separate domestic sphere for women that accompanied their increased responsibility within the family. Others focus on the efforts of prominent feminists, and argue that a turning point was reached because of a July 1848 women's rights convention in Seneca Falls, New York, which lobbied for improvements in the legal standing of married women. However, by this time New York had already passed its legislation of March 1848, that extended separate property rights

---

[9] For instance, in Maryland, "the policy of the (pre-1860) legislation...was, not to take from the husband the ownership which the common law gave him; but to protect from his creditors what came to him from her, leaving the ownership with him as before" (Bishop, *Commentaries*, vol. 2, p. 521; see also John F. Kelly, *A Treatise on the Law of Contracts of Married Women*, Jersey City, N.J.: D Linn & Co, 1882, p. 526; and Susan Lebsock, "Radical Reconstruction and the Property Rights of Southern Women," *Journal of Southern History*, vol. 43 (2) 1977: 195–216, p. 207.) One may speculate that downturns prior to the panic of 1837 did not lead to such widespread debtor protection laws because they were different in character, agrarian-based, and more localized in effect.

[10] Some twelve thousand lawsuits between 1800 and 1995 relate to married women. A search of cases by time period indicates that 40 percent of married women's cases before 1830 involved fraud and creditors, compared to 21.8 percent between 1830 and 1879, and 5 percent after 1920. South Carolina's 1744 "Act Concerning Feme Covert Sole Traders" illustrates the expansion of married women's rights to counter such problems: "whereas feme coverts in this province who are sole traders do sometimes contract debts in this province, with design to defraud the persons with whom they contract such debts, by sheltering and defending themselves from any suit brought against them by reason of their coverture," the colony therefore granted married sole traders the right to sue and correspondingly be sued.

to all married women.[11] Indeed, three years before feminists gathered for
the Seneca Falls Convention, and long before the passage of earnings laws,
the New York legislature enacted an 1845 statute that explicitly "secured to
every married woman who shall receive a patent for her own invention, the
right to hold and enjoy the same, and all the proceeds, benefits, and profits
as her separate property . . . as if unmarried."[12] Although no single explanation will suffice, the consensus from these studies appears to be that the laws
were changed by forces besides the increase in women's nonhousehold production. This conclusion is reinforced by the finding that "massive industrial
unemployment, particularly in the 1870s and 1890s, led many to question
women's right to labor," which implies that legal reforms during this period
were unlikely to have been caused by labor market pressures.[13]

Table 6.1 shows that statutory action progressed sequentially in terms
of three broad categories between 1830 and 1890. First, many jurisdictions
passed laws enabling married women to retain separate estates and property;
second, laws of the 1860s and 1870s granted such women the right to keep
their earned income; finally, the third wave of legislation permitted wives to
engage in business on the same basis as single women or as "sole traders."
However, distinct regional differences were evident. One can detect a "frontier effect," for example, in the finding that by 1890 all midwestern and
82 percent of the western states had approved separate estates for women.[14]
Moreover, 91 percent of western states had dissolved trading restrictions,
and 73 percent passed earnings acts by this period. A number of western
and midwestern states, including Kansas, Nevada, and Oregon, protected
women's property rights in their constitutions. Community property states

---

[11] The 1848 N.Y. Statute extended separate property rights to all married women. The Act
of 1860 stated that "A married woman may bargain, sell, assign and transfer her separate
personal property, and carry on any trade or business, and perform any labor or service
on her sole and separate account, and the earnings of any married woman, from her trade,
business, labor or services, shall be her sole and separate property." This statute was the
model for similar legislation in a number of other states.

[12] See Kelly, p. 456. Up to this period, only ten patents had been granted to women residing in
New York. Connecticut (1856) and West Virginia (1868) passed similar legislation.

[13] Claudia Goldin, *Understanding the Gender Gap: An Economic History of American Women*,
New York: Oxford University Press, 1990, p. 53.

[14] Mari J. Matsuda, "The West and the Legal Status of Women: Explanations of Frontier
Feminism," *Journal of the West*, vol. 24 (1) 1985: 47–56, argues that the "frontier effect"
was because of a number of factors including the relative scarcity of women. This notion
is reinforced by the record on women's suffrage, which several Western states granted in
the nineteenth century: Wyoming, 1869; Utah, 1870; Washington, 1883; Colorado, 1896;
Idaho, 1896. Washington state's decision was overturned by the Supreme Court, and later
revoked in 1889. When the legislation was finally approved in 1910, the vote carried in every
county, ending with a two to one majority. California, Arizona, Kansas, Oregon, Montana,
and Nevada also allowed women the vote between 1910 and 1915. See Mari Jo Buhle and
Paul Buhle (eds.), *The Concise History of Woman Suffrage*, Urbana: University of Illinois
Press, 1978.

Table 6.1. *State Laws Regarding Married Women's Economic Rights, by Year of Enactment*

State	Property Laws	Earnings Laws	Sole Trader Laws
Northeast			
Connecticut	1856 (patents)	1877	1877
Maine	1844	1857	1844
Massachusetts	1845	1874	1860
New Hampshire	1867	–	1876
New Jersey	1852	1874	1874
New York	1845 (patents)	1860	1860
Pennsylvania	1848	1872	–
Rhode Island	1848	1874	–
Vermont	1881	–	1881
South			
Alabama	1867	–	–
Arkansas	1873	1873	1868
Delaware	1875	1873	–
District of Columbia	1869	–	1869
Florida	–	–	–
Georgia	1873	–	–
Kentucky	–	1873	1873
Louisiana	–	–	1894
Maryland	1860	1860	1860
Mississippi	1871	1871	1871
North Carolina	1868	1873	–
Oklahoma	–	–	–
South Carolina	1870	–	1870
Tennessee	1870	–	–
Texas	–	–	–
Virginia	1878	–	–
West Virginia	1868 (patents)	1893	1893
Midwest			
Dakotas	1877	1877	1877
Illinois	1861	1861	1874
Indiana	1879	1879	–
Iowa	1873	1870	1873
Kansas	1868	1868	1868
Michigan	1855	–	–
Minnesota	1869	–	1874
Missouri	1879	1879	–
Nebraska	1881	1881	1881
Ohio	1861	1861	–
Wisconsin	1850	1872	–

(*continued*)

Table 6.1 *(continued)*

State	Property Laws	Earnings Laws	Sole Trader Laws
West			
Arizona	1871	–	1871
California	1872	1872	1872
Colorado	1874	1874	1874
Idaho	1887	–	1887
Montana	1872	1874	1874
Nevada	1873	1873	1873
New Mexico	–	–	–
Oregon	–	1880	1880
Utah	1895	1895	1895
Washington	1889	1889	1889
Wyoming	1876	1876	1876

Notes: The table includes those acts that granted separate control over property to married women (Property), the rights to their earnings without need of the husband's consent (Earnings), and the ability to engage in contracts and business without need of the husband's consent (Sole Trader). The table does not include legislation based on restrictions such as the right to trade only if abandoned by the husband, or if the husband were incapacitated or irresponsible, nor does it include legislation that was merely granted to afford relief from creditors. Married women's property right acts that were legislated primarily as debt relief include: Alabama, 1846, 1848; Arkansas, 1835, 1846; Florida, 1845; Georgia, 1868; Indiana, 1852; Kentucky, 1846; Maine, 1840, 1847; Maryland, 1841; Missouri, 1849; New York, 1849; North Carolina, 1849; Ohio, 1846; Oregon, 1857; South Carolina, 1868; Tennessee, 1825; Texas, 1845; Vermont, 1847; West Virginia, 1868. Kelly notes that debt relief legislation did not create a truly separate estate for women because control was still vested in the husband. Other acts that incorporated caveats such as the requirement that husbands were irresponsible, imprisoned or incapacitated, or appointed as trustees of their wives, include: Alabama, 1849; Arkansas, 1875; Connecticut, 1849, 1853, 1875; Delaware, 1865, 1873; Florida, 1881; Georgia, 1873; Idaho (no year mentioned); Illinois, 1874; Indiana, 1853, 1857, 1861; Kentucky, 1843, 1873; Louisiana, 1866; Maine, 1821; Massachusetts, 1835; Michigan, 1846; Minnesota, 1866; Mississippi, 1839; Missouri, 1865; Nebraska, 1881; New Hampshire, 1842, 1846; North Carolina, 1868, 1872, 1873; Ohio, 1868; Oregon, 1857; Pennsylvania, 1718, 1855, 1872; Rhode Island, 1880; Tennessee, 1835, 1858; Texas, 1865; Vermont, 1862, 1881; Virginia, 1876, 1877; West Virginia, 1868; Wisconsin, 1850, 1878. The 1845 act of New York (Chap. 11), the 1856 Act of Connecticut, and the 1868 Act of West Virginia explicitly accorded women the right to "hold a patent for an invention, as if she were unmarried" (W.Va. Code of 1868, Sec. 4).
Sources: Wells, *Separate Property*; Kelly, *Treatise*; Bishop, *Commentaries*, 2 vols.

such as California, Texas and Arizona inherited a civil law tradition that nominally granted joint ownership to husbands and wives.[15] In California and other southwestern areas, a married woman who carried on a business was considered to have the same rights as a single woman. This can be compared to the 71 percent of the southern states that had separate estates laws,

[15] Community property states were Arizona, California, Idaho, Louisiana, Nevada, New Mexico, Texas, and Washington.

the 47 percent that had sole trader laws, and the 41 percent that passed earnings laws. Southern states also tended to interpret the statutes more conservatively; for instance, Alabama and Virginia passed statutes whose ambit was severely limited to special cases. Florida and Texas passed no effective women's rights legislation in the nineteenth century, and in 1887 South Carolina went out of its way to formally bar married women from business partnerships.[16]

Legal historians have for the most part asserted that the consequences of married women's legislation were minimal. They argue that the antebellum property rights reforms increased the responsibility of women for the welfare of their families, without improving their economic status or their standing in the labor market.[17] For example, earnings laws were initially narrowly circumscribed in scope, with the main intention of protecting women who were burdened with profligate and irresponsible husbands. Courts also typically interpreted the statutes as exempting any work that was conducted in the home or for the benefit of the family, because they feared the transformation of the family relationship into a market relationship.[18] More generally, "the married women's acts themselves did not legitimate any radical shifts in the economic status of women."[19] A study of the New York statutes similarly opines that "full legal equality for married women loomed as a threat to the entire economic structure. Consequently the changes created by the statutes were either limited or illusory."[20] These assertions were not,

[16] Indeed, in many Southern states reforms occurred in the twentieth century. According to Susan Lebsock, "Radical Reconstruction and the Property Rights of Southern Women," *Journal of Southern History*, vol. 43(2) 1977: 195–216, p. 215, "major statutory changes in the married women's property laws in Alabama, Mississippi, Florida, Louisiana and Texas awaited the 1880s and beyond." Although Georgia passed separate estates legislation in 1873, it declared at the same time that the general contracts of married women were void. It was not until 1943 that Georgia allowed women the right to separate earnings. John C. Wells, *A Treatise on the Separate Property of Married Women*, Cincinnati: R Clarke & Co, 1878, p. 15, points out: "The first movement of the Florida legislature . . . was the ungracious extending of the criminal code so as to provide that a married woman may be convicted of the crime of arson, by burning her husband's property . . . it seems that here the whole business of legislating for married women stopped." Southern courts reinforced this tendency; as *Allen-West v. Grumbles*, 161 F. Cas. 461 (1908) acknowledged: "the Supreme Court of Arkansas has constantly and rigidly held to the rule of the common law in construing the married women's statute."

[17] See Elizabeth Warbasse, *The Changing Legal Rights of Married Women, 1800–1861*, New York: Garland, 1987; Basch, *In the Eyes*; and Chused, "Married Women's Property Law."

[18] Amy Dru Stanley, "Conjugal Bonds and Wage Labor: Rights of Contract in the Age of Emancipation," *Journal of American History*, vol. 75 (2) 1988: 471–99; and Reva B. Siegel, "The Modernization of Marital Status Law: Adjudicating Wives' Rights to Earnings, 1860–1930," *Georgetown Law Journal*, vol. 82 (7) 1994: 2127–211; Reva B. Siegel, "Home as Work: The First Woman's Rights Claims Concerning Wives' Household Labor, 1850–1880," *Yale Law Journal*, vol. 103(5) 1994: 1073–1217.

[19] Richard Chused, "Married Women's Property Law," p. 1362.

[20] Norma Basch, *In the Eyes*, p. 4.

however, tested for consistency with the evidence. Furthermore, quite apart from encompassing issues such as the economic status of women, the narrower question remains as to whether the existence of laws protecting individual property served as an incentive for women to alter their patenting behavior.

PATENTING AND CHANGES IN THE LAW

The previous chapters showed that the United States patent system is administered at the federal level, and places no restrictions on the race, gender, or citizenship of inventors eligible to hold patent property. Appropriate federal legal and property rights institutions functioned as "enabling factors," which are prerequisites for market expansion. Their absence would have retarded participation in the market economy; however, the presence of such institutions was not sufficient for inducing economic progress. For instance, states have domain over allied rights such as the ability to contract to convey the patent, to sue for compensation or deterrence in the event of infringement, and retain income or profits from commercialization of the invention. State holdings on torts, contracts, and other legal doctrines can strengthen or unravel property rights even if the latter are protected by the Constitution. Thus, the progress of women's property rights (broadly defined) was necessarily affected by policies at the state level. The Civil War heralded significant changes in women's property rights that increased the potential profits from their commercial efforts. The same period also marked a dramatic increase in their patenting activity. Some have argued that the timing was not coincidental, but causal. If so, the changes in the married women's laws would have stimulated an increase in women's investments in inventive activity and promoted greater efforts to obtain patent property.

The tables presented here show the association between per capita patenting (at the state level) and the different women's rights acts that were instituted in a particular state. Ideally one would want to compare the patenting record of married women patentees to unmarried patentees within the region, in terms of changes before and after the laws. The chapter on women inventors included data from city directories that were too limited and biased to yield reliable conclusions, but the figures seemed sufficiently distinct to warrant speculation that location-specific factors such as legal status might indeed account for some of the differences. Moreover, western and midwestern states were among the first to eliminate restrictions on the rights of women, and these frontier states were also prominent in per capita patenting.

Table 6.2 relates average per capita patenting by women to the timing of legal reforms. The results support the view that legal reforms caused women to increase their investments in inventive activity. Average per capita patenting did increase over time in states that had yet to pass any women's

Table 6.2. *Average per Capita Patenting by Women in Relation to Legal Reforms (weighted by female population)*

	1860s	1870s	1880s	1890s
Married Women's Property Rights				
Yet to Pass Law	6.0	16.1	36.2	53.1
	(31)	(13)	(9)	(9)
Passed in Current Decade	11.1	22.4	48.1	–
	(9)	(18)	(4)	(0)
Law Passed Before	28.5	59.1	54.9	80.0
	(9)	(18)	(37)	(40)
Sole Trader Laws				
Yet to Pass Law	9.9	24.2	34.1	48.8
	(43)	(25)	(20)	(20)
Passed in Current Decade	40.0	39.3	50.1	–
	(5)	(18)	(5)	(0)
Law Passed Before	3.2	100.4	72.3	103.7
	(1)	(6)	(25)	(29)
Earnings Laws				
Yet to Pass Law	12.4	18.9	29.6	44.6
	(43)	(23)	(20)	(20)
Passed in Current Decade	24.1	47.7	52.6	–
	(5)	(20)	(3)	(0)
Law Passed Before	3.2	65.0	63.1	91.0
	(1)	(6)	(27)	(29)
National Average	15.5	43.3	52.1	75.9

Notes: States that passed laws in the 1890s decade are included in the first category. Per capita patenting figures are weighted by population. The number of states in each category is included in parentheses. The 1890s patenting rates comprise those for the period up to March 1895, inflated by a factor of 1.9355.

rights laws but, in all instances where more than one state was involved, areas that had recently granted such rights experienced higher patenting rates. States that had previously enacted married women's statutes sustained rates of patenting that surpassed both of the other categories. However, the patterns are dominated by a strong upward trend, and might also reflect other features specific to a particular region. The table does not control for an array of variables that might affect the relationship between passage of married women's laws and female commercial activity. For instance, other laws might have been passed, or judicial decisions and cultural attitudes might have prevailed, that modified the married women's property laws, including changes in "marriage bars" or social sanctions against women inventors. Commercially developed areas that were rich in factors conducive to patenting, such as higher literacy rates and access to capital, might also have been more likely to pass laws protecting women's economic rights.

Systematic time series do not exist for these variables across states, but the level of urbanization (defined in terms of the presence or absence of cities within a county) is likely to be a good proxy.

Chapter 5 showed that women's patenting was affected by the degree of urbanization, but the direction of influence was unexpected: in the period before property rights laws were passed, women in rural areas (counties without a town of more than 25,000 residents) achieved higher patenting rates than women in urban and metropolitan regions. Moreover, even after adjusting for population, the distribution of women's patenting was far more concentrated in rural areas than was the case for male patentees, especially in the Midwest and West. The implication is that the typical urban advantages – access to education, information, social networks, financial capital – were not critical to female inventive activity. However, Table 6.3 shows that patenting in metropolitan counties (containing a city of over 100,000 residents) increased significantly after changes in laws granting property rights to women, and to a far greater extent than in rural areas. This increase might have occurred because concern about property rights was stronger in more developed markets, or perhaps because the property rights laws facilitated women's access to the advantages in urban areas that had promoted men's patenting.

Similarly, in rural and urban areas the passage of sole trader laws is associated with increases in patenting that are roughly comparable to the effects of the property laws. An exception occurs in metropolitan areas, where laws that granted women the right to independent businesses and contracts lead to higher patenting rates than is the case for property rights laws. For example, metropolitan areas in states that had legislated property rights laws in the 1870s experienced patenting rates of 2.9, whereas patenting in areas that had already passed property rights laws amounted to 42.4 per million women. The comparable figures for metropolitan counties in states that legislated sole trader laws in the 1870s were 13.6 in the current decade and 76.0 in states that had previously passed such laws. The higher rates in metropolitan areas after the passage of sole trader laws possibly reflect the greater potential for commercial activity and higher market demand in populous counties.

The rural/metropolitan differences also shed some light on the relationship between law and culture and, in particular, on the view that laws merely reflect prevailing norms or attitudes. Some scholars might feel that married women's laws were a function of cultural changes that were also favorable to inventive or commercial activity by women. That is, both the observed increase in patenting and changes in the laws towards married women could have been caused by changes in the omitted cultural variable. However, the results in Table 6.3 do not provide strong support for this proposition. If cultural norms indeed influenced both legal change and patenting, they could perhaps explain the divergence between rural and

Table 6.3. *Per Capita Patenting, Legal Reforms and Urbanization,*
1790–1895

	Married Women's Property Laws			
	1860s	1870s	1880s	1890s
Rural				
Yet to pass law	5.0	7.2	23.5	23.9
Passed in current decade	6.3	14.7	36.5	–
Law passed before	9.1	19.2	21.2	29.2
Total	7.6	17.7	21.4	28.9
Urban				
Yet to pass law	0.1	0.2	0.4	6.6
Passed in current decade	1.7	2.4	4.8	–
Law passed before	4.4	6.5	6.3	13.5
Total	2.8	5.3	5.9	13.0
Metropolitan				
Yet to pass law	1.0	1.8	2.7	3.3
Passed in current decade	3.5	2.9	0.0	–
Law passed before	17.3	42.4	40.8	52.9
Total	10.5	32.6	38.3	49.5

	Sole Trader Laws			
	1860s	1870s	1880s	1890s
Rural				
Yet to pass law	6.3	14.1	17.8	25.1
Passed in current decade	10.2	23.7	38.2	–
Law passed before	3.2	20.3	24.7	32.5
Total	7.6	17.7	21.4	28.9
Urban				
Yet to pass law	0.9	3.0	3.6	9.8
Passed in current decade	6.5	3.0	4.5	–
Law passed before	0.0	10.5	8.6	16.4
Total	2.8	5.3	5.9	13.0
Metropolitan				
Yet to pass law	3.9	12.7	17.7	20.3
Passed in current decade	23.5	13.6	0.0	–
Law passed before	0.0	76.0	60.5	78.9
Total	10.5	32.6	38.3	49.5

Notes: Rural refers to a location with fewer than twenty-five thousand inhabitants; urban, between twenty-five thousand and one hundred thousand; metropolitan, one hundred thousand and above. The figures are computed by dividing the number of patents within that urbanization category by total female state population. The 1890 patenting rates comprise those for the period up to March of 1895, inflated by a factor of 1.9355.

Sources: See the text and footnotes.

metropolitan patenting behavior in terms of cultural differences between rural and metropolitan areas; but it seems unlikely that attitudes in urban and metropolitan areas would differ sufficiently to account for patenting rates in urban counties that lagged behind both rural areas and metropolitan centers. Moreover, adverse views about married women's market participation were still in evidence even in the twentieth century, suggesting that cultural factors may have lagged behind women's commercial activity and legal change.[21]

The experience of women patentees also allows us to consider a number of additional issues, such as whether laws were associated with higher patenting rates after controlling for other factors; the differences between community property states and common law jurisdictions; and whether legal reforms were correlated with greater investments in inventive activity and the likelihood of trade in patent rights. Table 6.4 examines factors that influenced variation in the log of per capita patenting at the state level within each decade. The regressions, which are weighted by state population, show a statistically significant relationship between patenting by women inventors and legal changes affecting their property rights. A necessary condition for proving causality is that legal reforms preceded increases in patenting rates. The dummy variables *Prelaw* and *Postlaw*, respectively, represent states that had yet to pass married women's legislation, and those that had enacted laws previously. The omitted category refers to states that passed laws in the current period. Regressions 1 and 2, which are unweighted, show a statistically significant association between per capita patenting rates by women inventors and legal changes affecting their property rights, even after controlling for the strong upward trend. The negative and significant coefficients on *Prelaw*, combined with positive coefficients on the *Postlaw* dummy, imply that per capita patenting was lower in states that had not yet passed any laws, then increased markedly afterwards.

The analysis also examines the experience of the southern states (excluding the District of Columbia) and community property states (Regression 2.) Southern states recorded lower per capita patenting rates than other areas, and the difference persists even after controlling for time trends and changes in the law. Community property laws have been claimed to function in the same way as legal systems based on marital disability, because control of the common property was invariably vested in the husband.[22] The results support this interpretation, for they indicate that community property

[21] See Claudia Goldin, "Marriage Bars: Discrimination against Married Women Workers from the 1920s to the 1950s," in Patrice Higonnet, David S. Landes, and Henry Rosovsky (eds.), *Favorites of Fortune: Technology, Growth, and Economic Development Since the Industrial Revolution*, Cambridge, Mass.: Harvard University Press, 1991.

[22] See Donna C. Schuele, "Community Property Law and the Politics of Married Women's Rights in Nineteenth Century California," *Western Legal History* vol. 7(2) 1994: 244–81; and Lebsock, "Radical Reconstruction."

Table 6.4. *Regressions of Patenting in Relation to Legal Reforms (Dependent Variable: Log of Patenting Per Capita Within State, by Decade)*

	(1) (unweighted) Property Laws	(2) (unweighted) Property Laws	(3) (weighted) Property Laws
Constant	−2.35***	−1.80***	0.93**
	(3.86)	(2.71)	(2.05)
Time Trend			
1870s Decade	3.17***	3.27***	2.42***
	(4.00)	(4.13)	(4.36)
1880s Decade	3.34***	3.61***	2.71***
	(3.74)	(4.00)	(4.61)
1890s Decade	3.85***	4.17***	2.97***
	(4.26)	(4.54)	(5.17)
Legal Reforms			
Prelaw	−1.87***	−1.91***	−2.48***
	(2.34)	(2.40)	(3.94)
Postlaw	1.82***	1.36*	0.65
	(2.67)	(1.89)	(1.29)
Region			
South		−1.01*	−2.35***
		(1.82)	(6.30)
Community Property States		−0.85	0.45
		(1.16)	(0.64)
N	223	223	223
R^2	0.31	0.36	0.52

* significant at 7 percent level
** significant at 5 percent level
*** significant at 1 percent level

Notes: The regressions exclude the District of Columbia, in which the Patent Office was located. The female population weights comprise the decadal midpoint, computed by exponential interpolation. Community property states in the nineteenth century were: Arizona, California, Idaho, Louisiana, Nevada, New Mexico, Texas, and Washington.

Sources: See the text and footnotes.

jurisdictions had no special advantage in promoting patenting. Legal reforms clearly did not account for all of the variation in patenting at the state level over time, but it is difficult to control for regional factors because they are correlated with the changes in the laws. The issue of causation would be more effectively approached by considering the record for individual states within each region. In any event, the results are consistent with the view that the married women's property rights laws encouraged women's patenting activity.

The third regression is weighted by female population at the state level. A comparison of the weighted and unweighted regressions highlights the experience of the frontier states, where typically the female population was small. Per capita patenting was higher in the East North Central and western states after legal reform, contributing to the significantly positive coefficient on the *Postlaw* dummy in the unweighted regressions. However, when the state-level observations are adjusted for population, as in the weighted regression, the western states are overwhelmed by areas where population was larger and per capita patenting rates were lower. As a result, even though the *Prelaw* dummy remains significantly negative, the smaller weighting of the frontier states causes that *Postaw* dummy to become only marginally significant.

The effects from the earnings acts and sole trader legislation – which typically passed after the property acts – are dominated by the strong trends evident in the later periods. These statutes were associated with increases in patenting during, and after, the decade in which they were passed. Regressions which simultaneously control for property rights legislation show no additional influence from earnings and sole trader laws on per capita patenting. After including regional effects, the sole trader law is marginally significant, whereas the earnings acts are not. Judicial decisions restricted the earnings legislation to apply only to married women's labor that was not performed within the home, or for the benefit of the family, broadly defined. The results imply that courts rendered these laws ineffective by excluding occupations such as millinery outwork and assistance in the husband's trade or business.

Table 6.5 examines whether women increased their commitment to inventive activity after the laws changed. As stated before, a professional approach to invention is often linked with multiple patenting, and the attempt to extract profits and income from one's discoveries. Multiple patenting also helps to identify whether changes in the laws influenced women to increase and sustain their commitments to inventive activity. According to the regression results, the earnings and separate estates legislation seem to have been unrelated to the number of patents each woman filed. Thus, the property rights laws may have affected whether women in a state engaged in patenting at all, but not whether they chose to increase their investments in inventive activity. A relationship does exist between sole trader laws and the number of patents filed, but the exact nature of the link is unclear. One possibility is that women who were granted the right to secure contracts or to own businesses had a greater incentive to obtain multiple patents. More typically, women inventors of valuable patents formed businesses to exploit their inventions.

Two cases illustrate how the laws protected the property, both tangible and intangible, of women attempting to profit from patent rights. Mrs. Bonesteel, the defendant in *Voorhees v. Bonesteel*, 83 U.S. 16 (1872), owned an interest

Table 6.5. *Regressions of Total Career Patents in Relation to Legal Reforms (Dependent Variable: Log of Number of Patents per Person)*

	(1) Property Laws	(2) Sole Trader Laws
Constant	0.31	0.30
	(4.65)	(4.69)
Region		
New England	0.27	0.26
	(5.60)	(6.00)
Middle Atlantic	0.24	0.25
	(6.36)	(6.66)
West North Central	0.06	0.07
	(1.23)	(1.44)
East North Central	0.03	0.06
	(0.70)	(1.42)
West	0.01	−0.01
	(0.16)	(0.12)
Time Trend		
1870–1874	0.08	0.08
	(1.39)	(1.45)
1875–1879	−0.03	−0.04
	(0.50)	(0.68)
1880–1884	−0.07	−0.08
	(1.40)	(1.61)
1885–1889	−0.13	−0.14
	(2.67)	(2.95)
1890–1894	−0.16	−0.17
	(3.28)	(3.55)
Log (Per Capita Patenting)	0.02	0.02
	(4.42)	(3.87)
Patent Assigned	−0.16	−0.16
	(3.92)	(3.95)
Industry		
Industrial Machines	0.27	0.27
	(8.46)	(8.47)
Household Machines	0.09	0.09
	(2.26)	(2.27)
Apparel and Textiles	0.15	(0.15)
	(5.42)	(5.38)
Legal Reforms		
Property Rights Laws	0.02	–
	(0.58)	–
Sole Trader Laws	–	0.07
	–	(2.75)
	N = 4000	N = 4000
	$R^2 = 0.1$	$R^2 = 0.1$

Sources: See the text and footnotes. Assignments refer to patents sold at time of issue. Patents were categorized in terms of industry of final use.

in a patent license for making pavements. She also owned 1,145 shares in the Nicholson Pavement Company, which was formed to exploit the patent in Brooklyn. Her husband's creditors tried to attach the property to pay for his debts. After ascertaining that no fraud was involved, the courts protected the rights of Sophia Bonesteel, pointing out that the statutes permitted married women the rights to separate property and allowed them to retain the profits from mercantile business. However, as late as 1883, litigants in a New York case attempted to build a defense against a charge of violation of patent property, based on the grounds of marital disability. In *Fetter v. Newhall*, 17 F. Cas. 841 (1883), the defendant infringed a patent for drive screws, and tried to overturn the case by arguing that Mary Fetter, a married woman, had no right to assign the patent to the Fetter Drive Screw Company, nor to sue for infringement, for "at common law a patent-right granted or assigned to a married woman would be such personal property that her husband could, by virtue of his marital right, reduce it to possession and make it his own." Judge Wheeler refused this plea in deciding for the plaintiff and issued an injunction: "The laws of congress, however, of which patents are creatures, give the right to a patent to the inventor, whether sui juris or under disability, and to the assigns of the inventor. . . . This is the whole requirement. A married woman, and infant, or a person under guardianship, might be an inventor, or the assignee of an inventor . . . but . . . *the ability to make the instrument, or the aids to the disability, must be found in the laws of the states where all such rights are regulated.* The laws of New York free married women from disability to make such instruments, and make their property distinctly their own. . . . She could make the instrument in writing by the laws of the state, and when she had made it, it fulfilled the requirements of the laws of the United States. Thus the drive screw company took by her assignment what she attempted to assign to them; and she could sue in her own name in this forum, for infringement of her rights."

In short, the poor record for antebellum patenting by women appears to have been partially because of legal limitations on the rights of women to own property, and (to a lesser extent) trade on their own account. One would expect that women who gained business or work experience would find greater opportunities of detecting and satisfying market demand, as well as skills that might enhance their inventive abilities. When these legal constraints were removed or relaxed, inventive activity surged because women directed their efforts to devise and promote patented inventions with the objective of obtaining "fair compensation." Women inventors thus appear to have benefited from legal reforms that were directed to different ends than the protection of women who wished to pursue the profits that they expected to gain from inventive activity.

Lawsuits reinforce the suggestion that legal reforms enabled and encouraged married women, patentees and nonpatentees alike, to increase their commercial activity through several conduits. First, the maintenance of separate property and income afforded a measure of independence and control that mitigated uncertainty about the future.[23] Second, the ability to enter into partnerships, sign contracts, or to sue and be sued decreased the riskiness of independent ventures. It is significant that creditors, ever wary of "female pirates" who avoided liability behind the shelter of coverture, required the assurance of the statutes before providing funding. For instance, Mrs. Jennie Bornstein obtained a loan from Ellis Silberstein, a pawnbroker, and became a shopkeeper in Philadelphia in accordance with the 1872 statutes: "After making the necessary inquiry and satisfying himself that her purpose was commendable, and that, under the law, her separate earnings were secured to herself, so that they could not be taken and applied to her husband's debts, [Ellis Silberstein] loaned her $1500, with which she, in good faith, purchased a stock of goods and embarked in business on her own account." Silberstein's testimony in the case stressed the importance of the laws: "I asked her before I gave her the money if she had made application to the court. I said to her I knew she was not entitled under the law to her separate earnings unless she was a feme sole trader.... I saw the lawyer, Mr. Moyer, before I loaned the money, to see if it was all right. He said yes, he had drawn up the papers. Her husband had nothing to do with it. I would not have given him the money."[24] In contrast, decisions such as *De Graum v. Jones*, 23 Fla. 83 (1887), declaring that "a married woman has no contractual capacity and cannot bind herself personally," indubitably tended to deter women who wished to market and benefit from their inventions.

---

[23] Evidence of women attempting to attain financial and economic independence under the married women's laws is abundant: for separate bank accounts, see *Fullam v. Rose*, 160 Pa. 47 (1894), and *Hinkle v. Landis*, 131 Pa. 573 (1890). In *Stickney v. Stickney*, 131 U.S. 227 (1889), Mrs. Stickney's "repeated and express directions to invest the moneys for her benefit in her own name" were permissible only because of the separate estates laws of the District of Columbia. Profits from Mrs. F. B. Conway's Brooklyn Theatre were to be shared equally between husband and wife according to a contract they drew up with each other, *Scott v. Conway*, 58 N.Y. 619 (1874). Earnings from nursing were held as a wife's separate property in *Wren v. Wren*, 100 Ca. 276 (1893). Jane Anderson supported her twelve children from her separate earnings as a seamstress. Her claim that "she became and was entitled, under the Act of May 4, 1855, to all the rights and privileges of a feme sole trader" was supported by the courts, *Ellison v. Anderson*, 110 Pa. 486 (1885). In March 1881, Louisa Spering "presented her petition to the Court of Common Pleas of said county, under the Act of 3d April, 1872, entitled 'An Act securing to married women their separate earnings'... [to] be under her control independently of her husband." Despite her husband's insolvency later on, her business was able to expand to an establishment worth $14,000, *Spering v. Laughlin*, 113 Pa. 209 (1886).

[24] (*Orr & Lindsley v. Bornstein*, 124 Pa. 311 (1889) Sup. Ct. of PA.) Similarly, in the New Jersey case, *Aldridge v. Muirhead*, 101 U.S. 397 (1879), the Supreme Court pointed out that the loan would never have been made "had it not been supposed that the money was to be used for the benefit of Mrs. Aldridge.... The wife and her separate estate furnished the only security the parties supposed they had for the money which was loaned."

CONCLUSION

Married women increased their participation in commercial activity in general during the past century, but it was not clear whether these patterns were affected by the removal of legal restraints on their market-related economic activity. Some scholars support the view that married women's property rights laws exerted an independent influence and induced greater female participation in the market economy. A contemporary observer even equated the impact of these reforms to that of the abolition of slavery.[25] Others argue that the law merely provided an index of cultural change and such attitudes evolved over the course of the nineteenth century, affecting both the law and women's propensity to venture beyond traditional spheres. Legal historians have generally concluded that reforms in married women's property and sole trading rights were ineffectual because the laws failed to improve the economic status of women. The issue is obviously complex and unlikely to be settled definitely, both for conceptual reasons and because of the paucity of relevant data. However, the experience of nineteenth-century women inventors does seem to have been influenced by legal reforms.

Women inventors faced greater obstacles than men, but their patenting appears to have been motivated by similar influences; their efforts responded to market incentives and many attempted to gain income from their inventions. An important distinction exists when one compares patenting by men and women according to the degree of urbanization, for women in rural counties were far more likely to patent relative to women in cities, than their male counterparts. Indeed, before legal reforms, per capita patenting in rural areas exceeded both urban and metropolitan centers. Female patentees in western states were responsible for significantly higher per capita patenting rates, a result that coincided with more liberal laws toward women in the frontier areas.

Did legislation matter? Or did reforms that granted women separate property rights, the ability to act as sole traders, and the capacity to retain earnings from their nonhousehold labor prove to be merely nominal changes in irrelevant statutes? This chapter explored the possibility of a causal relationship between changes in married women's laws and patenting at the state level by considering per capita patenting rates before and after. When legal reforms protected their individual property rights, inventive activity surged because women were directing their efforts to devise and promote patented inventions with the objective of obtaining "fair compensation." Patenting in metropolitan areas in particular rose dramatically after the passage of legislation that granted women the rights to separate property and to

---

[25] "Excepting the abolition of slavery, no laws have wrought such a revolution in society, or whose influence in the future will be so deep and so far reaching," Jonathan Smith, *The Married Women's Statutes, and Their Results Upon Divorce and Society*, Clinton, Mass.: Clinton Printing Co., 1884, p. 29.

conduct business as sole traders. These patterns suggest that women initially had limited access to urban advantages that encouraged patenting by men, but property rights laws either removed those constraints or accompanied improvements in access. Thus, contrary to the view of a number of legal historians, statutory changes were influential because they provided incentives to which women responded by increasing their inventive activity. Women inventors were concerned about the extent and security of their separate claims on household income, rather than with the overall welfare of the household irrespective of their individual well-being.

In general, the experience of women patentees supports the arguments of economists who emphasize the role of institutions such as legal and property rights systems in eliciting and encouraging economic growth. Women were sensitive to the opportunities and incentives that legal and patent institutions offered, and the rules and standards to which they were subjected significantly affected their behavior. Patent grants to all true inventors were carefully protected at the federal level, but the efforts of women inventors were deterred by state restrictions on usufruct. Women responded to the reforms in state laws regarding married women's property and sole trader status that removed restrictions on their ability to take advantage of patent rights that allowed them to hold property and engage in commercial activity. The earnings acts confirm the importance of enforcement mechanisms because, although they might have had broader influence, such laws were nullified by judicial decisions. An assessment of changes in married women's property rights laws not only adds to our understanding of women's inventive activity; it also illustrates the impact of an inclusive patent system that was open to women and other under-appreciated classes of society. Patent institutions that gave women (and other disadvantaged groups) property rights to their technological ideas induced them to make potentially valuable contributions to the market for inventions. The issue of the impact of nineteenth-century legal reforms in these and other dimensions thus deserves further attention from students of democracy, because it raises fundamental questions about the long-term consequences of arbitrarily excluding groups from participation in the market economy.

7

Great Inventors and Democratic Invention

"The patent system added the fuel of interest to the fire of genius."
– Abraham Lincoln (1859)

Americans early on evinced a national distaste for inherited privilege and for rewards that were not commensurate with individual contributions to society. Their democratic ideal comprised rules and standards that ensured equal access to the opportunities associated with expanding markets and social progress, although few held illusions about the inequality in the distribution of resources and abilities that might ultimately result from such policies. This was especially true of institutions for the promotion of progress through technological change. The tendency to democratization was manifested in unique features of the U.S. patent system such as the examination of patent applications by technically qualified Patent Office employees, the award of property rights only to the first and true inventor, low fees, and few restrictions on the ability of patentees to exploit their inventions in the marketplace. This monograph contrasts the attributes of patent institutions in Europe and America, and holds that the American patent system for the most part achieved its aims. Patentees in the United States were drawn from the entire spectrum of the population, and inventive activity and productivity gains were more broadly distributed than in Europe. Some scholars might agree with this characterization but still contend that the "democratization of invention" was related to trivial improvements that were only of marginal relevance to the sources of productivity growth and technological change. A common argument is that economic growth depended on discrete advances vested in such "great inventions" as the telegraph, the railroad, and the steam engine, and that such ideas were generated through a different process.

  This chapter therefore examines the historical record on "great inventions" and "great inventors" to see how consistent the findings are with the claim of the democratization of invention. Consider Thomas Edison, one of the most recognizable and productive inventors in American history.[1] His

---

[1] Frank Lewis Dyer and Thomas Commerford Martin, *Edison, His Life and Inventions*, New York: Harper Bros., 1929.

inventive career spanned the period from the end of the Civil War through 1931, an era during which technological advances transformed everyday life. He is noted as the most prolific U.S. patentee, with a total of 1093 U.S. patents to his credit, including improvements in telegraphy, incandescent light bulbs, the stock ticker, storage batteries, movies, the phonograph, automobiles, and flying machines. Edison did not receive any formal schooling, and was untrained in modern science and mathematics. His methods were empirical and based on thousands of meticulously documented experiments. In his famous statement, Edison attributed much of his lifelong productivity to application rather than inspiration, and further noted that "genius is hard work, stick-to-it-iveness, and common sense." Despite his humble background, Edison became a celebrated inventor, as well as a successful and wealthy entrepreneur, who founded numerous companies throughout the world to exploit his inventions.

What were the factors that contributed to Edison's success and how typical was he relative to other inventors of his time? The public has long been fascinated with heroic inventors, and their experience forms a staple of children's books and popular biographies. The topic also has absorbed the interest of historians and economists. Joseph Schumpeter distinguished between invention and innovation on the grounds that the two might involve quite separate abilities: the inventor conceives of the idea, whereas the entrepreneur is responsible for ensuring its practical application, or its innovation.[2] This distinction is important in the study of economic growth, because it addresses the central and long-standing question of the extent to which technical change responds to market forces or is otherwise endogenously determined by the "fuel of interest." If technical change is indeed induced, it implies that inventors will tend to employ entrepreneurship in the pursuit of profits, increasing their investments in inventive activity under conditions that enhance expected net returns, such as the characteristic expansion of markets during the initial stages of industrialization.

In contrast, other scholars have argued that inventions were independent of industrial demand. One approach considers technology to be exogenous, as a *deus ex machina* – or perhaps more appropriately, as a *machina ex deo*. Fluctuations in output are sometimes modeled as responding to random technology shocks to the system. A separate school instead proposes that supply factors are predominant in explaining technological change. In this view, the existing stock of knowledge permits only a limited amount of

---

[2] Joseph Schumpeter, *The Theory of Economic Development*. Cambridge, Mass.: Harvard University Press, 1961; and J. Schumpeter, *Business Cycles*. New York: McGraw-Hill, 1939. Schumpeter's (1939) argument was hedged about with a number of caveats that should not be overlooked. For instance, he avoided explicit appeal to the notion of autonomous inventions, "although this seems to carry a connotation more relevant to our argument."

additional inventions; thus, technological bottlenecks need to be resolved by major discoveries before output can be affected. Both of these approaches deny or downplay the responsiveness of inventive activity to financial returns and to demand-related factors. Joel Mokyr, for example, distinguished between "microinventions" that are responsive to prices and incentives, and "macroinventions" that primarily result from "strokes of genius, luck, or serendipity," and emerge *"ab nihilo"* to transform the state of technology and the economy.[3] Mokyr's recent work argues that there is "a great deal of autonomy to [useful knowledge], which cannot be explained in terms of demand or factor endowments."[4] Moreover, the democratic free market process is no safeguard, and indeed under some circumstances may serve to enshrine inefficient technologies to a greater degree than other less desirable political systems.

According to this tradition, truly significant advances are either exogenous with respect to economic returns, or else related to individual characteristics including education, scientific and technical knowledge, whether the inventor is an outsider to the affected industry, or psychological traits such as a "need for achievement." In short, even if they concede that marginal improvements are related to market demand, some scholars regard important inventions as haphazard rather than systematic, and largely unresponsive to material incentives. Great inventors are typically depicted as less than entrepreneurial, if not naive eccentrics, uninterested in or incompetent at appropriating the returns to their efforts because of a lack of simple business sense. The Commissioner of Patents' *Annual Report* for 1887 called attention to "the fact that quite as much genius is required to develop a property or manufacture and place it advantageously upon the market... and it is the misfortune of inventors, as a general rule, not to be possessed of business habits and business faculties." *The National Cyclopaedia*, while praising Jason Osgood's business acumen, felt that he "was exceptional among inventors from the fact that he was a shrewd, careful, and successful business manager, efficient in promoting his inventions."[5]

These perspectives are based on assumptions about the motivations of individuals who prove to be successful inventors, or about the conditions that lead to successful invention. In particular, they tend to be more skeptical of the similarities between inventors and entrepreneurs than Jacob Schmookler, who emphasized the responsiveness of inventive activity to the market, or Abbott Usher, who proposed a "cumulative synthesis" between invention

[3] The implication is that "regardless of where they came from, genuinely important and new ideas were neither cheap nor elastically supplied. Technology was, as I have argued repeatedly, constrained by supply." Joel Mokyr, *The Lever of Riches: Technological Creativity and Economic Growth*, New York: Oxford University Press, 1990.

[4] Joel Mokyr, *The Gifts of Athena: Historical Origins of the Knowledge Economy*. Princeton, N.J.: Princeton University Press, 2002, p. 293.

[5] *The National Cyclopaedia of American Biography*, New York: James T. White & Co., 1926.

and innovation.[6] Before these claims can be assessed, a more clearly specified model of entrepreneurship is needed. In particular, three propositions need to hold if invention during this era is to be deemed entrepreneurial. First, inventors were responsive to perceived demand and market incentives. Second, there was systematic investment in inventive activity rather than "noneconomically oriented tinkering." Finally, inventors secured property rights to their discoveries and sought to appropriate the returns from their efforts. It also is relevant to consider how such processes changed over time, especially with the advent of the "new economy" of the 1920s.

This chapter discusses where and under what conditions important technological knowledge was generated, based on a quantitative analysis of "great inventors" and their patenting that Kenneth Sokoloff and I conducted.[7] We examined the experience of important American inventors, drawn from biographical dictionaries and the histories of some 420 "great inventors" and their inventions between 1790 and 1930.[8] The data set also incorporated

---

[6] A. P. Usher, "Technical Change and Capital Formation." In *Capital Formation and Economic Growth*. Cambridge, Mass.: NBER, 1955: 523–58. Schmookler (1962) provides a similar perspective to ours: "This suggests that the inferences drawn from the behavior of the minor inventions . . . probably apply in considerable measure to major inventions, too." Jacob Schmookler, "Economic Sources of Inventive Activity." *Journal of Economic History* 22, 1962: 1–20. See also J. Schmookler, *Invention and Economic Growth*. Cambridge, Mass.: Harvard University Press, 1966.

[7] B. Zorina Khan and Kenneth L. Sokoloff, "'Schemes of Practical Utility': Entrepreneurship and Innovation among 'Great Inventors' During Early American Industrialization, 1790–1865," *Journal of Economic History*, vol. 53 (2) 1993: 289–307; B. Zorina Khan and Kenneth L. Sokoloff, "Entrepreneurship and Technological Change in Historical Perspective: A Study of Great Inventors During Early Industrialization," *Advances in the Study of Entrepreneurship, Innovation, and Economic Growth*, vol. 6 (1993): 37–66; B. Zorina Khan and Kenneth L. Sokoloff, "Institutions and Democratic Invention in 19th Century America," *American Economic Review*, vol. 94 (May, Pap. And Proc.), 2004: 395–401.

[8] The main source of the sample is the *Dictionary of American Biography* [*DAB*], New York: Charles Scribner's Sons, 1928–1936. The *DAB* list of inventors is based on a consensus among distinguished specialists that the individual had "made some significant contribution to American life." This is supplemented by *Who Was Who in America: Historical Volume, 1607–1896*, Chicago: Marquis (1963) and *The National Cyclopaedia of American Biography*, various volumes, New York : J. T. White, 1926–1984.) Additional details were obtained from a number of sources including articles and book-length biographies. The 420 "great inventors" were selected from the *DAB* if their date of birth was before 1885. The information includes their date and place of birth, father's trade, schooling, and age at first major invention. The classifications also cover inventive specialization (if any), occupations before and after the first major invention, whether the first major invention was related to prior occupation, and if their subsequent trade was related to the invention. Their occupational status was categorized as merchants and white-collar professionals (including a poet, missionary, and a university president); machinists, engineers, and full-time inventors (mechanics were classed as machinists); artisans (such as typesetters and goldsmiths); manufacturers; farmers, and others. The first major invention was determined by the *DAB*'s account, whereas the inventive career of the inventor was measured as the difference between his first and last patent plus one. The *DAB* also provided details on the inventor's source of income, which were grouped

material from other sources, including inventors' patents, citations from patent specifications, commercialization activities, and litigation. The sample comprised approximately 4,500 patents out of the total 16,900 patents that the great inventors who were born between 1740 and 1885 obtained over their lifetime. It should be noted that the overwhelming majority of the great inventors were prolific patentees: thirteen failed to patent their inventions, and only three of these nonpatentees were from the birth cohorts after 1820.

A valid concern is the extent to which the biographical sample and their patents capture truly important inventions. Patents are the most conveniently available measure, but not all inventions are patented, and the propensity to patent may vary across time or industry; patents differ significantly in terms of commercial and technical value; and patents may be filed as part of an aggressive corporate strategy rather than because of their intrinsic value.

in terms of assignments and royalties from licensing, commercialization, both of these, or no income. The sample used for the comparison with "ordinary patentees" is described in Chapter 4. The *DAB* assessment of importance of inventors can arguably be proxied by the space allotted to each individual, ranging from 444 column lines to Alexander G. Bell, to 40 lines for Moses S. Beach. I created an index that measured the space for a particular inventor relative to the average of ninety-four lines.

Patent citations and assignments at issue relate to patents in the sample. Inventor citations comprise a count of patents that mentioned a great inventor's name in the patent specifications, and the majority of these "inventor citations" are contemporary with the great inventor's own cohort. The second metric counts the number of citations that a specific patent in the sample received from patentees who filed patents after 1975. These "long-term patent citations" in part indicate inventions that patentees today regard as still germane to their technical field. Patent citations explained 32 percent of the variation in historical importance for technical inventors, but only 14 percent for nontechnical inventors. Correlations of career patents with inventor citations (0.7) and patent citations (0.7) imply that the quantity of patents is related to quality. Commercially valuable inventions (as measured by patent assignments) are significantly correlated with technically valuable inventions, as shown by the correlation coefficients of assignments with inventor citations (0.6), and patent citations (0.7.) Assignments are positively related to career patents (0.8.) Inventor citation has a lower but significant correlation with patent citation (0.5), which in part reflects changes in technological fields of interest over time. The *DAB* index is correlated with number of patents (0.5), inventor citations (0.5), patent citations (0.4), and assignments (0.3.) The relatively low correlation for assignments suggests that entry in the *DAB* was not simply due to commercial success. [All correlation coefficients are statistically significant (p < .0001.)] Litigation information was obtained from U.S. federal court records through searches by inventors' names.

The *Annual Reports of the Commissioner of Patents* for various years provided data on patents filed between 1790 and 1930 at both the aggregate and individual level. Patent records include date of issue, city of residence at time of patenting, and the subject matter of each patent. Each patent was classified according to its sector of final use, and county of patentee residence. The regions include Northern New England (Maine, New Hampshire, and Vermont); Southern New England (Connecticut, Massachusetts, and Rhode Island); Southern Middle Atlantic (Delaware, New Jersey, Maryland, and the District of Columbia); and the South (Alabama, Arkansas, Florida, Georgia, Kentucky, Louisiana, Mississippi, North Carolina, South Carolina, Tennessee, Texas, and Virginia.)

Moreover, inventors differ in terms of eminence. Their reputations vary over time, so a biographical dictionary of the early twentieth century might reflect the views of the period of publication rather than those of posterity. Several metrics were constructed to ensure that the results were representative. First, the relative importance of inventors in the biographical dictionary was measured in terms of space allotted. Second, inventor citations provided an index of technically valuable inventions during the inventors' lifetime, whereas patent citations gauged the technical relevance of great inventors' patents to the modern period (since 1976.) Taken together, these measures allow us to follow Galton's 1869 definition of genius in terms of "the opinion of contemporaries, revised by posterity." Third, assignments of patent rights comprised an index of commercial value.

The report here is presented in two parts, the first of which deals with the record up to the end of the Civil War, whereas the second part summarizes the findings for the postbellum period through 1930. The conclusion highlights the role of patent institutions in aiding inventors from ordinary backgrounds to make important technological contributions during the early nineteenth century through the "second industrial revolution." Moreover, by facilitating the creation of tradeable assets in inventive ideas, patent institutions disproportionately helped relatively disadvantaged individuals to extract returns from their efforts.

GREAT INVENTORS BEFORE 1865

The description of important technological change in the antebellum era centers on the experience of 160 great inventors who filed their first patent by 1846. The roster of these early "great inventors" includes popularly recognized technological heroes such as James Eads, John Roebling, Samuel Morse, Charles Goodyear, Paul Moody, Thomas Blanchard, Robert Fulton, and Eli Whitney. The 150 "great inventors" who were also patentees received 1,178 patents, or slightly less than 2 percent of the total awarded over the period.

Overall, one is impressed by the democratic nature of significant inventions in the early nineteenth century. The majority were inventors who had little or no formal education, whose experience on the job revealed problems or bottlenecks that they generally solved through persistent trial and error experimentation.[9] Their patterns of patenting were procyclical and similar

---

[9] This is not to say that some discoveries were not haphazard. However, although it is true that many inventors proceeded by trial and error, this merely describes the *method* of discovery; it does not imply that their *objective* was random or haphazard. Charles Goodyear's process was the outcome of an unforeseen combination, but it was his specific intent to discover such a process. Several other inventors, such as Nathaniel Hayward, Horace Day, and E. M. Chaffee were making similar experiments, induced by the large market for durable rubber products.

to those of "ordinary inventors." Great inventors effectively responded to expected profit opportunities that appeared when markets expanded. They were clustered in major centers of manufacturing and invention, for those who were born elsewhere migrated to the more extensive markets in New York, Southern New England, and the rapidly growing Midwest. The majority were noted for their entrepreneurial activities, for they rapidly changed locations and occupations to take advantage of their discoveries, founded enterprises to produce their inventions, and sold or licensed the rights to their patents.

One of the salient features of the great inventors is how similar some of their patterns of inventive activity were to those of ordinary patentees. Most significant, perhaps, is the finding that important inventions resembled patents in being strongly and positively associated with the extent of markets. Like patentees in general, the great inventors were disproportionately concentrated in the Northeast, and especially in Southern New England and New York, where low-cost transportation networks had facilitated a rapid expansion of commerce early in the antebellum period. This geographic distribution was characteristic of where they filed their patents (Table 7.1) as well as where they were born (Figure 7.1.) The correspondence holds not only at the state level but also at the county level, where great inventors were even more concentrated than the ordinary patentees in counties with high rates of general patenting.

The procyclicality of both great inventor patents and overall patents during the antebellum period provides further support for the thesis that inventive activity responded to market conditions. Figure 7.2 shows that the two annual series track each other closely, with rapid growth during the years of interruptions in foreign trade prior to the War of 1812, as well as during the economic expansions from the early 1820s to the mid-1830s, and in the 1850s. Moreover, they both exhibit periods of stagnation or slight decline during the protracted economic downturns following the War of 1812 and the Panic of 1837. In short, far from being exogenous, inventive activity by great inventors was influenced by much the same market-related forces as invention by ordinary patentees.

Both the procyclicality and the geographic clustering of patenting in areas with low-cost access to major economic centers are consistent with the responsiveness of inventive activity to market conditions during early industrialization. It is of course possible that the clustering of important inventions was partially because of geographic variation in population characteristics related to inventive potential, such as the level of education or the distribution of technical skills. Judging from the experience of the great inventors, however, such supply-side variables are not persuasive as an explanation of important inventions in the antebellum period. Table 7.2 supports the finding that higher education was hardly a necessary prerequisite for important technological discovery during the antebellum period. Nearly half of

Table 7.1. *Regional Shares of Total Patents, Great Inventor Patents, and Population (1790–1930)*

Region	1790–1829	1830–1845	1846–1865	1866–1885	1886–1905	1906–1930
New England						
Patents	34.4%	30.1%	24.7%	19.7%	16.7%	11.4%
G.I. Patents	55.1	34.1	29.6	29.1	29.1	18.3
Population	21.0	13.2	10.1	9.1	7.6	7.2
Middle Atlantic						
Patents	54.5	52.3	48.3	40.6	37.6	30.8
G.I. Patents	35.5	57.7	55.7	51.5	41.1	62.0
Population	34.4	30.0	26.5	23.1	20.5	21.1
Midwest						
Patents	3.0	8.3	20.8	30.3	34.5	36.8
G.I. Patents	1.9	3.2	13.3	13.6	22.9	14.5
Population	3.3	17.3	29.2	34.0	36.0	32.6
South						
Patents	8.1	9.2	5.1	6.0	6.8	10.8
G.I. Patents	7.5	5.0	1.4	1.5	2.3	3.6
Population	41.3	39.7	32.9	31.9	31.5	31.7
West						
Patents	–	–	1.0	3.4	4.6	10.2
G.I. Patents	–	–	0.0	2.9	2.7	1.6
Population	–	–	1.4	1.9	4.5	7.5

Notes and Sources: The population figures are interpolated from the decadal U.S. Census of Population. The regional distribution of total patents were computed from the U.S. Patent Office Annual Reports. The great inventor patents for the period before 1865 include all patents filed by great inventors to that date; after 1865 the distribution of great inventor patents refers to a sample of patents. See text and notes to Table 7.2.

the sample had little or no formal schooling, whereas less than a quarter attended college.[10] The technically qualified college graduates were certainly overrepresented relative to the general population, but the shares of great inventor patents were even more weighted toward those with limited schooling, since they produced larger numbers of patents on average than their more erudite peers. This qualitative pattern held over time through 1865, in all sectors and for virtually all subregions. Patentees from the South and foreign countries – such as John Roebling, who emigrated from Germany

[10] The terms "primary" and "secondary" are arbitrary, as the former refers to little or no schooling, whereas the latter encompasses more years of schooling but no college. Inventors for whom the extent of schooling is unknown seem likely to have had low levels of educational attainment. Rossman (1935) similarly concluded from a study of inventors in 1927–1929 that little relationship existed between the amount of formal schooling and inventiveness in terms of numbers of patents. (J. Rossman, "A Study of the Childhood, Education and Age of 710 Inventors." *Journal of the Patent Office Society* vol. 17 (1935): 411–21.)

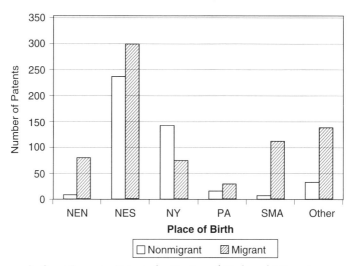

Figure 7.1. A. Great Inventor Patents by Region of Birth and Migration, 1790–1865
Notes and Sources: Counts of number of patents filed by inventors born in a specific region (Northern New England, Southern New England, New York, Pennsylvania, Southern Middle Atlantic, Other.) A patent is attributed to a migrant if the patentee was born in a state other than the one listed as his residence on the patent records.

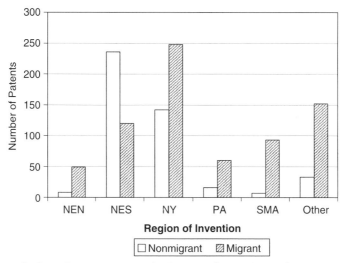

Figure 7.1. B. Great Inventor Patents by Region of Invention and Migration, 1790–1865
Notes and Sources: Counts of number of patents filed by residents of a specific region. See above.

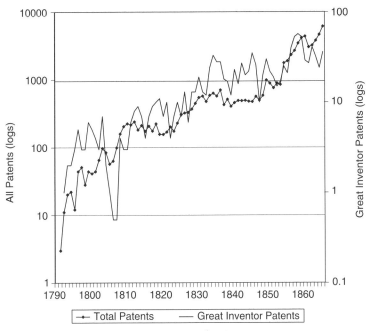

Figure 7.2. Total and Great Inventor Patents (logs), 1790–1865
Notes and Sources: U.S. Patent Office *Annual Reports*, and sample of great inventors from *Dictionary of American Biography* (see text and notes to tables.)

with specialized training in civil engineering – provided the only exceptions.[11] As such, in the antebellum period the distribution of technical skills does not explain geographic differences in rates of invention.

Rather than formal schooling, practical training through apprenticeships and informal influences, including family background, were of greater importance to potential inventors. For instance, eight of the nine machinist/inventor fathers had sons in the same profession, as was the case with the McCormicks, the Stevens family, and the Goodyears. However, less than a third of the great inventors followed their father's profession, because of the occupational mobility to be expected from a society undergoing the shift from agriculture to industrialization. Table 7.2 does suggest a link between commercial backgrounds and great inventors, as a quarter of the antebellum inventors came from merchant/professional families, whereas the fathers of a

---

[11] Most of the great inventors of the period noted for their formal engineering skills were of foreign birth, including Henry Burden, John Allen, Christian Detmold, Robert Dunbar, John Ericsson, and William Hudson. Little or no trend is apparent in the patent shares of those with formal schooling over the period. College-educated individuals were least important in manufacturing and agriculture but were relatively more important in transportation. Although dominant in the South, they comprised a distinct minority in all regions, and were least evident among Southern New England inventors.

Table 7.2A. *Personal Characteristics, Patenting and Productivity of Great Inventors by Birth Cohort*

	2.A. Personal Characteristics of Great Inventors			
	Pre-1820 Birth Cohort ("Antebellum inventors")		Post-1820 Birth Cohort ("Postbellum inventors")	
	Number of Inventors	Percent	Number of Inventors	Percent
Educational Background				
Primary	76	47.5%	65	25.0%
Secondary	22	13.8	53	20.4
College	15	9.4	80	30.8
Technical	23	14.4	51	19.6
Unknown	24	15.0	12	4.6
Father's Occupation				
Artisan	17	15.6%	36	20.3%
Farmer	43	39.4	37	20.9
Engineer/Machinist/ Full-time Inventor	9	8.3	25	14.1
Merchant/Professional	28	25.7	61	34.7
Manufacturer	8	7.3	18	10.2
Other	4	3.7	–	–
Occupational Class at First Invention				
Artisan	24	15.0%	16	6.2%
Farmer	8	5.0	9	3.5
Engineer/Machinist/ Full-time Inventor	53	33.1	114	43.8
Merchant/Professional	36	22.5	68	26.2
Manufacturer	37	23.1	48	18.5
Other/Missing	2	1.3	1	0.4

	Number of Inventors	Percent	Average Patents	Number of Inventors	Percent	Average Patents
Length of Career, by Average Patents Granted						
0–5 years	37	23.1%	1.7	22	8.5%	2.2
6–10 years	8	5.0	3.5	8	3.1	6.0
11–20 years	21	13.1	6.1	36	13.8	25.9
21–30 years	35	21.9	9.3	45	17.3	30.6
>30 years	59	36.9	23.9	149	57.3	83.5

	Patents	Pct. by Migrants	Patents	Pct. by Migrants
Lifetime Patents Filed, by place of birth and outmigration				
Northern New England	92	87.0%	1697	97.6%
Southern New England	537	55.5	1785	64.7
New York	213	34.7	1872	58.3
Pennsylvania	45	64.4	459	67.6
Southern Middle Atlantic	118	91.5	266	60.3
South	48	64.4	260	87.0
Midwest/West	34	44.1	2670	89.6
Foreign	91	100.0	3483	100.0

(continued)

Table 7.2A *(continued)*

	Pre-1820 Birth Cohort ("Antebellum inventors")				Post-1820 Birth Cohort ("Postbellum inventors")			
	Number of Inventors	Pct.	Average Patents	Average Citations	Number of Inventors	Pct.	Average Patents	Average Citations
Age at First Invention								
< 20 years	8	5.0%	9.6	1.0	3	1.2%	117.7	8.7
20–24 years	22	13.8	19.0	0.7	35	13.7	130.6	9.9
25–29 years	22	13.8	20.0	0.4	72	27.7	64.8	4.2
30–34 years	35	21.9	17.8	0.5	66	25.4	46.1	4.0
35–39 years	28	17.5	7.3	0.3	33	12.7	39.0	2.5
40–44 years	14	8.8	5.9	0.1	18	6.9	28.7	2.8
45–55 years	21	13.1	4.6	0.0	18	6.9	19.9	1.4
> 55 years	10	6.3	1.9	0.1	7	2.7	7.0	0.0

further 23 percent were artisans and manufacturers. However, this was more true of native New York and Pennsylvania inventors, half of whom were born to merchant/professional families. Southern New England inventors came predominantly from farming communities, suggesting that family-related influences might have been stronger in urban areas and related more to innovation than invention.

In the first decade of the Republic, patenting by both ordinary and great inventors was dominated by merchants and professionals from cities like Philadelphia, an advantage that steadily declined over time. However, this did not imply that technological change was then dependent on a technically adept cadre of machinists and engineers. Among ordinary patentees, machinists and engineers were overrepresented relative to the general population, but they were outnumbered by those from commercial, artisanal, professional, and other less technical occupations. One is reminded that a strikingly broad spectrum of the population was participating in invention at all levels during the first half of the nineteenth century. Technical backgrounds and skills were clearly an advantage, especially in the transportation sector, but the nature of technology at the time was such that they were far from indispensable even for "great inventions." The occupational distribution for the antebellum great inventors in Table 7.2 exhibits a similar pattern to the experience of ordinary patentees. Roughly one third of the sample comprised machinists, engineers, and full-time inventors. The majority, however, consisted of manufacturers, merchants, and white-collar professionals, farmers, and others whose jobs did not require technical skills; artisans from traditional crafts such as silversmiths and engravers accounted for the remainder. Skepticism about the idea that such population characteristics account for regional patterns is reinforced by the observation that great inventors in Southern New England were markedly less well educated and less inclined toward technical

Table 7.2B. Patenting and Productivity of Great Inventors By Birth Cohort (Percent)

	1820–1839			1840–1859			1860–1885		
	Patents	Assignments	Citations	Patents	Assignments	Citations	Patents	Assignments	Citations
Age when first patent granted									
Below 20	1.1	0.0	0.0	0.9	0.0	0.0	0.2	0.0	0.0
20–29	3.0	1.5	2.3	5.9	5.0	2.2	4.0	3.1	3.4
30–39	14.5	8.4	12.4	26.6	23.3	18.7	23.5	23.7	20.3
40–49	26.7	25.3	17.1	27.1	24.5	26.0	39.0	45.4	31.1
50–59	24.4	27.1	28.1	19.6	23.5	18.4	24.6	21.5	26.2
60–69	21.7	26.0	29.5	13.2	14.2	18.4	8.7	6.4	19.1
70+	8.7	11.7	10.6	6.7	9.5	6.2	0.0	0.0	0.0
Industry									
Agriculture/Food	9.5	9.5	3.7	4.1	4.5	1.2	0.9	0.2	1.7
Electric/ Energy	8.7	6.6	7.8	34.3	33.2	31.9	34.5	33.3	20.5
Eng/Construction	10.0	11.7	12.4	4.8	2.3	8.1	7.3	4.4	20.5
Manufacturing	53.2	61.5	53.0	31.1	29.0	38.8	29.9	29.1	28.6
Transportation	6.9	5.9	8.3	15.7	14.0	8.1	18.7	18.8	9.8
Miscellaneous	12.3	4.8	14.7	12.4	17.1	11.8	12.1	14.2	19.1
Education									
Primary or None	39.5	50.5	44.7	31.8	30.4	34.9	9.9	4.0	13.4
Secondary	23.0	27.5	20.8	27.8	32.3	24.6	11.5	15.0	3.0
College (nontechnical)	27.8	18.0	26.7	22.1	18.1	23.8	35.7	30.5	47.6
Technical	9.7	4.0	7.8	18.3	19.2	16.7	42.9	50.5	36.0

(continued)

Table 7.2B (continued)

	1820–1839			1840–1859			1860–1885		
	Patents	Assignments	Citations	Patents	Assignments	Citations	Patents	Assignments	Citations
Primary Method of Appropriating Returns									
Proprietor	57.2	37.8	65.4	59.1	58.6	58.1	52.3	43.8	65.2
Employee	15.7	24.1	16.1	12.5	15.5	6.8	28.6	39.6	15.9
Direct Exploitation	5.5	5.4	7.4	8.0	2.8	8.2	12.0	12.0	10.8
Sell/License patent	17.6	30.3	9.0	13.1	11.1	16.5	7.1	4.6	8.1
None	3.9	2.5	2.1	7.3	12.0	10.4	0.0	0.0	0.0
Total Number of Patents	960	308	239	1352	579	407	919	591	407

Notes and Sources: The estimates have been computed over 4,325 patents awarded to the sample of "great inventors" who were born through 1885, drawn from the *Dictionary of American Biography (DAB.)* Both the classifications of educational status and the way in which inventors extracted income from their inventions were based on a close reading of the biographies in the *DAB.* The classification of the way income was extracted refers to the overall career of the inventor, as many inventors employed different methods over different inventions, and over different stages of their careers. Thus, classification was made at the inventor level, and applied to all of his or her (there is one woman among the great inventors) patents. The categories include: inventors who frequently sold or licensed the rights to the technologies they patented; those who sought to directly extract the returns by being a principal in a firm that used the technology in production or produced a patented product; and those who were employees of such a firm. The overall sample of "great inventors" consists of two waves. The first wave (160 inventors, comprising the "antebellum sample"), consists primarily of great inventors born through 1820. This sample included the information for each of the patents inventors received through 1865, as well as a tally of total patents they received over their respective careers. Some of the information here regarding the antebellum great inventors differs from earlier published results because previous work was based on patenting information through 1865, whereas this table is based on total lifetime patents. The second wave (260 inventors, comprising the "postbellum sample") included patents from every fifth year through 1930, and thus omits the patents received late in the careers of inventors who were born in the 1870s and 1880s. Although the sample of individual patent records is thus incomplete, I obtained information on the total number of patents ever awarded so the average figures for total patents and for career length do indeed reflect the complete record. Career length is defined as the number of years between the first and last patent, plus one. Thus inventors with no patents (thirteen in the entire period), or only one patent, are estimated to have a career length of one year. Great inventor patents on average are likely to be more valuable than patents of other patentees, but I constructed several additional metrics for valuable inventions. These include patent assignments (an index of commercial value); references to the inventor that appeared in subsequent patent specifications ("inventor citations"); and the number of citations to patents in the great inventor sample that appear in patents filed after 1975. The latter "long-term patent citations" in part indicate inventions that patentees today regard as still germane to their technical fields, and proxy for the inventor's impact on technologies in the modern period.

Table 7.2C. *Top Twenty Great Inventors: Ranked in Terms of Lifetime Patents, Citations, and Historical Importance*

	Lifetime Patents			Patent Citations			Inventor Citations	
Inventor	No.	Index	Inventor	No.	Index	Inventor	No.	Index
Thomas Edison	1093	453	Henry A. Wood	63	104	Charles Kettering	85	183
Carleton Ellis	753	138	Elihu Thomson	52	252	Thomas Edison	79	453
Elihu Thomson	696	252	Thomas E. Murray	51	105	George Westinghouse	76	272
Henry A. Wood	440	104	Carleton Ellis	44	138	John M. Browning	53	107
Walter Turner	343	78	Thomas Edison	43	453	Elihu Thomson	44	252
George Westinghouse	306	272	Howard R. Hughes	37	95	Carleton Ellis	37	138
Thomas E. Murray	270	105	Simon Lake	33	192	Elmer Sperry	36	210
Elmer Sperry	264	209	Elmer Sperry	29	210	John R. Rogers	35	80
John W. Hyatt	205	130	Hiram P. Maxim	27	106	Walter Turner	30	78
Luther Crowell	192	140	Peter Hewitt	24	82	Louis Goddu	27	73
Peter Hewitt	187	82	Henry A. House	24	78	Ottmar Mergenthaler	27	127
Horace Wyman	179	81	Marshall B. Lloyd	24	89	Nikola Tesla	27	285
Rudolf Eickemeyer	169	85	George Westinghouse	23	272	Benjamin Lamme	26	121
Reginald Fessenden	168	103	Thomas Midgley	22	198	Peter Hewitt	22	82
John Thomson	165	78	Charles VanDepoele	22	142	Henry E. Warren	19	103
Charles Fortescue	157	113	James H. Rand	21	125	Elisha Gray	17	98
Sidney H. Short	153	91	Nikola Tesla	21	303	Alexander G. Bell	16	472
Charles Kettering	147	183	James E. Emerson	16	72	John Appleby	15	73
James J. Wood	139	87	Richard H. Rice	15	79	Charles Fortescue	15	113
Patrick Delany	136	98	John M. Browning	14	107	Herman Hollerith	14	103

Notes and Sources: For a description of the data set, see the text and notes to other tables. The index of historical importance comprises a measure of the space allocated to the inventor in the *Dictionary of American Biography*, relative to the average (index = 100) for all inventors.

196

occupations than their counterparts in areas with lower inventive activity such as the Southern Middle Atlantic and the South. The evidence on great inventors conforms well with the view that high regional inventiveness was associated with a wider segment of the population directing their resources toward invention and innovation, in response to the opportunities and challenges presented by expanding markets.

Some might argue that successful invention was largely a matter of individual genius or fortune, which made it unlikely that technological discoveries could be endogenous with respect to demand. The first problem for this perspective comes from the clustering in patenting, as well as in the origins, of these inventors. If successful invention were driven by randomly distributed factors like genius or luck, one would not expect the manifest extent of geographic concentration. Greater doubt is fostered by an examination of the life cycles of the great inventors. Thomas Blanchard notwithstanding, less than a third of the sample made their first significant invention before the age of thirty. The distribution of age at first patent indicates that middle-aged and older men were predominant in inventive activity. Great inventors on average were older than the general working population, and more than 25 percent were in their forties or fifties. Because a dominant role for genius would presumably be reflected in an age distribution more skewed toward youth, these data suggest that experienced and committed, rather than uniquely gifted, individuals were the principal source of important inventions.

Few of the great inventors are eligible cases of serendipity or a single lucky finding. The career of a great inventor from first to last patent typically spanned an extended period: only 23.1 percent had careers of less than five years, and nearly 60 percent were active more than twenty years. Even among the few whose inventive careers (as gauged from their patenting records) were limited to one year or to one invention – including Eli Whitney, Samuel Batchelder, William Crompton, and Pliny Earle – there seem to be few good candidates for the lucky strike hypothesis. William Crompton, for instance, was a textile worker who identified two defects in the structure of looms. The cams restricted the number of warp harnesses that could be used, and had to be changed every time a new pattern was woven. Crompton solved these problems by using an endless-chain feature in his widely adopted loom, and also incorporated a motion of the warp that put less strain on the threads. An example of the two thirds who made useful discoveries for more than a decade is James Bogardus, who patented over a twenty-year period successful inventions for a clock, ring-flyer, sugar mill, banknote plates, gas-meter, cast-iron supports for buildings, and an automatic pencil.

In short, far from being haphazard or unsystematic, great inventions appear to have been the outcome of investments in inventive activity directed at salient needs manifested through the market. One illustration of this is through the relation between occupations and inventions. Part of the

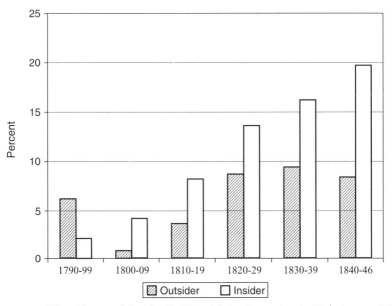

Figure 7.3. "Outsiders and Insiders": First Major Invention in Relation to Prior Occupation

Notes and Sources: The first major invention of an insider was related to previous occupation; outsiders represent inventions that were unrelated to previous or current occupation. See text and notes to tables for sources.

argument that significant inventions are unrelated to demand is frequently expressed in terms of an insider/outsider dichotomy: "reflective students of economic and engineering history must be struck by the curious circumstance that revolutionary inventions are usually conceived not within but without an industry."[12] However, when one considers the relationship between first major invention (because inventors frequently switched their occupation afterward) and previous occupation among the great inventors, this "outsider hypothesis" is not sustained (Figure 7.3.) It is true that the majority of the great inventors active in the 1790s could be deemed outsiders, but this was largely because of the predominance of merchants and white-collar professionals in the early cohorts of inventors. With the decline in the prevalence of this class that soon occurred, the pattern shifted. Over the entire period from 1790 through 1846, 64 percent of the first major inventions were produced by men within the respective industry. In contrast to the paradigm of the technically adept outsider revolutionizing an industry, the majority of great inventors appear to be entrepreneurial individuals who contrived

---

[12] W. Kaempffert, p. 2010, "Systematic Invention." *The Forum* vol. 70, 1923: 2010–2018 and 2116–2122. See also S. C. Gilfillan, *Sociology of Invention*. Cambridge, Mass.: Follett, 1935.

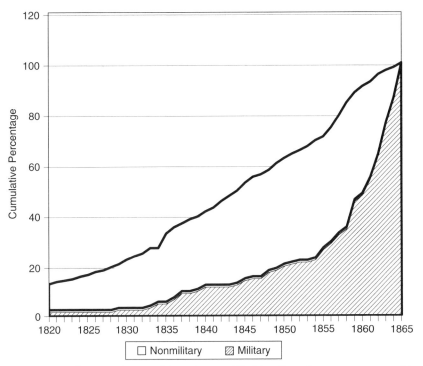

Figure 7.4. Military and Nonmilitary Patents Filed by Great Inventors, 1790–1865
Notes and Sources: See text and notes to tables. Military patents include patents for firearms, weapons, explosives, and cartridges, among others.

"schemes of practical utility."[13] Insiders, who perhaps had stronger incentives to invest in inventive activity, and better information about the state of the market, were the norm. A typical experience was that of Michael Simpson, an importer of wool that frequently arrived imbedded with burrs. Simpson recognized the need for a cost-reducing method of combing out these foreign particles, and patented a successful device in 1837, the British rights to which were sold for £10,000. He then turned to the manufacture of this and other textile-related machinery.

This entrepreneurial response to perceived need was dramatically demonstrated during wars and interruptions in foreign trade. For example, Daniel Treadwell, an erstwhile silversmith, took advantage of the shortage of screws brought about by the embargo of 1807, and invented a screw-machine that he operated in Saugus, Massachusetts, until the peace following the War of 1812. Figure 7.4 highlights the record of invention during the Civil War and provides a further case study of entrepreneurial flexibility. Here the

[13] J. L. Bishop, vol. 2, p. 512, *A History of American Manufactures, from 1608–1860*, 3 vols. Philadelphia: Edward Young & Co, 1868.

magnitude of the stimulus to invention was so great that insiders increased their investments in inventive activity, and outsiders were induced to redirect their efforts. Military-related invention engaged the abilities of one third of the forty-one inventors in the sample who were still active in 1861, involving a major shift in focus for the majority. No one demonstrates this point better than the legendary Richard Gatling, who had previously specialized in farm machinery, but others such as Benjamin Babbitt, whose most recent invention had been in the medical area, were equally adaptable. The great inventors produced some eighty-one inventions relating to cartridges, guns, ordnance, and war vessels from 1858 to 1865 – or 70 percent of all their military-related inventions since 1790.

The substantial shift in the direction of inventive activity among great inventors paralleled the change in orientation by ordinary patentees. *The New York Times* noted in 1861 that "the war has so stimulated the inventive Yankee brain that the Office at Washington fairly groans . . . under the weight of instruments of destruction."[14] Whereas a total of forty-seven patents for firearms were granted from 1790 through 1846 (or less than one a year), eighty-four firearm patents were recorded in 1865 alone. Applications for inventions relating to weaponry, knapsacks, ambulances, artificial limbs, and tents rose precipitously from 1860 to 1865, totalling 1,752 during the war years. Less obviously, inventive activity by both categories of patentees was directed toward solving problems such as the need for improvements in baling machines to compact hay and straw during army transportation. Inventions for machines to process substitute fibres such as flax and hemp increased as cotton became scarce. After the war, applications for baling devices fell and attentions were subsequently focused on cotton presses for use by free labor on the Southern plantations. As in the other instances examined earlier, during the war years the inventive patterns of great inventors and ordinary patentees varied congruently.

Another method of gauging entrepreneurial orientation among great inventors is to examine two additional dimensions of flexibility: occupational and geographical mobility. If the great inventors were entrepreneurial, then one would expect to observe considerable mobility directed at promoting the commercial exploitation of their inventions. This was indeed the case. As discussed earlier, nearly two thirds of the great inventors produced inventions that were related to their trade, and were accordingly already in a position to appropriate some of the returns. Furthermore, roughly 42 percent changed their occupation afterward to one that would allow a more ready pursuit of economic advantage. When other methods of extracting returns

---

[14] New York Times, December 6, 1861, p. 4. Overall patenting fell during the war, but the percentage of military inventions soared to the highest proportion in American history to date. See B. Zorina Khan, "Creative Destruction: Technological Change and Resource Reallocation during the American Civil War," (unpublished paper, 2005.)

from the invention are taken into account, such as royalties, licensing fees, and sales of patent rights, the overwhelming majority of great inventors were actively seeking income derived from their inventive activity. Some individuals showed such a remarkable degree of flexibility in their pursuit of material gain that it might be termed fluidity. Josiah Warren was originally a music teacher and orchestra leader who invented a lard-burning lamp, which he marketed profitably for a while. After he devised a printing press, however, he started a journal, and received other patents related to printing. Inventors such as William Mason were not as mobile across trades, but were no less inclined to adapt to changing circumstances. An apprentice in a cotton factory, Mason manufactured power looms from 1832 to 1833 after obtaining a patent. He then proceeded to invent and produce a ring frame and his famous self-acting mule, before shifting his focus to the manufacture of textile goods rather than capital equipment. His firm eventually employed some one thousand workers, and produced furnaces, rifles, printing presses, and locomotives.

Entrepreneurial flexibility was no less evident in terms of willingness to migrate to more promising markets, or geographical mobility in general. The antebellum period witnessed the rise of new centers of manufacturing and invention in townships such as Troy, Lowell, Waterbury, and Trenton, as well as the national expansion westward. Among an extremely mobile population, the great inventors stood out as especially inclined to take advantage of opportunities by moving – with the most mobile tending to be the most prolific in terms of numbers of patents. Individuals such as Jacob Perkins, Richard Gatling, and Cyrus McCormick readily relocated when it was useful for the commercial development of their ideas. Overall, 70 percent of all great inventors migrated to two or more states over their career. More than 80 percent at some point filed a patent in a state other than that of their birth, with over 10 percent filing in three states. A number of inventors, including Samuel Colt, Joseph Saxton, William Crompton, John Howe, and Samuel Slocum, even traveled to Europe to take advantage of the opportunities there. Because only a quarter of the 1860 population were recorded as inter-state migrants, the great inventors were markedly more geographically mobile than the general population.[15]

Their exceptional mobility also was manifest in terms of the distribution of great inventor patents by region and migratory status (where the latter refers to whether the patent was filed in a state other than the inventor's state of birth.) Migrants clearly dominated patenting in all regions except Southern New England. The record for Southern New England is particularly interesting, because it implies that the technological leadership of this region

---

[15] According to the 1860 Population Census (pp. xxxiii–xxxiv), 24.8 percent of the population had emigrated from their state of birth, typically relocating in states adjacent to their state of origin.

was based on natives to the area, as opposed to centers like New York City whose dominance was largely because of in-migrant inventors. The net flow of great inventors was from areas with less commercial development and economic opportunity (such as Northern New England or the Southern Middle Atlantic), toward more extensive markets (such as New York or Southern New England), or regions undergoing rapid expansion (such as the Midwest.)

The building of the Erie Canal (1817–1825) is frequently pinpointed as a pivotal event in American technological history.[16] As Kenneth Sokoloff showed, total patents in New York received a one-time boost during this period, which allowed the subregion briefly to overtake Southern New England as the leader in per capita terms.[17] At the same time, among the great inventors the locational inventive advantage shifted away from Southern New England toward New York, a shift that was sustained for the rest of the antebellum period. Improved access to markets, and New York's subsequent preeminence as a center of manufacturing and commerce served to attract migrant inventors from distant and disparate locations. The transferral of geographical inventive advantage may be partially explained by the substantial share of great inventors, born in Southern New England, who ultimately migrated to New York. This movement from one highly commercialized locality to another is consistent with the interpretation that migrant inventors were acting entrepreneurially to increase the returns to their inventive activity. These systematic patterns at the county level militate against the idea that important inventions were random, and instead suggest a common motive factor. Location-specific variables may have stimulated individuals to make great inventions, but men of great inventive potential also were inclined to migrate to particular centers of activity. These loci were possibly attractive for the array of commercial opportunities available, as well as for other conditions conducive to invention.

Insight into the entrepreneurial inclinations of inventors also can be discerned from their attempts to appropriate returns from their inventions (Table 7.3.) Such efforts encompassed a variety of methods, including direct use of the invention in production, assignment or sale of rights, licensing, and litigation. The typical great inventor combined ingenuity in both invention and commercial exploitation, proving to be a shrewd entrepreneur who

---

[16] By the middle of the century, the cadre of engineers and machinists who had received practical training during the Erie Canal construction would incorporate a more technical approach to patenting and invention in the United States. Most prominent American-born engineers during the first half of the nineteenth century learned their profession either on the Erie Canal itself or from an engineer who had been there, so that, according to F Gies and J. Gies, *The Ingenious Yankees*, New York: Crowell, 1976, p. 148, the canal "proved to be a magnificent engineering school."

[17] Kenneth L. Sokoloff, "Inventive Activity in Early Industrial America: Evidence from Patent Records, 1790–1846," *Journal of Economic History*, vol. 48 (Dec.) 1988: 813–50.

Table 7.3. *Great Inventor Patents By Source of Income and Occupation (1790–1865)*

Occupation	Source of Income from Invention				Total
	Royalties	Mfg	Both	None	
Artisan					
N	5	73	55	2	135
%	3.7	54.1	40.7	1.5	
Full-time Inventor					
N	45	27	69	19	160
%	28.1	16.9	43.1	11.9	
Engineer/Machinist					
N	60	150	110	26	346
%	17.3	43.4	31.8	7.5	
Merchant/Professional					
N	22	35	135	56	248
%	8.9	14.1	54.4	22.6	
Manufacturer					
N	5	39	165	7	216
%	2.3	18.1	76.4	3.2	
Farmer/Other					
N	5	27	30	0	62
%	8.1	43.6	48.4	0.0	
Total	142	351	564	110	1167

Notes and Sources: This table reports the distribution of great inventor patents by the principal occupation of the inventor over his career as well as by the methods he employed to secure returns to his inventions. Hence, each of the inventor's patents is classified in the same manner. Methods by which inventors realized returns are royalties (inclusive of the licensing or sales of the rights to the patented invention), manufacturing (inclusive of manufacturing that used the invention), both royalties and manufacturing, and none. In general, no information was available on those inventors whose patents were classified in the "none" category. See the text for further information.

efficiently promoted his inventions, motivated by a desire for profit. Few failed to secure rewards from their inventions.[18] These exceptions generally support the thesis that returns were related to entrepreneurial abilities, because they tended overwhelmingly to be college-educated professionals, had brief inventive careers, and produced inventions that were unrelated to

---

[18] Only two antebellum inventors are recorded as receiving no benefits. However, as it is likely that many of the twenty inventors for whom no record of income exists also did not receive substantive returns, I chose the conservative route of including them in the 'no income' category. All told, an upper estimate is therefore that 14 percent of inventors, who accounted for less than 10 percent of all great inventor patents, gained minimal returns.

their occupations; they further did not try to commercialize their ideas, and were apt to be nonmigratory.

The assignment or sale of patent rights could prove to be profitable when the invention was demonstrably useful, and when the inventor had reputational capital to draw on. Some inventors maintained long-term relationships with enterprises, such as Henry Burden's with the Troy Iron and Nail Factory, which paid him a retainer of $10,000 per year for the rights to his spike machine. Alternatively, the decision to license involved the patentee in a measure of risk-taking, but the difference in payoff could be significant. Christopher Latham Scholes assigned his typewriter patent rights to the Remington Company for $12,000, whereas his partner James Densmore opted for royalties and subsequently received over $1,500,000. Almost 40 percent of those who simply assigned the rights or licensed the patent were from the merchant/professional class. Those who chose this strategy were in the minority, as 85 percent of the inventors for whom information is available were directly involved in commercial exploitation of their invention through manufacture, or both manufacture and licensing.

Entrepreneurs are normally credited with transforming the invention into a usable product, and such innovation is often associated with the greatest potential return. For instance, Cyrus McCormick received $20 to $35 in patent royalties per reaper, but gained an estimated unit profit of $80 through manufacturing.[19] Before 1825, half of all great inventor patents accrued to individuals who manufactured the product in question and were presumably directly affected by the growth of markets. Subsequent to the rapid industrial expansion of the 1820s and 1830s, it became increasingly common for these inventors to license as well as manufacture. Because patent assignments or licenses could be restricted to specific locations, such practices often made it possible to exploit a larger market than if the inventor chose the manufacturing strategy alone. According to Table 7.3, 76.4 percent of all patents granted to manufacturers were filed by those who chose this dual route to appropriating returns, as compared to 42 percent of patents by great inventors in other occupations. Although the joint strategy was preferred, unless inventors benefited from learning by doing, their licensees and assignees could become competitors on the expiration of the patent. This may be one reason why multiple patenting was so prevalent among inventors who had their own manufacturing enterprises. Their establishments tended to incorporate the latest technology, including developments by other inventors. Inventor-manufacturers such as Hiram Pitts, Cyrus McCormick, Horace Day, Richard Hoe, George Esterly, and George Bruce aggressively acquired the assignment rights to patents and designs that they employed in their operations. Many of their

---

[19] W. T. Hutchinson, *Cyrus Hall McCormick* (2 vols.), New York and London: Century Co., 1930 (vol. 1, pp. 278 and 292.)

companies became virtual monopolies because of their superior policies of innovation.

Great inventors in the antebellum period thus attempted to appropriate returns from their inventions, and for the most part succeeded. If the propensity to patent typifies economic men motivated by expected profit, then virtually all of the great inventors fall within this category: only 10 of 160 failed to secure patents for their discoveries. Like Benjamin Franklin and Thomas Jefferson, some made their inventions freely available for the public good. Robert Dalzell belongs to this category of "Patent Dissenter," who apparently objected to the individual accumulation of profit on ethical grounds. However, although he reportedly limited his gains to 7 percent, he amassed a considerable income from employing his unpatented devices in manufacturing elevator systems for grain storage. The same is true of Thomas Rogers, who provided specifications of his improvements in locomotives to the Patent Office, but did not obtain a patent "to ensure their being public property." Rogers was amply remunerated by producing locomotives for which a ready demand existed, based on his excellent reputation among railroad owners. Three machinists – Gridley Bryant, Sylvanus Brown, and Isaac Dripps – made unpatented improvements that transformed the productivity of their firms. Thomas Kingsford relied on secrecy rather than patents to protect his discovery of a process for making starch from corn, which he manufactured in his Oswego, New York, factory in partnership with his son. However, when it became apparent that others were replicating his results, Kingsford switched to edible cornstarch, for which he obtained a patent in 1863. It is noteworthy that all of the above individuals produced inventions that were job-related. Most were able to obtain some return from their efforts, either through enhanced reputations that led to greater remuneration, or through manufacturing. Although these inventors were able to extract returns without patents, the vast majority of great inventors did not. That only a few individuals chose to bypass the patent system was because of the readily duplicable nature of technology, and to the degree of competition in antebellum product markets.[20]

Although a valid patent was helpful, it was no guarantee that an inventor would be able to appropriate the return to his invention. That ability depended, among other factors, on aspects of the legal system such as the attitudes of the judiciary. Influenced by the frustrations of Eli Whitney, Charles Goodyear, and Oliver Evans in the courtroom, some observers have questioned whether important inventions could be protected. Although there is some truth to the idea that the more significant the discovery, the greater the incentive for infringement, this did not imply that inventors were

---

[20] Just one example is provided by Mathias Baldwin, who was a pioneer in locomotive production in the United States; he gained access for half an hour to a locomotive that was imported, then returned to his workshop and reproduced it (see Bishop, vol. 2; p. 538.)

unable to realize substantial returns. Ithiel Town, an engineer and architect whose design simplified bridge structures, was readily able to identify infringers, whom he charged double the price collected from more honest users. Nathaniel Wyeth filed over fourteen patents dealing with cutting and shipping ice, but he ignored infringers in the domestic market because he was gaining large returns from shipping overseas.

Thirty, or less than one fifth of all great inventors were actually involved in litigation, whereas only forty, or 3 percent of their patents, were at issue. For the 80 percent who never appeared in the courts, it is likely that their patent rights and reputation were sufficient to ensure out-of-court settlements, or that patent infringement was not critical because the inventors could appropriate returns through other means. At the same time, the per patent rate of litigation for great inventor patents was three times as high as the rate for ordinary patents, indicating that important patents had a higher probability of being litigated. One reason for this is that inventors employed litigation as a strategy to maintain market share and preempt rivals, both actual and potential. An example is Cyrus McCormick, who maintained a phalanx of lawyers full time on his payroll. William Woodworth's wood-planing machine was similarly litigated in over seventy-five lawsuits throughout the country, resulting in a virtual monopoly over the industry. The proportion of *cases* in relation to total patents filed was 2 percent for all inventors, but amounted to 10 percent for great inventors. Because precedent was established in the first successful outcome, these plaintiffs may have been more interested in suppressing competitors than in defending the patent *per se*. The litigation records are thus consistent with the evidence presented above in suggesting a strong concern with extracting an economic return.

Even if they agreed that marginal inventions might be market induced, many economic historians studying the sources of early inventive activity have viewed important inventions as largely haphazard and unresponsive to the prospect of material gain; such inventions tend to be attributed instead to idiosyncratic factors such as innate ability, technical expertise or serendipity. However, information from the antebellum sample of 160 great inventors does not support this perspective and instead is more consistent with the process of democratization. Ordinary individuals were stimulated by higher perceived returns or demand-side incentives in general to make long-term commitments to inventive activity. These individuals were not especially distinctive in terms of age, education, or occupational background. Instead, great inventors were typically entrepreneurial and responded systematically to changes effected by the remarkable expansion of the antebellum era. Far from being random, patenting by great inventors corresponded closely to the procyclical patterns observed for general patentees and, like ordinary patentees, the great inventors were highly concentrated in districts with access to broad markets. Moreover, the great inventors took advantage of expanding opportunities by migrating in disproportionate numbers to areas with ready access to markets, as well as by changing occupations to exploit their inventions. In sum, the experience of the antebellum great inventors seems to be entirely consistent with the thesis that technological change during early industrialization was because of a process of democratization. History has distinguished the most successful from an array of inventors and elevated them above their peers, but one is more struck by their similarity to other patentees in the common quest to employ patent institutions to benefit from their efforts.

THE SECOND INDUSTRIAL REVOLUTION

How relevant is the experience of such antebellum inventors in explaining technological change after the Civil War? After all, the common perception is that major inventions during the Second Industrial Revolution tended to originate from corporate research and development laboratories whose employees were formally trained in science and engineering. Again, the canonical figure from this perspective is Thomas Alva Edison. Ironically, Edison's biography itself highlights the ambiguities of such a model in explaining inventive activity at the turn of the century. From 1876 through 1886 Edison worked at Menlo Park, and in 1887 he established the West Orange Laboratory, both of which are regarded as the precursors to the modern age of team invention in research and development laboratories. The Edison laboratory undoubtedly attracted a large number of brilliant young scientists, machinists, and engineers; however, Edison's own background and the methods of his assistants were more in keeping with the earlier cohort of experimentalists. These trial-and-error methods were only gradually updated after inventors began to apply their formal theoretical knowledge to test scientifically based hypotheses. For instance, Charles Kettering's lab at General Motors tested over thirty-three thousand different chemical compounds in the quest to create an antiknock fuel, before Thomas Midgley decided to use his knowledge of the periodic arrangement of chemical elements to predict the correct solution to the problem. The Maxim family illustrates such changes in the nature of invention over time. Maine-born inventors Hiram Stevens Maxim and his brother Hudson came from a poor family in a remote rural area, and both were predominantly self-trained. Hiram worked as an apprentice at an early age. He later achieved fame and fortune as the creator and producer of the world-famous Maxim gun, and was knighted for his efforts. Unlike his father, Hiram Percy Maxim was formally educated as an 1886 graduate of the Massachusetts Institute of Technology. Hiram Percy was himself a great inventor, and applied his technical knowledge to devices such as the silencer (an invention that caused much outcry in its day as a "menace to public safety.")

We can address such issues by studying a sample of "postbellum inventors," some of whom were still active in the 1950s. The assessment of great

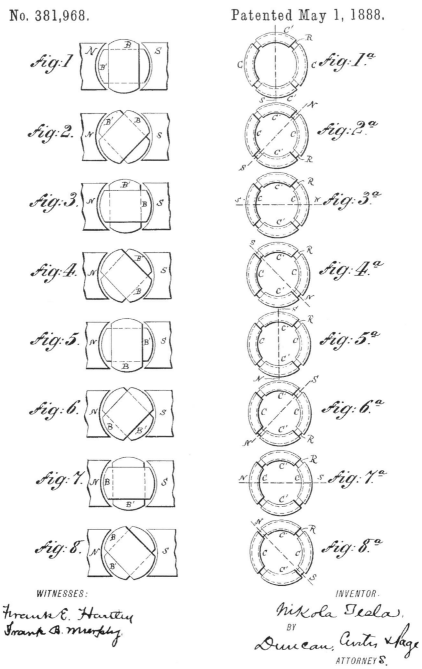

Illustration 9. Electromagnetic motor patent granted to Nikola Tesla in 1888. (*Source*: U.S. Patent Office.)

inventors in the second industrial revolution is primarily based on the patenting activity of 260 inventors born between 1820 and 1885, many of whom contributed to the "new economy" of the early twentieth century, and to the rapid economy-wide growth in productivity. Even the most cursory statistics (Table 7.2) reveal the evolving nature of patenting and invention in this period. The propensity to patent was even higher than before the Civil War, because only three inventors in this cohort failed to obtain patent protection. The great inventors in the postbellum sample were drawn from a narrower range of the population than their earlier counterparts. Engineers, machinists, and full-time inventors comprised 43.8 percent of the later sample, and only sixteen were employed as artisans. Half of the postbellum inventors had completed college-level education, and fifty-one of these were trained in science or engineering.

In the antebellum period, the average great inventor filed his first patent at the age of thirty-four, and was granted twelve patents over a twenty-five-year period; whereas the average postbellum inventor filed his first patent at the age of thirty-two, and obtained fifty-seven patents during a a career that lasted almost thirty-two years. As might be expected, more specialization in invention and lengthier careers were reflected in higher numbers of patents granted. Edison, of course, is an outlier, but others such as Carleton Ellis (753 patents), Elihu Thomson (696), Henry A. W. Wood (440), Walter V. Turner (343) and George Westinghouse (306), were close rivals. Although inventors with longer careers and larger numbers of patents were probably more likely to be viewed as great inventors, this was not necessarily the case. For instance, John F. O'Connor, a Chicago engineer who received over eight hundred patents for railroad draft gearing inventions, remained obscure and is not part of the sample. O'Connor, an Irish immigrant who was born in 1864, went to work as a boy and gained an education through night school and correspondence courses. His last patent was filed in 1935, and the majority were assigned to his employer, the William H. Miner Company of Illinois.

A number of scholars have used biographical information to gauge the relationship between age and the productivity of innovators, and compared theorists to experimentalists.[21] Indeed, Carleton Ellis fits the profile of the theoretically motivated genius. Ellis, who was born in 1876, a generation

---

[21] David Galenson proposes a life cycle approach to creativity, and discerns two different types of innovators: "conceptual artists" are theorists who primarily make their most significant discoveries early in their careers; whereas "experimental artists" or empiricists are those whose "genius" emerges later in the life-cycle after a long gestation period of trial and error. See David W. Galenson, *Painting outside the Lines: Patterns of Creativity in Modern Art*, Cambridge, Mass.: Harvard University Press, 2001. See also Hans Eysenck, *Genius: The Natural History of Creativity*, Cambridge: Cambridge University Press, 1995; the many monographs by Dean Keith Simonton, including *Genius and Creativity: Selected Papers*, Greenwich, Conn.: Ablex Publishing Corp, 1997; and Harvey. C. Lehman, *Age and Achievement*, Princeton, N.J.: Princeton University Press, 1953.

after Edison, created his first invention while still a chemistry student at the Massachusetts Institute of Technology. Approximately half of Ellis's 753 patents were granted during the first two decades of his career, and more than 70 percent of the citations that other patentees made to his inventions refer to work he completed before he was forty. Similarly, Elihu Thomson, whose company later merged with Edison General to form General Electric, obtained more than 40 percent of his patents while still in his thirties. Thomson had taught physics and chemistry and his work was based on a thorough understanding of the scientific principles underlying his inventions. Scientist William Channing, while collaborating with Moses G. Farmer to improve electric telegraphs, was able to achieve enough within a space of three years to earn himself an entry as a great inventor.

However, the most productive great inventors were not conceptual in approach, and their overall patenting behaviour reflected the work of experimentalists with long careers. Similarly, the numbers of long-term citations are disproportionately higher for older inventors. Thomas Edison was more typical in being "a trial-and-error inventor . . . [who] scorned scientific theory and mathematical study which might have saved him time." He and his assistants found a solution to the problem of producing a durable filament for the incandescent lightbulb only after months of trying thousands of different fibers. Edison obtained his first patent in 1869 at the age of twenty-two, and his last patent was granted some sixty-three years later. Edison's productivity in terms of patenting peaked during his forties, but most of his patents were filed well after this period. During this lengthy career, some have argued, his best work was completed in his first decade of invention, and the rest of his career lacked the creativity of his youth. However, the evidence suggests that he was a productive inventor for most of his life: the majority of his long-term patent citations relate to patents that he filed after he was fifty, and this is similar to the pattern for his inventor citations.

Economic models of human creativity focus on the acquisition and accumulation of human capital that lead to increases in productivity. Sources of additions to human capital include experience, apprenticeship, and formal schooling. The economic model further predicts that inventors with specialized formal education in technical subjects will tend to exhibit a different pattern over the life cycle, relative to experimentalists who primarily benefit from experience or untutored empiricism. Experience comprises continuous additions to human capital that would be associated with a life-cycle pattern of productivity that peaks later in the individual's career. Thus, if genius is indeed only "1 percent inspiration" we should observe a life-cycle pattern rather similar to Edison's. By contrast, if formal science or engineering education enhanced the initial rate of capital accumulation we should observe an earlier peak in inventive productivity among the technically qualified inventors, as this type of creativity would likely dissipate more rapidly over the life cycle. The age distribution of both patents and citations awarded to inventors who received formal technical training in science or engineering, relative to all other inventors is quite distinct, because they are skewed toward earlier ages, and fall off more rapidly, as the human capital model predicts.

Creativity is commonly regarded as an intensely individualistic factor, and has been explained in terms of age and psychological, cognitive or genetic characteristics. The results for the great inventors are consistent with a range of other studies that find similar life-cycle patterns for exceptional individuals in art, music, literature, and the sciences. However, this simple stratification by age may be misleading, because it fails to control for a number of potentially significant factors, such as those related to time, opportunity, industry, and location. The tendency for older individuals in the sample to be productive in terms of patenting and citations also may result from selection biases, because unproductive inventors might be more likely to leave the sample at younger ages, whereas older inventors may be able to draw on more resources such as funding, independently of their productivity. The effects of age are likely to be commingled with those of birth cohort and differences in career profiles. For instance, Theodore R. Timby and Henry A. Wood both filed their first patent at age twenty, but Timby was born in 1822 and Wood was born in 1866. By contrast, individuals from the same birth cohort may enter the sample at different ages. Thomas Seavey Hall was born in 1827, but only became an inventor after he had retired from wool manufacturing, and a train accident gave him the idea of a means to prevent future mishaps; whereas his contemporary, James Eads, received his first patent twenty years earlier.

The stock of an inventor's human capital is likely to reflect cohort effects because new discoveries will have been made or old ones discredited, industries differ in attractiveness or potential value over time, and the opportunities for exploiting knowledge or obtaining support also will vary across cohorts. Inventors with minimal education who were born before the Civil War were able to contribute to commercially valuable technologies. The relatively uneducated inventors or those from rural areas were no less likely to produce valuable inventions, as gauged by commercial value as well as technical value, and their patents were just as likely to be assigned and cited in the patent specifications of other inventors. However, Figure 7.5 illustrates how, as technology advanced and became more complex and capital intensive, would-be inventors needed more formal training to operate on the frontiers of the most advanced industries. College-educated individuals were increasingly characteristic of great inventors over time, and by the final cohort formal education was clearly a prerequisite for technologically significant inventions. Inventors with formal technical qualifications accounted for only 8 percent of all patents filed by those who were born between 1820 and 1839, and even fewer assignments and citations. For the 1860–1885 birth cohort, formal scientific and engineering education led to higher productivity,

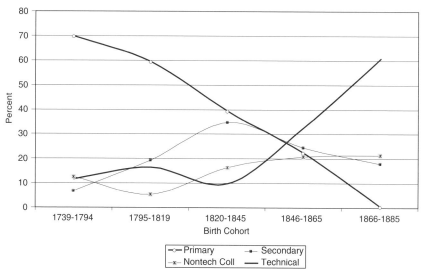

Figure 7.5. Great Inventor Patents by Formal Schooling and Birth Cohort (Percent distribution)

Notes and Sources: Primary schooling includes no formal education and a few years of schooling; technical includes patents filed to inventors with college-level science and engineering qualifications. See text and notes to tables.

as shown by the disproportionate fraction of great inventor patents, assignments, and citations that graduates in these fields generated. The change in the nature of inventive productivity was already becoming evident by the 1840 birth cohort, and consolidated with the postbellum cohort. Although only 37 percent of the inventors who were born in the 1860–1885 cohort were technically qualified, they produced 45.1 percent of patents, 52.1 percent of assignments, 40.4 percent of all long-term citations, and 60.9 percent of inventor citations.

The distribution of inventors and their activities allows us to trace the relative significance of location and industrial factors. Throughout the nineteenth century, agriculture was still the largest source of employment in the economy, but it attracted little attention from patentees in general and less from great inventors. Important inventors were clustered in the newest industries that promised the highest marginal rate of return. Inventors who were born before 1839 directed most of their efforts toward crucial discoveries in manufacturing and transportation. For instance, Allen B. Wilson was born in 1824 and produced a number of significant improvements to sewing machines. Wilson was able to finance the costs for his first patent application by selling off a share of his potential patent rights. He later formed a partnership with another great inventor, Nathaniel Wheeler, which became the fourth largest sewing-machine manufacturing establishment in the country.

Another famous great inventor partnership of the time, George Babcock and Stephen Wilcox, focused on the manufacture of steam engines and boilers, and several other inventors of the same cohort also directed their attention to steam ships.

By the following birth cohort, inventive and commercial interests were shifting toward the burgeoning electrical industry. The early twentieth century was renowned as "a new era" of the most rapid aggregate productivity growth in American economic history, in part because of the diffusion of innovations related to electricity.[22] The foundations for the surge in productivity were laid in the second half of the nineteenth century, with the development of telegraphy, underwater cables, arc lights, and long-distance power generation. These enterprises involved numerous great inventors, including George Westinghouse, Moses G. Farmer, Stephen D. Field and Nikola Tesla. Electrical innovations comprised the single largest industry in terms of patents and assignments. Other measures of productivity similarly highlight the pivotal role of the electrical industry in the late nineteenth and early twentieth centuries. Inventors who belonged to the 1840–1859 birth cohort and contributed to innovations in electricity and energy garnered an average of fourteen long-term patent citations, relative to the overall average of 5.2. Similarly, electricity inventors were mentioned more frequently in contemporary patent specifications (12.8) relative to the average number of inventor citations (4.5.)

The role of cohort-specific leading sectors, especially in the case of electricity, is reminiscent of the modern dot-com phenomenon. As inventive activity became more technically demanding, it is likely that the returns to technological specialization increased, as well as the benefits to belonging to the cohort with the most current training. The technological contributions to the "new economy" were disproportionately by specialized younger inventors, whereas the patentees of miscellaneous inventions were likely to be older. Like the modern dot-com boom, inventions and investments in electrical discoveries generated large fortunes, and readily attracted flows of venture capital for speculative research and development. Patentees were able to parlay their property rights in promising inventions into part ownership in numerous companies on advantageous terms that included retention of patent rights and royalties. Edison attracted capital from prominent investment bankers and financiers, as well as from investors in the several hundred public corporations with which he was associated. Just one of these enterprises, the Edison Electric Light Company, along with Drexel,

[22] According to the *Wall Street Journal*, March 14, 1918, p. 6, "the electro-chemical industry of the United States has been so developed as now to make it certain that it is about to establish a new era in American industry." In 1884, the American Institute of Electrical Engineers was founded, and included three great inventors (Alexander G. Bell, Thomas A. Edison, and Franklin L. Pope) among its six vice presidents.

Morgan & Co., provided $150,000 to finance his R&D in the electrical field between October 1878 and March 1881. However, he was not atypical in this regard, for investors were eager to fund start-ups allied with other great inventors.

Throughout the nineteenth century, patenting among the population of "ordinary inventors" clustered in specific regions and in urban areas. The antebellum period featured growing market demand and an increase in per capita patenting in Southern New England and New York state.[23] Metropolitan counties registered higher rates of inventive activity, but much of the rise in early patenting was because of residents of rural counties that had recently gained access to markets. Even though national markets emerged shortly after the Civil War, with the rapid extension of railroad networks, regional disparities in patents and assignments persisted throughout the nineteenth century. Patenting rates in the South and the West remained lower than in the Northeast while, within the Northeast, New England's importance decreased relative to the Middle Atlantic. Such locational factors were even more pronounced among the great inventors in the postbellum sample. Great inventors' patenting was strongly focused in the Middle Atlantic (51.2 percent of patents) and New England (25.9 percent), although a growing fraction was active in the Midwest. The Middle Atlantic dominated in part because inventors were attracted to the region from other parts of the country. Henry Sargent was an extreme example, who moved to twenty-six different cities between 1854 and 1893, but great inventors were exceptionally mobile and tended to migrate to more profitable markets.

The notion of patenting and inventive activity as a means of achieving eminence, especially for disadvantaged groups, is borne out by the experience of foreign-born inventors. Both ordinary patentees and great inventors were disproportionately of immigrant origin when compared to the general population. Between 1870 and 1930, the foreign-born accounted for 10–14 percent of the total population, whereas 21 percent of all patentees were foreign-born, as were 23 percent of the individuals who made important technological contributions to the second industrial revolution. Only two of the postbellum great inventors were black, and neither was native-born. Elijah McCoy, a black engineer, was born in Canada in 1844; and Jan Earnst Matzeliger was an immigrant from Dutch Guiana. Matzeliger, a poor factory worker in Lynn, Massachusetts, was lauded for making the most outstanding contribution to the mechanization of shoe production: his patented machine was able to last shoes of any type of leather quality, and led to enormous increases in the productivity of shoe manufacturing.[24] Immigrant great inventors from the postbellum cohorts were more likely

[23] Sokoloff, "Inventive Activity," 1988.
[24] Sidney Kaplan, "Jan Earnst Matzeliger and the Making of the Shoe," *Journal of Negro History*, vol. 40 (1) 1955: 8–33 and *DAB*.

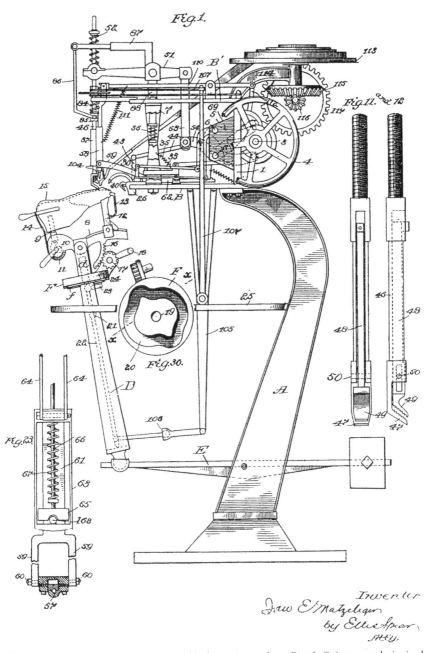

Illustration 10. Jan Matzeliger, a poor black immigrant from Dutch Guiana, revolutionized shoe manufacturing with his lasting machine. (*Source*: U.S. Patent Office.)

to be from humble backgrounds than their U.S.-born counterparts. Forty percent had minimal education, relative to 25 percent of native-born great inventors; and only 10 percent of the immigrant great inventors were technically trained, compared to 22 percent of the American inventors. Many came to the United States with no resources, but through their inventions built up large enterprises: one such was Akiba Horowitz, a Russian who changed his name to Conrad Hubert on arrival in the United States, and whose patents formed the foundation for the Eveready® Company. Nikola Tesla, although technically trained, arrived in the United States with little money, and at one point earned his living through digging ditches.

Migration to centers of invention and assignment occurred in part because of institutional factors such as ready access to venture capital, patent agents, legal counsel, and networks of other inventors working on similar issues. The role of networks can be seen from the numerous partnerships or loose alliances that inventors formed with each other, ranging from consultation, co-invention, employment, and assignment contracts, to long-term joint ownership of enterprises. These networks were especially evident among inventors in the electric industry, where younger inventors briefly worked for established entrepreneurs such as Edison or Westinghouse, before breaking off to form their own enterprises. Westinghouse, when he wished to switch from air-brake inventions to the electrical field, purchased key patents to work on, and hired the services of then-junior inventors William Stanley and Nikola Tesla. Stanley had been an assistant to Hiram Stevens Maxim and after leaving Westinghouse Electric he established his own firm, the Stanley Electric Manufacturing Company, which he eventually sold to General Electric. Such network effects accumulated, resulting in growing industrial specialization within and across regions. For the birth cohort from 1840 to 1859, 68 percent of New England patents and 39 percent of Middle Atlantic patents were in the electric industry. By the next cohort, important electrical discoveries were clustered primarily in the Middle Atlantic region. The electrical industry's share of New England patents had fallen to 24 percent; whereas the share for the Middle Atlantic increased to 42 percent, and this region accounted for 75 percent of all electrical patents. Similarly, great inventors in the Midwest region increasingly specialized in transportation-related patents.

Technical schooling and professional contacts may have been more important after the Civil War period, but informal networks still played a role in promoting inventiveness, as shown by the increase in the percent who belonged to families of inventors or technically trained professionals. These networks in part explain geographical clustering of important inventions independently of the availability of educational institutions. Ralph Emerson, a resident of Rockford, Illinois, was distinguished for improvements in agricultural inventions and knitting machines. We can trace at least three generations of inventors in the Emerson family, as well as complex intragenerational networks of innovation and invention.[25] Emerson organized his own firm, Emerson & Company, which manufactured products that incorporated inventions of his own as well as those of other Rockford great inventors, John Manny and William Worth Burson. Ralph Emerson's neighbor, John Manny, was a wealthy patentee and manufacturer of reapers, whereas William Burson not only founded the Burson Knitting Company but also established the knitting industry for which Rockford, Illinois became famous. Ralph Emerson married Adeline Talcott, and was a coinventor with Adeline's father Waitt Talcott, and another relative, William A. Talcott. The Emersons' daughter, "Hattie" Elizabeth Emerson, applied for a tubular knitting machine patent when she was only twenty years old, and another four years later. Hattie married William E. Hinchliff, who also patented two knitting machine inventions, and their son Ralph Hinchliff obtained eleven patents, some of which were related to knitting machines and assigned to the Burson Knitting Company. Several other relatives living in the area were manufacturers or patentees or both. Beyond the family, Rockford residents included numerous successful knitting and agricultural implement manufacturers, patentees, and patent dealers. Such geographical and familial networks to some extent substituted for formal technical training, and enabled specialization outside of engineering or science programmes.

Specialization, education, and other forms of human capital accumulation not only yielded higher productivity in inventive activity, they also contributed to entrepreneurship and the ability to profit from innovation. College-educated inventors were better able to marshall resources and efforts that allowed them to extract returns from their inventions. Almost a third of all great inventor patents in the postbellum sample went to owners or principals of firms who had acquired higher education. Undoubtedly, college education in some instances simply proxied for wealthier backgrounds or social connections, rather than heightened organizational skills or entrepreneurial insight. However, a more privileged background did not explain all of the advantage of higher education: formal technical education was not necessarily an aid to entrepreneurship, and inventors with science or engineering backgrounds obtained 27 percent of patents to owners and principals – about the same as inventors with only secondary education. At the same time, almost none of the technically qualified patentees chose the route of simply selling off their rights. Instead, inventors without a technical background or with minimal education were more likely than other inventors to sell the rights to their patents, or to license those rights to others. This tendency appeared early in the nineteenth century, and increased over time. It illustrates the changing nature of invention and innovation as markets expanded. By the early twentieth century, successful innovators needed to be

[25] This information was obtained by matching patent records and information from the manuscript censuses.

able to mobilize large-scale financing beyond family resources, manage complex organizations, attract and train skilled employees, and rapidly respond to economic signals across the nation or even internationally.

Changes also occurred in the competitive environment, especially in terms of litigation. Lawsuits tended to relate more to aggressive commercial strategies rather than to weak intellectual property rights. Less than 20 percent of the antebellum great inventors had been involved in lawsuits, but almost half (47.2 percent) of the postbellum great inventors were party to litigation. Owners or principals in firms were implicated in over two thirds of all lawsuits, and only 7.8 percent of litigation involved employees. The most competitive areas in terms of technology also were the most litigious. The inventors of electrical discoveries were four times more likely than the average inventor to be involved in federal litigation, and were responsible for 41.4 percent of all cases. Inventors such as Edison were alert to new discoveries made elsewhere, and used the patent records as a source of information to locate promising areas for technological profit opportunities. Edison made contributions to the quadriplex telegraph, shifted to the telephone after the breakthrough patents of Alexander G. Bell and Elisha Gray, and similarly changed direction throughout his career whenever a new field opened. His companies purchased the rights to numerous patents by other inventors, in order to work on that area of innovation. Edison's companies were noted for their rapid absorption of competing technologies and technologists. It is therefore not surprising that he was involved in 11.3 percent of the total 533 great inventor cases litigated during this period.

The Supreme Court had long upheld the dictum that "In the construction of patents and the patent laws, inventors shall be fairly, even liberally, treated," so patentees had great leeway in formulating strategies. Former competitors, such as Gordon McKay and Charles Goodyear, pooled patents to resolve overlapping claims. A combination even paid John Good $150,000 a year *not* to operate his rope manufacturing factory that featured the superior technology of his key 1885 patent. Nevertheless, a number of these lawsuits involved antitrust charges of attempts to monopolize the industry based on the advantage of patent ownership. George Eastman's firm, Eastman Kodak, was charged several times with antitrust violations. In 1912 he controlled more than 75 percent of the entire photography market and earned 171 percent in profits. Eastman Kodak bought out competing patents, filed lawsuits against competitors, stipulated exclusive contracts with suppliers, and required principals in acquired companies to sign agreements not to re-enter the industry. The firm joined with Edison Manufacturing Company and eight others to organize the Motion Picture Patents Company, which controlled over 70 percent of that industry. Although the holding company managed the pooled patents, its major function allegedly was to bring hundreds of patent lawsuits. Then, as now, courts at times found it difficult to disentangle the legitimate entrepreneurial exploitation of patent rights from welfare-reducing monopolistic strategies, especially when the defendant was universally regarded as a public benefactor.

CONCLUSION

The nineteenth century proved to be the age of patented inventions. The economic expansion that began early in the 1820s was pervasive and associated with widely ranging inventive efforts. The first railroad in the country was built in Massachusetts, as well as the first locomotive in New York; and steam engines were vastly improved, exciting interest even in Europe. Productivity increased in mining, and scientific advances such as improved seed culture raised agricultural efficiency while the sector became increasingly mechanized. Manufactured commodities produced by new methods and machines were not only more numerous and varied – ranging from porcelain and silk to varnish, dyes, chemicals, canned goods and paper – but also of improved quality. The American patent system played a central role in influencing the nature of inventors and inventions during this formative era.

Other economists have pointed out that patents allow the appropriation of returns from new ideas, and aid in the diffusion of information. They also have debated the parameters of the optimal patent grant in terms of the costs and benefits of patent length, breadth and height. My analysis returns to a perspective that was regarded as a commonplace truism in the nineteenth century: it underlines the democratic nature of the patent institutions that the United States introduced. The U.S. patent system, with the support of the legal system, facilitated the entry of relatively disadvantaged individuals into the field of technology, enabled them to specialize in invention, mobilize resources to fund patenting, and commercialize their discoveries. Specific features included the relatively low patent fees; the award of patents to the first and true inventor (which protected poor inventors who needed to raise money to obtain a patent); the centralized examination system after 1836 (which led to economies of scale; the training of a large cadre of patent examiners unlike any other in the world; the provision of a signal that facilitated capital mobilization for risky projects; the acquisition of information not just about prior arts but also about potential new areas of interest; and increased security of property rights.) There were minimal restrictions on the rights of patentees after the patent was granted (no compulsory licenses, "Crown use," or working requirements.) The patent office itself was self-financing and independent and thus less susceptible to political corruption or arbitrary political dictates. Patent rights comprised secure assets that were extensively traded, and gave inventors with only modest resources the opportunity to appropriate private returns as well as to make valuable contributions to society. These conclusions were true for undistinguished patentees as

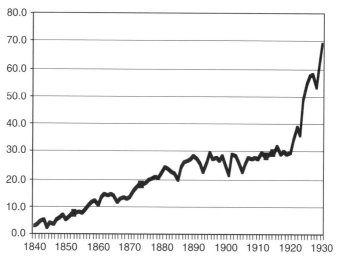

Figure 7.6. Patent Assignments at Issue as a Percent of Total Patents Granted, 1840–1930
Notes and Sources: U.S. Patent Office *Annual Reports* and LexisNexis.®

for the so-called great inventors. Throughout the nineteenth century, important inventions were generated by patentees from ordinary backgrounds, as gauged by their family origins, rural residences, occupations, and educational level.

However, by the first few decades of the twentieth century, the nature of invention and innovation was in the process of being transformed. Inventors such as Thomas Alva Edison had acquired valuable training and insights through apprenticeship and experimentation. By the era of the "new economy" of the 1920s, Elihu Thomson, an immigrant scientist, was a more typical great American inventor than Thomas Edison. The acquisition of human capital through formal education increased the inventor's ability to resolve problems more quickly, although this type of capital also depreciated more rapidly. Formal education also was associated with organizational skills and the ability to mobilize large-scale capital, which allowed "conceptual" inventors to capture longer term returns through corporate equity ownership. Skills in innovation became more valuable as markets became national or even international, the scale and scope of enterprise expanded, and both firms and inventors competed to acquire and sustain first mover advantages. By the 1930s a significant number of great inventors were employees with specialized skills in science or engineering that they directed toward expanding the frontiers of the newest complex technologies of the day. Figure 7.6 reflects the general transformation that occurred in the 1920s, for the percent of patents assigned at the time the patent was issued changed from a relatively stable 20–30 percent between 1880 and 1920, to more than 60 percent by 1930.[26] Such changes may have been associated with a decline in the role of the independent inventor, or the increased participation of firms in the market for technology.[27] Nevertheless, through all of these changes, the U.S. patent system comprised a key institution in the progress of technology, and it also stood out as a conduit for creativity and achievement among otherwise disadvantaged groups.

[26] Assignments indicate the sale or transfer of patent rights. An unsystematic examination of the patent records during the 1920s suggests that many of these patent rights were being transferred to companies. Part of the change in the patterns of assignments in the 1920s and 1930s may have occurred because of enforcement of the work for hire doctrine, which facilitated the assignment of employee patents to the employer.

[27] For a summary report on their extensive project on the market for invention, see Naomi R. Lamoreaux and Kenneth L. Sokoloff, "Long Term Change in the Organization of Inventive Activity," *Proc. Nat. Acad. Sci.*, vol. 93 (November) 1996: 12686–92.

8

Copyright in Europe and the United States

"Our literary workmen...ask simply for markets"

– G. H. Putnam (1879)

Joseph Whitworth, one of the mid-nineteenth-century British observers of the American system of manufactures, linked American economic success to its scheme of inventions, the prevalence of printed materials, and the consequences of a democratic system. He attributed inventiveness in the United States to its favorable laws, widespread literacy and education, and to a free and universal press. He noted that "the humblest labourer can indulge in the luxury of his daily paper, everybody reads, and thought and intelligence penetrate through the lowest grades of society."[1] These observations provoke more specific inquiries about the relationship of copyright policies to the democratic process and economic growth. Despite the common source of patent and copyright statutes in the intellectual property clause of the U.S. Constitution, policies toward authors and inventors differed significantly.

A review that jointly considers copyrights and patents allows us to understand the rationale for both to a greater extent than is possible through a survey of either in isolation. The chapters on patents pointed to the unique character of American patent policies and their success in promoting flourishing markets in patented inventions and patent rights. Unlike their European counterparts, in the United States patentees were drawn from diverse backgrounds, and their inventions improved productivity across all sectors and industries. To an even greater extent than was true of patents, U.S. copyright doctrines also diverged significantly from those in France and England. This chapter first describes the evolution of the French and British copyright systems, which allows us to better gauge American exceptionalism in this area of intellectual property rights.

International copyright laws today recognize a moral right in authorship that is incompatible with the American approach to intellectual

---

[1] Nathan Rosenberg (ed.), *The American System of Manufactures*: The Report of the Committee on the Machinery of the United States 1855 and the Special Reports of George Wallis and Joseph Whitworth, Edinburgh: Edinburgh University Press, 1969, p. 389.

property.[2] The moral rights regime does not weigh costs and benefits; instead, it unilaterally favors authors over public users (holding the incentive effect constant), and as a corollary also limits diffusion and the operation of markets in protected works. Indeed, to the extent that some moral rights are viewed as inalienable, the existence of a market in those rights is prohibited. The moral rights approach is often coupled with an elitist notion of authorship as a creative act of individual genius. In contrast, American statutes were directed toward the pragmatic ends of learning, and the holders of copyright were not elevated above users. The U.S. Constitution favored the public, facilitated the operation of markets, and protected economic rights rather than moral rights. Like patents, American copyright policies and patterns were utilitarian and pragmatic, with features and doctrines that were carefully calibrated to further the ends of a market-oriented democracy.

Copyright policy has pervasive implications for the nature of democracy in society, to a greater extent than is true of patents. Economists such as Milton Friedman and Friedrich Hayek have argued that free markets and economic democracy promote political democracy, and early American policies toward copyrights are consistent with this perspective. Democratic values emphasized equal and widespread access to learning, and the importance of information flows for maintaining political freedom, as well as active participation by all citizens, whereas strong copyrights impinged on the fullest attainment of these objectives. American legal and political analysts frequently pointed to the potential for conflict between copyright and the First Amendment (protecting freedom of speech.)[3] Indeed, during the debates about the U.S. Constitution concerns were voiced about whether copyright grants were compatible with a free society. During the Pennsylvania

[2] Moral rights include the right to be identified as the creator; the right to publish or withhold from publication; the right to prevent distortion or changes in the work; and the right to withdraw an item from circulation.

[3] See Nimmer, "In some degree [copyright] encroaches upon freedom of speech in that it abridges the right to reproduce the "expression' of others, but this is justified by the greater public good in the copyright encouragement of creative works. In some degree it encroaches upon the author's right to control his work in that it renders his "ideas' per se unprotectible, but this is justified by the greater public need for free access to ideas as part of the democratic dialogue." David Nimmer, "Does Copyright Abridge the First Amendment Guarantees of Free Speech and Press?", 17 *U.C.L.A.L.Rev.* 1180, 1192–93 (1970). Also see *Triangle Pubs. v. Knight-Ridder Newspapers*, 445 F. Supp. 875 (1978):

"When the Copyright Act and the First Amendment both seek the same objective, their future coexistence is easily assured. However, when they operate at cross-purposes, the primacy of the First Amendment mandates that the Copyright Act be deprived of effectuation. Rather than strike down an entire act as overbroad in such a situation, the judiciary prefers to interpret such a statute as narrowly as needed to preserve it for the effectuation of those of its purposes deemed consistent with the Constitution."

deliberations, Robert Whitehill was concerned that "Tho it is not declared that Congress have a power to destroy the liberty of the press; yet, in effect they will have it. For they will have the powers of self-preservation. They have a power to secure to authors the right of their writings. Under this, they may license the press, no doubt; and under licensing the press, they may suppress it."[4]

The ultimate objective of U.S. intellectual property policies was to promote the public good and it was for this very reason that copyrights were distinguished from property rights in inventions. As the Supreme Court noted in *Wheaton v. Peters*, 33 U.S. 591 (1834): "It has been argued at the bar, that as the promotion of the progress of science and the useful arts is here united in the same clause in the constitution, the rights of the authors and inventors were considered as standing on the same footing; but this, I think, is a non sequitur . . . for when congress came to execute this power by legislation, the subjects are kept distinct, and very different provisions are made respecting them." Patent policies in the "Republic of Technology" were predicated on the belief that strong incentives for individual initiative would benefit social welfare because they elicited new and useful improvements on the existing arts. Copyright policy, in direct contrast, recognized that the rights of "authors" might conflict with those of society to a greater extent. When that happened, the rights of the public were to be primary.[5]

A strategy that offered strong patents and weak copyrights accorded with the objective of the Constitution. First, potential consumers of cultural goods included most members of society, and restricted access had important and deleterious implications for the degree of education, equality, and democracy. Second, the rationale also held in terms of the incentive effects on authors and artists. For, even if internalization were less effective than for patents and the appropriation of private benefits was more limited, the statutory grant of weaker rights was likely to have a smaller disincentive effect for creators of copyrighted materials than for creators of useful inventions. Alternative methods of compensating contributions to cultural goods might be more efficient than the grant of exclusive rights. The elasticity of supply was high for many cultural goods whose providers were motivated by the prospect of celebrity or reputation, in which case no additional monetary incentive might be needed. "Authors" could obtain returns from complementary goods,

---

[4] See Merrill Jensen (ed.), *The Documentary History of the Ratification of the Constitution: Ratification of the Constitution by the States, Pennsylvania*, Madison: State Historical Society of Wisconsin, vol. 2, 1976, p. 454.

[5] See *Myers v. Callaghan*, 5 F. 726 (1881): "It is the object of the act of congress to 'secure' the right which thus primarily exists. Indeed, the statutes of copyright seem to imply the existence of a natural right of the author to the product of his brain. . . . It seems to me, on the contrary, that these various provisions of law in relation to copyright should have a liberal construction, in order to give effect to what may be considered the inherent right of the author to his own work."

rather than from the copyright monopoly per se. Third, when significant network externalities were present, as in the case of knowledge and cultural inputs, weaker enforcement of property rights might lead to improvements in social welfare.[6] As one supporter of weaker copyrights put it: "Books, it is quaintly said, sell one another. Every book that is read makes a market for more even of the same character."[7] Finally, copyright-like outcomes could be created through private bilateral means such as contract or price discrimination, or through alternative doctrines such as unfair competition, which led to lower social costs than a tax on all users. Thus, economic considerations meshed with a democratic focus on users rather than the elite class of producers of cultural goods.

This calculus led to the conclusion that copyright grants should be more narrowly interpreted and enforced less strictly than patent rights in order to promote social welfare. Senator Ruggles, who had been instrumental in reforming and strengthening patent rights in 1836, regarded copyright as a tax on the public. He declared that "The multiplication of cheap editions of useful books, brought within the reach of all classes, serves to promote that general diffusion of knowledge and intelligence, on which depends so essentially the preservation and support of our free institutions."[8] As will be seen in the next chapter, during the period when the United States was itself a developing country, it regarded widespread copyright "piracy" of foreign materials as international fair use. In the realm of domestic copyrights, free access to copyright material was also upheld when it qualified as fair use, to a greater extent than was true in any other country. In more recent years the social consequence of copyright reemerged as a key policy issue, and some scholars (one of whom is now a Supreme Court Justice) have proposed that only an "uneasy case" can be made for copyright protection.[9] European commentators are much less likely to question the award of strong property rights to authors, for reasons that have their origins in economic history.

THE EVOLUTION OF EUROPEAN COPYRIGHT POLICIES

Like patents, the grant of book privileges originated in the Republic of Venice in the fifteenth century, a practice that was soon prevalent in a number of

---

[6] Lisa Takeyama, "The Welfare Implications of Unauthorized Reproduction of Intellectual Property in the Presence of Demand Network Externalities," *Journal of Industrial Economics*, vol. 42 (2) 1994: 155–66, finds that unauthorized reproduction of intellectual property in the presence of demand network externalities can increase social welfare, by allowing firms to price discriminate across consumers.

[7] Report to accompany Senate Bill No. 32, 25th Congress, 2d Session, June 25, 1838.

[8] Report to accompany Senate Bill No. 32, 25th Congress, 2d Session, June 25, 1838.

[9] Stephen Breyer, "The Uneasy Case for Copyright," 84 *Harvard Law Review* 281 (1970), p. 284, concludes that "although we should hesitate to abolish copyright protection, we should equally hesitate to extend or strengthen it."

other European countries. In the early years of printing, books and other written matter became part of the public domain when they were published. Donatus Bossius, a Milan author, petitioned the duke in 1492 for an exclusive privilege for his book, and successfully argued that he would be unjustly deprived of the benefits from his efforts if others were able to freely copy his work. He was given the privilege for a term of ten years. However, authorship was not required for the grant of a privilege, and printers and publishers obtained monopolies over existing books as well as new works. For instance, in 1479 three printers were given the exclusive right to print the breviary of the diocese of Wurzburg. Since privileges were granted on a case-by-case basis, they varied in geographical scope, duration, and breadth of coverage, as well as in terms of the attendant penalties for their violation. Grantors included religious orders and authorities, universities, political figures, and the representatives of the Crown.[10]

The French privilege system was introduced in 1498 and was well-developed by the end of the sixteenth century. Privileges were granted under the auspices of the monarch, generally for a brief period of two to three years, although the term could be as much as ten years. Applicants could obtain protection for new books or translations, maps, type designs, engravings, and artwork. Petitioners paid formal fees and informal gratuities to the officials concerned. Because applications could only be sealed if the king were present, petitions had to be carefully timed to take advantage of his route or his return from trips and campaigns. It became somewhat more convenient when the courts of appeal such as the Parlement de Paris began to issue grants that were privileges in all but name, although this could lead to conflicting rights if another authority had already allocated the monopoly elsewhere. The courts sometimes imposed limits on the rights conferred, in the form of stipulations about the prices that could be charged. Privileges were alienable property that could be bequeathed, assigned, or licensed to another party, and their infringement was punished by a fine and at times confiscation of the output of "pirates."

After 1566, the Edict of Moulins required that all new books had to be approved and licensed by the Crown. Favored parties were able to get renewals of their monopolies that also allowed them to lay claim to works that were already in the public domain. By the late eighteenth century an extensive administrative procedure was in place that was designed to restrict the number of presses and to engage in surveillance and censorship of the publishing and printing industry.[11] Manuscripts first had to be read by a

[10] This discussion of the early system of privileges follows Elizabeth Armstrong, *Before Copyright: The French Book-Privilege System, 1498–1526*, Cambridge: Cambridge University Press, 1990.

[11] See Raymond Birn, "The profits of ideas: Privilèges en librairie in eighteenth century France," *Eighteenth-Century Studies*, vol. 4 (2) 1970–1971:131–68; and Robert L. Dawson, *The*

censor, and only after a permit was requested and granted could the book be printed, although the permit could later be revoked if complaints were lodged by sufficiently influential individuals. Decrees in 1777 established that authors who did not alienate their property were entitled to exclusive rights in perpetuity. Because few authors had the will or resources to publish and distribute books, their privileges were likely to be sold outright to professional publishers. However, the law distinguished between authors' rights and the rights granted to publishers: if the copyright was sold, the privilege was only accorded a limited duration of at least ten years, the exact term to be determined in accordance with the value of the work, and once the publisher's term expired the work passed into the public domain. The fee for a privilege was 36 livres. Approvals to print a work, or a "permission simple" that did not entail exclusive rights also could be obtained after payment of a substantial fee. Between 1700 and 1789, a total of 2,586 petitions for exclusive privileges were filed, and about two thirds were granted.[12] The result was a system that resulted in "odious monopolies," higher prices and greater scarcity, large transfers to officials of the Crown and their allies, and pervasive censorship. It likewise disadvantaged smaller book producers, provincial publishers, and the academic and broader community.

Privileges also were prevalent in other areas of cultural output such as plays, musical compositions, and operas. In some instances, wealthy patrons subsidized the efforts of composers and artists, without constraining their rights or limiting access for competitors, but in others patronage was associated with monopolistic restrictions that extended through generations. Henri II granted a privilege for printing music in 1551 to Robert Ballard and his cousin Adrian le Roy. This family was able to maintain its monopoly rights even after the death of Henri II, and their hold on the music printing industry lasted for almost two hundred years.[13] An illustration of the way in which privileges limited cultural diffusion occurred in the realm of opera during seventeenth-century France. Louis XIV, in authorizing the founding of the Académie d'Opéra (Paris Opera) in 1669, initially gave the poet Pierre Perrin a twelve-year monopoly over all operatic performances in France. The director of the Paris Opera, Jean-Baptiste Lully, parlayed his influence with Louis XIV to extend the monopoly in perpetuity, and further garnered additional privileges such as limits on the number of musicians who could perform outside the Académie d'Opéra. He vigorously maneuvered to suppress competitors such as the Comédie Française. In 1672 Lully obtained sole publication rights for the opera's librettos and sold shares in the right

French Booktrade and the "permission simple" of 1777: Copyright and the Public Domain, Oxford: Voltaire Foundation, 1992.

[12] See Birn, p. 149.

[13] See Samuel F. Pogue's entry on "Ballard," in D. W. Krummel and Stanley Sadie (eds.), *Music Printing and Publishing*, New York: W. W. Norton & Company, 1990: pp. 159–63.

to several printers including the Ballards. Lully was able to make a fortune and bequeathed the privileges to his heirs.[14]

The French Revolutionary decrees of 1791 and 1793 ostensibly replaced privileges with uniform statutory claims to literary and cultural property, based on the principle that "the most sacred, the most unassailable and the most personal of possessions is the fruit of a writer's thought."[15] The subject matter of copyrights covered books, dramatic productions, and the output of the "beaux arts" broadly defined to include designs and sculpture. However, in the realm of inventions, the Revolution did not eliminate the institutional inequities of the *ancien régime*, and this also was true of restrictions in such other dimensions as the Paris Opera. In 1790, a Revolutionary commentator referred to the "privileges, so broad, so despotic, and in consequence so odious" of the Académie d'Opéra.[16] Nevertheless, far from abolishing the privileged position of the Paris Opera, after the Revolution the city councillors turned the Opera over to two businessmen and provided them with public subsidies to keep the enterprise solvent.

Just as the former trappings of monarchical privileges were replicated in opera and inventions, so, too, were privileges in books perpetuated, despite the Revolution. Some observers felt that early copyrights in Revolutionary France were "of all property rights the most humble and the least protected," because they were enforced with a care to protecting the public domain and social welfare.[17] For instance, authors were required to deposit two copies of their books with the Bibliothèque nationale or risk losing their copyright. Although France is associated with a strong author's rights approach to copyright and proclamations of the "droit d'auteur," these ideas evolved slowly and hesitatingly, mainly in order to meet the self-interest of the various

---

[14] See Henry Prunières, *La vie illustre et libertine de Jean-Baptiste Lully*, Paris: Librairie Plon, 1929. Ironically, although Lully is considered the father of French opera and influenced its movement away from the Italian style, he was himself an Italian immigrant who took French citizenship.

[15] Jane Ginsburg, p. 996, "A Tale of Two Copyrights: Literary Property in Revolutionary France and America," 64 *Tul. L. Rev.* 991 May, 1990, argues that "the principles and goals underlying the revolutionary French copyright regime were far closer to their U.S. counterparts than most comparative law treatments (or most domestic French law discussions) generally acknowledge. The first framers of copyright laws, both in France and in the U.S., sought primarily to encourage the creation of and investment in the production of works furthering national social goals. This study stops at the end of the Napoleonic era, substantially before the development of personalist doctrines, such as moral rights, by French copyright scholars and courts. These doctrines did provoke theoretical and practical divergences between the French and U.S. copyright regimes."

[16] This quotation is taken from Victoria Johnson, "Founding Culture: Art, Politics and Organization at the Paris Opera, 1669–1792." (Ph.D. Dissertation, Department of Sociology, Columbia University, 2002). Johnson's thesis supports the idea of institutional persistence, which she terms "imprinting," and examines the specific experience of the Paris Opera in the eighteenth century.

[17] E. Laboulaye, 1858, cited in Ginsburg, 1990, p. 1012.

members of the book trade.[18] During the *ancien régime*, the rhetoric of authors' rights had been promoted by French owners of book privileges as a way of deflecting criticism of monopoly grants and of protecting their profits, and by their critics as a means of attacking the same monopolies and profits. This language was retained in the statutes after the Revolution, so the changes in interpretation and enforcement may not have been universally evident.

By the middle of the nineteenth century, French jurisprudence and philosophy tended to explicate copyrights in terms of innate rights of personality. However, the idea of the moral claim of authors to property rights was not incorporated in the law until early in the twentieth century. The droit d'auteur first officially appeared in a statute of April 1910, which declared that "the alienation of a work of art does not imply, unless the contrary is explicitly stated, the alienation of the right of reproduction." In 1920 visual artists were granted "droits de suite" or a claim to a portion of the revenues from the resale of their works. Subsequent evolution of French copyright laws led to the recognition of the right of disclosure, the right of retraction, the right of attribution, and the right of integrity.[19] These "moral rights" were (at least in theory) perpetual and inalienable. Thus, they could be bequeathed to the heirs of the author or artist, regardless of whether or not the work was sold to someone else. The self-interested rhetoric of the owners of monopoly privileges had now emerged as the keystone of the "French system of literary property" that would shape international copyright laws in the twenty-first century. History has its ironies!

In England, as in France, copyright law began as a monopoly grant to benefit and regulate rights of guilds, and as a form of surveillance and censorship over public opinion on behalf of the Crown.[20] England similarly experienced a period during which privileges were granted, such as a seven-year grant from the Chancellor of Oxford University for an 1518 work. Charles II bestowed royal patents on Thomas Killigrew and Sir William Davenant in 1660, which allowed them to establish two theater companies, the King's Players and the Duke's Players. The Licensing Act of 1737 gave the Lord Chamberlain the right to censor all plays and to protect the monopoly of these two London theaters. A royal charter in 1557 created a more notorious and inveterate privilege that was bestowed on the Worshipful

---

[18] Russell J. DaSilva outlines the two systems in "Droit Moral and the Amoral Copyright: A Comparison of Artists' Rights in France and the United States," 28 *Bulletin of the Copyright Society* 1 (1980).

[19] The "droit de divulgation" or a publication right; "droit de retrait ou de répentir" or a right to retract or modify the work; the right of integrity or "droit au respect de l'oeuvre" is the right to prevent alteration of the work; and the "droit de la paternité" is the right to be known as the creator.

[20] See Michael Rushton, "The Moral Rights of Artists: Droit Moral ou Droit Pécuniaire?" *Journal of Cultural Economics*, 22 (1) 1998:15–32.

Company of Stationers. This publishers' guild controlled the book trade for the next two hundred years. The Stationers' Company created and regulated the right of their constituent members to make copies, so in effect their "copy right" was a private property right that existed in perpetuity, independently of state or statutory rights. The Company itself monitored and enforced its monopoly through its Court of Assistants, which maintained a register of books, issued licences, and sanctioned individuals who violated their regulations.

The English system of privileges was replaced in 1710 by a copyright statute (the "Statute of Anne"). This statute "wholly ignored the authors of books, and certainly was not intended to confer any additional rights on them."[21] Rather, the legislation was a response to concerns about the power wielded by the Stationers' Company. Its promoters intended to restrain the power of the publishing industry and destroy its monopolistic structure. According to the law, the grant of copyright should be made available to anyone, not just to the Stationers' Company. Instead of a perpetual right, the copyright term was limited to fourteen years, with a right of renewal, after which the work would enter the public domain. Price control measures could be introduced if any person brought a complaint that prices were "too high and unreasonable." The statute also stipulated that copies of publications would have to be lodged with nine specific libraries. These institutions were hardly accessible to the masses, including the royal library, and the universities of Edinburgh, Oxford and Cambridge. During the eighteenth century, Cambridge University tended to promptly sell off many of the books that the publishers deposited with them, because the university librarians viewed the titles as insufficiently learned and too crassly commercial.[22]

Subsequent litigation and judicial interpretation added a new and fundamentally different dimension to copyright. Just as in France, publishers tried to promote the idea that copyright was based on the natural rights of authors or creative individuals, in order to protect their perpetual copyright. The recognition of natural rights benefited the publishers because, as the agent of the author, those rights would devolve to the publisher. Moreover, if copyrights derived from these innate rights, they represented property that existed independently of statutory provisions and could be protected at common law in perpetuity. The booksellers engaged in a series of strategic litigation that

[21] John Feather offers an extensive account in *Publishing, Piracy, and Politics: An Historical Study of Copyright in Britain*, New York: Mansell, 1994, p. 64. The English copyright statute was entitled "An Act for the Encouragement of Learning, by Vesting the Copies of Printed Books in the Authors or Purchasers of Such Copies, During the Times Therein Mentioned," 1709-10, 8 Anne, ch. 19.

[22] The 1842 copyright statute stipulated that the British Museum would be given a copy of any book printed in the United Kingdom within three months of its publication, and within a year if published in the British dominions overseas.

culminated in their defeat in the landmark case, Donaldson v. Beckett, 98 Eng. Rep. 257 (1774). The court ruled that authors held a common-law right in their unpublished works, but on publication that right was extinguished by the statute, whose provisions determined the nature and scope of any copyright claims. This transition from publisher's monopoly rights to statutory author's rights implied that copyright had transmuted from a straightforward license to protect monopoly profits into a fully fledged property right whose boundaries would henceforth increase at the expense of the public domain.

Between 1735 and 1875 fourteen Acts of Parliament amended the initial copyright legislation. Copyrights extended to sheet music, maps, charts, books, sculptures, paintings, photographs, dramatic works and songs delivered in a dramatic fashion, and lectures outside of educational institutions. Copyright owners had no remedies at law unless they complied with a number of stipulations that included registration, the payment of fees, the delivery of free copies of every edition to the British Museum (delinquents were fined), as well as complimentary copies for four libraries, including the Bodleian and Trinity College. The ubiquitous Stationers' Company administered registration, and the registrar personally benefited from the monetary fees of 5 shillings when the book was registered and an equal amount for each assignment and each copy of an entry, along with 1 shilling for each entry searched. Foreigners could obtain copyrights if they presented themselves in a part of the British Empire at the time of publication. The book had to be published in the United Kingdom, and prior publication in a foreign country – even in a British colony – was an obstacle to copyright protection. Only the copyright holder and his agents were allowed to import the protected works into Britain. The term of the copyright in books was for the longer of forty-two years from publication or the lifetime of the author plus seven years, and after the death of the author a compulsory license could issue to ensure that works of sufficient public benefit would published. The "work for hire" doctrine, which vested in employers copyrights to items that employees created in the course of employment, was in force for books, reviews, newspapers, magazines, and essays, unless a distinct contractual clause specified that the copyright was to accrue to the author.

The statute of Anne did not succeed in creating a competitive publishing industry, so English judges used the common law to regulate the monopoly of copyright owners. One of the most notable doctrines allowed unauthorized use of a publication for the purposes of "fair use."[23] Fair use at this time related to whether or not the work of an author interfered with the copyright of a second author or publisher, and therefore was analogous to doctrines of unfair trade. Many of the early cases dealt with abridgments of an original

---

[23] The best treatise on the development of fair use is William Patry, *The Fair Use Privilege in Copyright Law*, Washington, D.C.: Bureau of National Affairs, 1995.

copyrighted work.[24] Fair abridgments involved books that were sufficiently distinct from the original work, and abridgers who did not merely intend to evade the copyright statutes by introducing cosmetic changes to the original. Courts drew on different rationales for deciding whether or not a case involved fair use, including intent, productive or transformative use, whether the alleged infringement would substitute for the original, and socially valuable purposes such as for reviews or criticism. However, it was impossible to draw up bright line rules that determined whether a subsequent work drew fairly on a prior publication.[25]

A British Commission that reported on the state of the copyright system in 1878 felt that the laws were "obscure, arbitrary, and piecemeal" and were compounded by the confused state of the common law.[26] The piecemeal nature of the numerous laws that were simultaneously in force led to conflicts and unintended defects in the system. For instance, it resulted in a peculiar version of the first sale doctrine in which, if a painting or photograph was sold without any written contractual allocation of the copyright, neither party retained the copyright and it was lost altogether. Some of the penalties were disproportionate when applied outside the context to which they had originally been levied. A £2 fine originally directed toward dramatic performances was applied to each dramatic song performed without permission at nonprofit events, giving some enterprising individuals the incentive to purchase powers of attorney from composers in order to make a profession of pursuing amateur performers and successfully collecting the fine on the spot. The report discussed, but did not recommend, an alternative to the grant of copyrights, in the form of a royalty system in which "any person would be entitled to copy or republish the work on paying or securing to owner a remuneration, taking the form of royalty or definite sum prescribed by law." The main benefit would be to allow the public to have early access to cheap editions, whereas the main cost would be to the publishers whose risk and return might be negatively affected.

The Commission noted that the implications of copyrights for the colonies were "anomalous and unsatisfactory." Publishers in England practised price

[24] "If I should extend the rule so far as to restrain all abridgments, it would be of mischievous consequence, for the books of the learned...would be brought within the meaning of this act ..." – *Giles v. Wilcox*, 2 Atk. 141, 3 (1740).

[25] "No certain line can be drawn, to distinguish a fair abridgment; but every case must depend upon its own circumstances" – *Dodsley v. Kinersley* (1761).

[26] According to a British Commission appointed in 1878, "The law is wholly destitute of any sort of arrangement, incomplete, often obscure, and even when it is intelligible upon long study, it is in many parts so ill-expressed that no one who does not give such study can expect to understand it...the piecemeal way in which the subject has been dealt with affords the only possible explanation of a number of apparently arbitrary distinctions between the provisions made upon matters which would seem to be of the same nature." G. H. Putnam (ed.), *Question of Copyright*, New York and London: G. P. Putnam's son, 1896, p. 213.

discrimination, modifying the initial high prices for copyrighted material through discounts given to reading clubs, circulating libraries and the like, benefits which were not available in the colonies. In 1846 the Colonial Office acknowledged "the injurious effects produced upon our more distant colonists" and passed the Foreign Reprints Act in the following year. This allowed the colonies that adopted the terms of British copyright legislation to impose a tariff of 12.5 percent on their imports of cheap reprints of British copyrighted material. The proceeds of the tariff would be distributed among the copyright owners. However, enforcement of the tariff seems to have been less than vigorous because, over the period from 1866 to 1876, only £1,155 was received from the nineteen colonies that took advantage of the legislation (£1,084 from Canada, which benefited significantly from the American reprint trade.) The Canadians argued that it was difficult to monitor imports, so it would be more effective to allow them to publish the reprints themselves and collect taxes for the benefit of the copyright owners. This proposal was rejected, but under the Canadian Copyright Act of 1875 British copyright owners could obtain Canadian copyrights for Canadian editions that were sold at much lower prices than in Britain or even in the United States.

The Commission made two recommendations regarding colonial copyrights. First, the bigger colonies with domestic publishing facilities should be allowed to reprint copyrighted material on payment of a license to be set by law. Second, the benefits to the smaller colonies of access to British literature should take precedence over lobbies to repeal the Foreign Reprints Act, which should be better enforced rather than removed entirely. Some had argued that the public interest required that Britain should allow the importation of cheap colonial reprints since the high prices of books "are altogether prohibitory to the great mass of the reading public" but the Commission felt that this should be adopted only if the copyright owner consented. They also devoted a great deal of attention to what was termed the "American Question" but took the "highest public ground" and recommended against retaliatory policies. As Chapter 9 shows, the "American Question" in the realm of copyright was to remain unresolved until late in the twentieth century.

This discussion of the French and English systems has distinguished between two major systems of copyright, both of which initially were primarily viewed in the context of the rights of publishers. The civil law system of privileges ultimately asserted that the author has a moral right or natural right in his artistic creation, which extends beyond the sale of the item, potentially in perpetuity.[27] This system of personal or innate rights, at least at the rhetorical level, highlighted the concerns of authors, and only

---

[27] For instance, see Jane Ginsburg, "A Tale of Two Copyrights: Literary Property in Revolutionary France and America," 64 *Tul. L. Rev. 991*, May 1990.

incidentally attended to the consequences for society in general. It provided a contrast to the Anglo-American system, which was more concerned with the economic bargain underlying the limited grant of a monopoly to authors and their assignees in exchange for the improvement of social welfare from the products of their efforts. The welfare aspect of copyright as a means to "encourage learning" was increasingly emphasized in eighteenth-century England at the expense of publishers' rights. However, technological changes improved the profits of printers and publishers, and encouraged authors to join the publishers' lobby for stronger recognition of their rights in order to gain a larger share of the commercial gains.[28] Ironically, the English system would find its most faithful representation in the United States, especially after Britain ratified the 1886 Berne Convention and amended its laws in line with copyright practices in Continental Europe.

COPYRIGHT IN THE UNITED STATES

In the period before the Declaration of Independence, individual American colonies recognized and promoted patenting activity, but copyright protection was not considered to be of equal importance, for a number of reasons. To a new country, pragmatic concerns were likely of greater importance than the arts, and markets were sufficiently narrow that an individual could saturate the market with a first run printing. Local publishers produced ephemera such as newspapers, almanacs, and bills, whereas the more substantial literary works were imported. Moreover, a significant fraction of output was devoted to works such as medical treatises and religious tracts whose authors wished simply to maximize the number of readers, rather than the amount of income they received. Protection was granted to literary works on an individual, ad hoc basis, until a few writers including John Ledyard and Noah Webster lobbied for the passage of general copyright statutes.

In 1783, Connecticut became the first colony to approve an "Act for the encouragement of literature and genius" because "it is perfectly agreeable to the principles of natural equity and justice, that every author should be secured in receiving the profits that may arise from the sale of his works, and such security may encourage men of learning and genius to publish their writings; which may do honor to their country, and service to mankind."[29] Although this preamble might seem to strongly favor author's natural rights,

---

[28] Technological changes in nineteenth-century printing included the use of stereotyping which lowered the costs of reprints, improvements in paper-making machinery, and the advent of steam powered printing presses. Graphic design also benefited from innovations, most notably the development of lithography and photography. The number of new products also expanded significantly by the end of the century, encompassing recorded music and moving pictures.

[29] *Copyright Enactments of the United States, 1783–1906*, compiled by Thorvald Solberg, Washington, D.C.: Library of Congress, 1906, p. 11.

the language merely reflected eighteenth-century rhetoric, for the rules and standards were certainly not consistent with the theory of natural rights. The statute specified that books were to be offered at reasonable prices and in sufficient quantities, or else a compulsory license would issue.[30] Moreover, rights were only extended to residents of Connecticut and to residents of other states that offered reciprocal benefits. Similar statutes were enacted in Massachusetts, Maryland, New Jersey, New Hampshire, and Rhode Island in the same year; Pennsylvania and South Carolina in the following year; Virginia and North Carolina in 1785; and Georgia and New York in 1786.

The first federal copyright statute was approved on May 31, 1790, "for the encouragement of learning, by securing the copies of maps, charts, and books to the authors and proprietors of such copies, during the times therein mentioned." The lack of distinction between authors and proprietors is a clear refutation of copyright as an innate right, and this is also true of the stipulations that had to be fulfilled before the copyright was regarded as valid. The copyright act required authors or proprietors to deposit a copy of the title of the work at the district court in the area where they lived, on payment of a nominal fee of 60¢. Registration secured the right to print, publish and sell maps, charts and books for a term of fourteen years, with the possibility of an extension for another fourteen year term.[31] It was felt that copyright protection would serve to increase the flow of learning and information: publication would contribute to democratic principles of free speech, and the diffusion of knowledge also would ensure broad-based access to the benefits of social and economic development for both producers and consumers.[32] In this spirit, the Act authorized piracy of foreign works, for it specified that "nothing in this act shall be construed to extend to prohibit the importation or vending, reprinting or publishing within the United States, of any map, chart, book or books . . . by any person not a citizen of the United States."[33]

[30] See *Patten v. Goodwin*, 1 Root 172 (1790), a case brought in the superior court of Hartford County, Connecticut against Goodwin, who owned the copyright for *Webster's institute of English grammar*. The charge was that he "did not supply the public with a sufficient number of said books; and that he sold them at an exorbitant price."

[31] In 1790 the duration of copyright protection comprised fourteen years from registration, with the possibility of renewal for a further fourteen years; after 1831 the maximum term was twenty-eight years from time of registration with the right of renewal for fourteen years, whereas the 1909 statute allowed twenty-eight years plus extension for a further twenty-eight years if the author were still alive. See appendix.

[32] Some senators argued that "literature and science are essential to the preservation of a free Constitution.'" U.S. Senate Journal, 1st Cong. 8–10; U.S. Annals of Congress, 1st. Cong. 935–36.

[33] Original Copyright Act, First Congress, Second Session, Chapter 15, May 31, 1790: "An Act for the encouragement of learning, by securing the copies of maps, charts, and books, to the authors and proprietors of such copies, during the times herein mentioned." See *Copyright Enactments of the United States, 1783–1906*, compiled by Thorvald Solberg, Washington, D.C.: Library of Congress, 1906.

Table 8.1. *U.S. Copyright Registrations, 1790–1800*

	1790–1795		1796–1800		1790–1800	
	No.	Percent	No.	Percent	No.	Percent
Identity of Filer						
Author	157	56.5%	240	51.5%	397	53.4%
Other Proprietor	121	43.5	226	48.5	347	46.6
Subject						
Atlases, maps, navigation	23	8.1	34	7.0	57	7.4
Biography	11	3.9	31	6.3	42	5.4
Commerce	16	5.7	24	4.9	40	5.2
Directories and dictionaries	20	7.1	14	2.9	34	4.4
Law	16	5.7	27	5.5	43	5.7
Music, poetry, and plays	25	8.8	61	12.5	86	11.1
Novels and fiction	4	1.4	15	3.1	19	2.5
Religion and philosophy	41	14.5	49	10.0	90	11.7
Science and medical	23	8.1	56	11.4	79	10.2
Social and political	43	15.2	62	12.7	105	13.6
Textbooks	45	15.9	89	18.2	134	17.4
Miscellaneous nonfiction	16	5.7	27	5.5	43	5.6
Total	283	100.0	489	100.0	772	100.0
Location						
Kentucky	0	0.0	1	0.2	1	0.1
Massachusetts	92	32.3	107	21.7	199	26.3
Maine	4	1.4	2	0.4	6	0.8
North Carolina	0	0.0	4	0.8	4	0.5
New Hampshire	13	4.6	17	3.4	30	4.0
New York	35	12.3	69	14.0	84	11.1
Pennsylvania	117	41.1	146	29.6	263	34.8
Rhode Island	9	3.2	6	1.2	15	2.0
South Carolina	4	1.4	10	2.0	14	1.8
Virginia	6	2.1	8	1.6	14	1.8
Vermont	3	1.1	0	0.0	3	0.4
Dept. of Interior/State Dept.	2	0.7	121	24.5	123	16.3
Total	285	100.0	494	100.0	756	100.0

Source: Federal Copyright Records, 1790–1800, James Gilreath (ed.), Library of Congress, Washington, D.C., 1987.

As Table 8.1 indicates, in the first decade after the enactment of the statute almost a half of all copyrights were issued to "proprietors" such as publishers, rather than authors. The first U.S. copyright was obtained in June 1790, when John Barry registered his spelling book in the District Court of Pennsylvania. The majority of copyrights similarly related to practical items such as textbooks, dictionaries, atlases, and tables for measuring longitude. Religious works rapidly fell in importance even during this early period. Mrs. Mercy Warren, the first woman to obtain a federal copyright, registered her *Poems dramatic & miscellaneous* in the District Court of Massachusetts in October 1790.[34] Charles Evans's bibliography included some thirteen thousand items that were published during the same period, indicating that the majority of early authors did not apply for copyright protection. However, filings increased at a rapid rate, from 2,212 between 1796 and 1831, to 10,073 in the following decade, and 40,000 in the period between 1841 and 1859.[35] By 1870, when registration was rationalized in one office at the Library of Congress, approximately 150,000 entries had been lodged. These included the first filings in Ohio (1806), Illinois (1821), Michigan (1824), California (1851), and Colorado (1864). Copyright records included icons in American literature such as Harriet Beecher Stowe's *Uncle Tom's Cabin*, which was registered in the District Court of Maine in May 1851. However, the majority of copyrights related to items other than books, such as plays, musical compositions, and maps.

Annual data on American copyright registrations begin in 1871, when 12,688 registrations were recorded by the Library of Congress. The annual count of items registered steadily increased. Part of the reason for the growth in registration was because of legislative policies, which continually expanded the scope of copyright protection. Amendments to the original statute (see appendix) extended protection to many other works including musical compositions, performances, engravings, and photographs. Thus, by the end of the century, "writings" were held to encompass dance choreography, sculpture, chromos, furniture designs, ticker tape, movies, and piano rolls on organettes. Legislators refused to grant perpetual terms to copyright holders, but the length of protection was extended in general revision of the laws in 1831 and 1909 and subsequently. In addition, the judiciary extended the scope of protection under alternative doctrines such as unfair competition and trade secret laws, to cover property such as stock market quotations. Nevertheless, American copyright grants remained among the least generous in the world in terms of scope, duration, and interpretation at law.

Statistics on copyrights per se contain little systematic information that is of interest to economic historians, for a number of reasons. First, because

---

34 District Court Ledger, 43 MA 3, October 23, 1790. The second woman to copyright a book under the Federal law was Mrs Hannah Adams; on July 6, 1791 she filed for the copyright on *A View of Religion in Two Parts* [District Court of Mass., 43 MA 6]. See *Federal Copyright Records, 1790–1800*.

35 *Annual Report of the Commissioner of Patents for 1861*, Washington, D.C.: Government Printing Office, 1862.

all copyrights are granted on application, the variance of copyright grants in terms of value is much greater than for patents. Second, the multiplicity and variety of items that may be copyrighted imply that a simple aggregation is unlikely to be meaningful. It is unclear what copyright counts signify because they encompass maps, songs, music, paintings, poems, illustrations, encyclopedias, plays, movies, and items such as lamp stands, altarpieces, and commemorative porcelain plates. Third, copyrights are not exclusive, because there is no presumption of novelty or incremental improvement over the state of the art. Instead, there is substantial duplication in copyrighted material, as any novel reader or movie fan realizes. Thus, empirical analysis of aggregate copyright registrations is likely to be of little value. Litigation records about copyrights, by contrast, can yield insights into the relationship between copyright, markets, and social welfare.

COPYRIGHT LITIGATION

Reported copyright disputes addressed questions about the balance between individual rights, economic objectives of increasing growth and welfare, and the prerequisites for political democracy. However, as Figure 8.1 shows, few conflicts were recorded in the formal legal system in the antebellum period. The writers of religious tracts, or newspapers, pamphlets, and other ephemera that predominated in early American society, had little incentive to pursue either copyrights or infringers.[36] Authors, publishers, and other copyright holders also might have had alternative, more effective means of sanctioning infringement of their rights. The early publishing industry was a small and close-knit community, in which infringement was easy to detect and prosecute privately. For instance, when the *Transcript*, a Boston newspaper, began to publish passages from Miss Kemble's *Journal*, her publishers wrote a stern letter in 1835 warning that such an action was "not only a violation of courtesy but of copyright," but did not pursue the matter to court.[37] As the chapter on patent litigation indicated, the attitude of the courts toward intellectual property was extremely favorable and enforcement was predictable. This may have created an environment that encouraged the private settlement of disputes. It is therefore not surprising that fewer than eight hundred copyright disputes were brought before the courts between 1790 and 1909. Although copyright litigation is by no means representative, it still gives us valuable insights into the market for copyrighted

[36] Piracy by traveling groups performing in small Midwestern towns provides another example of infringement that was unlikely to be prosecuted, since "it is difficult and in many cases impossible to serve him with injunctions and court orders, because of his migratory habits," and "he is in almost every instance entirely without attachable means" so the copyright owner was unlikely to benefit from legal action. H.R. No. 1191–53, p. 2 (1894).

[37] David Kaser, *Messrs. Carey & Lea of Philadelphia*, Philadelphia: University of Pennsylvania Press, 1957, p. 60.

material, the views of judicial policy makers, and changes in copyright over time.

Table 8.2 shows the distribution of a sample of copyright lawsuits decided between 1790 and 1909, excluding *ex parte* cases. The first panel illustrates the regional distribution of some 654 cases over this period. The majority of disputes were filed in New York and Pennsylvania, and New York accounted for a constant share of almost half of all copyright cases, reflecting its importance as a center of publishing and commerce. In the early period (the first column), it is instructive to compare the Midwest and the South, two regions of equivalent population size but different levels of literacy, freedom, and equality of income. As in many measures of technological capability, the South reveals a rather low prevalence of copyright registrations, book sales, and marketing of printed materials. Southerners preferred English authors, they purchased few (but expensively bound) books, and sales of printed matter were concentrated in urban regions – all indicators of the lack of a mass market. In 1856 the leading publishing company of Ticknor and Fields sold over $10,000 worth of books in Cincinnati, an amount that was equal to its sales in the entire South.[38] The company spent more on advertising in the city of New York in 1856 than it did for all of the states in the South.[39] Moreover, the latter region accounted for less than five percent of total U.S. book production in the same period. The evidence on litigation therefore reflects the absence of expansive markets and commerce in printed materials in the South, an absence that was likely to have been both a cause and effect of the corresponding inequality and lack of democracy.

As discussed earlier, a system of author's rights holds the product to be an extension of the creator's personality, and rights such as those of identity and paternity survive the alienation of the product. However, according to the American doctrines on copyright ownership, authors and artists were acknowledged to possess only a pecuniary or economic interest. Thus, the force and protection of copyright law were not linked to the identity of the authors or their creativity but simply directed toward the legal owner of the copyright regardless of who that person or institution might be. Table 8.2 shows the identity of plaintiffs, and here a notable result is that the fraction of copyright plaintiffs who were authors (broadly defined) was initially quite low, and fell continuously during the nineteenth century. By 1900–1909, only 8.6 percent of all plaintiffs in copyright cases were the creators of the item that was the subject of the litigation. Instead, by the same period,

---

[38] See Warren Tryon, "Publications of Ticknor and Fields in the South, 1840–1865," *Journal of Southern History*, vol. 14, no. 3. (August 1948): 305–30. Tryon argues that the state of the Southern book trade owed more to its lack of railroad transportation and manufacturing than to social factors such as slavery.

[39] Ronald Zboray, "The Transportation Revolution and Antebellum Book Distribution Reconsidered," *American Quarterly*, vol. 38, no. 1 (Spring, 1986): 53–71, especially p. 63.

Table 8.2. *Copyright Litigation: Percentage Distribution of Reported Cases,*
1790–1909

	1790–1879	1880–1889	1890–1899	1900–1909	All
A) Regional Distribution of Cases					
Mid-Atlantic	57.2	54.9	61.3	63.2	59.8
Midwest	15.1	14.8	11.3	10.4	12.5
New England	18.4	7.4	9.5	12.7	12.2
South	4.0	5.7	9.5	2.4	6.1
West	2.0	6.6	4.8	3.8	4.1
Supreme Court	3.3	10.7	3.6	7.6	6.1
Total					
Number	152	122	168	212	654
Percent	23.2	18.7	25.7	32.4	100
B) Identity of Plaintiffs					
"Authors"					
Book author	24.3	20.9	18.8	8.6	17.3
Artist	6.3	0	2.9	6.6	4.3
Composer	1.8	0	0.7	1.3	1.0
Mapmaker	4.5	2.3	1.5	0	1.9
Photographer	0	3.5	9.4	4.0	4.5
Playwright	9.9	2.3	6.5	3.3	5.6
Assignees					
Producer	7.2	7.0	5.1	7.3	6.6
Publisher	21.6	37.2	42.0	52.3	39.7
Other	9.9	11.6	2.9	2.0	5.6
Other					
Manufacturer	2.7	3.5	3.6	3.3	3.3
Financial	1.8	0	0.7	4.0	1.8
Other	9.9	11.6	5.8	7.3	8.2
Total					
Number	111	86	138	151	486
Percent	22.8	17.7	28.4	31.1	100
C) Subject of Copyright Infringement					
Printed Matter					
General books	26.2	23.1	24.1	27.1	25.9
Dictionaries and Directories	0.7	3.3	10.2	8.7	6.2
Law Books and Reports	12.6	8.2	3.6	8.2	8.1
Maps	10.6	2.5	1.2	1.4	3.7
Other	7.2	2.5	6.0	2.4	4.4
Music and Plays					
Music and Opera	4.6	11.6	3.6	6.2	6.2
Plays	18.5	6.6	12.6	8.6	11.6

(continued)

Table 8.2 (*continued*)

	1790–1879	1880–1889	1890–1899	1900–1909	All
C) Subject of Copyright Infringement (*continued*)					
Visual Arts					
Chromos and Pictures	7.3	10.0	4.8	6.7	7.0
Paintings	0.0	2.5	1.8	10.6	4.3
Photographs	0.0	9.1	10.2	6.7	6.5
Other					
Advertisements	2.7	5.0	1.8	0.5	2.2
Information and Ideas	0.7	3.3	5.4	7.2	4.4
Titles and Trademarks	2.0	9.1	3.0	3.4	4.0
Miscellaneous	6.0	3.3	8.4	3.4	5.3
Total					
Number	151	121	166	208	646
Percent	23.4	18.7	25.7	32.2	100
D) Issues Litigated					
Abandonment	10.7	10.5	6.4	7.2	8.3
Compliance with statutes	8.9	9.3	14.3	6.5	9.8
Contract	9.8	11.6	6.4	5.3	7.7
Ownership/Originality	34.8	24.4	22.1	25.5	26.5
Remedies	2.7	9.3	9.0	11.1	8.3
Scope of copyright protection	10.8	19.7	17.8	19.6	17.2
Unfair competition	9.8	10.5	12.9	7.8	10.2
Other	12.5	4.7	10.7	17.0	12.0
Total					
Number	112	86	140	153	491
Percent	22.8	17.5	28.5	31.2	100

Notes and Sources: Decisions of the United States Courts involving Copyright and Literary Property, 1789–1909 (4 vols.), Washington, D.C.: Library of Congress, 1980. The sample included all copyright cases in the first two volumes, which listed cases in alphabetical order. I excluded *ex parte* cases and those that dealt with peripheral issues such as customs claims and the price of postage for printed materials.

the majority of parties bringing cases were publishers and other assignees of copyrights. Copyright enforcement was therefore largely the concern of commercial interests, and not of the creative individual. Indeed, some courts repudiated the focus on authors, such as in the case involving two telegraph companies, *National Telegraph News Co. v. Western Union Telelegraph*

Co.: "Is the enterprise of the great news agencies, or the independent enterprise of the great newspapers, or of the great telegraph and cable lines, to be denied appeal to the courts, against the inroads of the parasite, for no other reason than that the law, fashioned hitherto to fit the relations of authors and the public, cannot be made to fit the relations of the public and this dissimilar class of servants? Are we to fail our plain duty for mere lack of precedent? We choose, rather, to make precedent – one from which is eliminated, as immaterial, the law grown up around authorship."[40] Assignees were granted the same rights as authors, ensuring that trade in copyrights was unhindered by uncertainty about enforcement.

The rejection of innate author's rights in American copyright laws shaped a number of important doctrines, which had the effect of facilitating market transactions and arguably increased public welfare relative to an author's rights system. It was evident to the judiciary that "the primary object of this provision is to promote the progress of science and useful arts, thereby benefiting the public, and as a means to that end, and as a secondary object, to secure exclusive rights to authors."[41] Markets become most efficient when trades are anonymous and divorced from the identities of the parties on either side of the bargain, and when prices reflect present discounted values adjusted for systematic risk rather than idiosyncratic factors. The American approach to copyrights was thus likely to enhance both of these aspects of market efficiency relative to a system that granted authors moral rights or rights of personality.

The market orientation of American copyright is illustrated in Table 8.2 by the subject matter of items that were the focus of litigation. American democracy has long been noted (or faulted) for promoting a public taste that values the practical over the decorative or artistic. Indeed, according to Gilreath, "The constitutional copyright provisions' emphasis on the useful arts sought not to bolster a professional literary establishment of novelists, poets, and critics such as the one that existed in England but rather to ensure that books with demonstrably practical benefits to American society would be available to the readers of the new Republic."[42] The copyright registration in Table 8.1 showed that nonfiction and items of low creativity predominated

---

[40] 119 F. 294 (1902).

[41] *Koppel v. Downing*, 24 Wash. L. Rep. 342 (1897). According to a report to Congress, "The enactment of copyright legislation by Congress under the terms of the Constitution is not based upon any natural right that the author has in his writings,... but upon the ground that the welfare of the public will be served and progress of science and useful arts will be promoted... not primarily for the benefit of the author, but primarily for the benefit of the public...." H.R. Report No. 2222, 60th Cong., 2d Sess. 7 (1909).

[42] James Gilreath, p. xxiii, *Federal Copyright Records*. See also *J. L. Mott Iron Works v. Clow et al.*, 82 F. 316 (1896): "The object of that provision was to promote the dissemination of learning, by inducing intellectual labor in works which would promote the general knowledge in science and useful arts.... It sought to stimulate original investigation, whether in literature, science, or art, for the betterment of the people, that they might be instructed and improved with respect to those subjects."

among early copyrighted material. Table 8.2 similarly indicates that litigation continually surrounded items with a low degree of creativity, such as maps, atlases, legal treatises, law reports, dictionaries, and directories, and compilations of data. Utility and learning, rather than creativity, comprised the central focus of property rights in cultural products. When judges ignored this fundamental principle of the practical, market-orientation of American copyright, it led to flawed conclusions, such as a decision that photographs could not be protected because they were not "creative" products.[43]

A number of cases addressed whether strictly commercial items such as advertisements, or "mere" compilations were worthy of copyright protection. For instance, the American Trotting Register Association kept a list of racehorses that had made times of two minutes, thirty seconds or better. The company brought an infringement suit against the publishers of a yearbook that had used their data, and Judge Ricks found that the defendant "availed himself of the industry of the complainant, and has used the tables which it compiled at great expense and labor. This the defendants certainly cannot do under the law."[44] In *Bleistein v. Donaldson Lithographing Co.*, the lower court held that show bills to advertise the circus were "merely frivolous . . . tawdry pictures," and as such did not deserve the protection of the statutes that were designed to promote science and the useful arts. On appeal to the Supreme Court, Justice Holmes delivered the majority opinion that "if they command the interest of any public, they have a commercial value . . . and the taste of the public is not to be treated with contempt." He also suggested that the very fact that property rights were infringed indicated that they had value and were worth protecting, a doctrine prevalent in early patent decisions.[45]

Table 8.2 also reveals an increase in the proportion of cases that dealt with property rights in information. In 1829, the Circuit Court for the district of New York considered a case brought by Edwin B. Clayton and his associates, for protection from infringement of their publication of current stock prices and a review of the state of the market.[46] They lost the case because it was felt that information on fluctuating daily prices did not contribute to learning and

---

43 *Wood v. Abbott*, 30 F. Cas. 424 (1866): Plaintiffs were photographers who made photographs of pictures. The judge rejected the plea against infringers because "that combination of creative or imitative power and mechanical skill by which the artist works out his own conception . . . the fruits of which the law was intended to protect, is not brought into play." However, in the same year, in *Rossiter v. Hall*, 20 F. Cas. 1253 (1866) it was ruled that a copy by photography did infringe, because "to hold otherwise, would work a substantial repeal of the copyright laws in many cases."

44 *American Trotting Register Assn v. Gocher et al.*, 70 F. 237 (1895).

45 *Bleistein v. Donaldson Lithographing Company*, 188 U.S. 239 (1903). Holmes also pointed out "it would be a dangerous undertaking for persons trained only to the law to constitute themselves final judges of the worth of pictorial illustrations." Justices Harlan and McKenna dissented on the grounds that the Constitution intended to promote the useful arts and that this did not extend to "a mere advertisement of a circus."

46 *Clayton et al. v. Stone et al.*, 5 F. Cas. 999 (1829).

science, and thus was not within the scope of the statute. However, by the end of the nineteenth century, specialized trades in information and publications with financial information had become routine and were clearly of value in the marketplace, which the courts took into account in their interpretation of the statutes. In *William B. Dana Company v. United States Investor* (1895), the plaintiff (publisher of the *Commercial and Financial Chronicle*) received a favorable judgment in its infringement lawsuit.[47] Even uncopyrighted price quotations were granted protection under common law because "legislatures and courts generally have recognized that the natural evolution of a complex society is to be touched only with a very cautious hand."[48]

Courts enforced a number of doctrines that were based on their interpretation of the Constitution as a mandate for furthering welfare through market transactions rather than by protecting personal rights of literary or artistic individuals. These included stipulations regarding first sale and work for hire, which were counter to the holdings under civil law or moral rights, and developed prior to statutes that have since encapsulated the same principles. Under the moral rights system, an artist or his heirs can claim remedies if subsequent owners alter or distort the work in a way that allegedly injures the artist's honour or reputation; according to the first sale doctrine, the copyright holder loses all rights after the work is sold.[49] Even at first glance, it is clear that the moral rights doctrine, if enforced, would tend to create market uncertainty, and increase transactions costs relative to a first sale system. Moreover, it is likely to decrease monetary returns to copyright holders, as the market price would fall to take into account the lower stream of benefits that purchasers derive from the product. Under the American system, if the copyright holder's welfare was increased by nonmonetary concerns, these individualized concerns could be addressed and enforced through contract law, rather than through a generic federal statutory clause that would affect all property holders.

Work for hire decisions also repudiated the right of personality, in favor of facilitating market transactions.[50] In the realm of copyright, such doctrines

---

[47] See *Publishers' Weekly*, Nov. 2 (1895) no. 1240, p. 772.

[48] *Board of Trade v. Christie Grain & Stock Co.*, 198 U.S. 236 (1905).

[49] See, for instance, *Harrison v. Maynard, Merrill & Co.*, 61 F. 689 (1894). Harrison was a secondhand-book dealer who bought from a third party a quantity of merchandise published by Maynard et al. that had been damaged in a fire. Harrison rebound and sold the books. The damaged books had been sold originally with the stipulation that they were only to be used for waste paper. The court argued that "the right to restrain the sale of a particular copy of the book by virtue of the copyright statutes has gone when the owner of the copyright and of that copy has parted with all his title to it, and has conferred an absolute title to the copy upon a purchaser."

[50] Work-for-hire doctrines allow intellectual property rights to be allocated to employers if the work falls within the normal scope of employment.

were upheld long before they were reservedly introduced in patent cases, and indeed were implied by the first copyright statutes.[51] As stated before, the 1790 copyright statute did not require authorship per se for it allowed grants to be issued to either authors or proprietors, and almost a half of the earliest copyrights were recorded to proprietors, including publishers, printers, and employers. The work for hire doctrine was reiterated in cases that dealt with copyrights in art, drama, maps, and printed matter. In *Lawrence v. Dana*, 15 F. Cas. 26 (1869), Justice Clifford noted that "Title to the notes or improvements prepared for a new edition of a book previously copyrighted may, in certain cases, be acquired by the proprietor of a book from an employee, by virtue of the contract of employment, without any written assignment.... Although the services were gratuitous, the contributions of the complainant became the property of the proprietor of the book, as the work was done, just as effectually as they would if the complainant had been paid daily an agreed price for his labor." In 1895 Thomas Donaldson filed a moral rights claim that Carroll D. Wright's editing of Donaldson's report for the Census Bureau served to "emasculate" his research and was therefore "damaging and injurious to the plaintiff, and to his reputation" as a scholar. The court rejected his claim, for "having been employed and paid for his work, he has no right of property of any kind in the bulletin." To do otherwise, it was further argued, would create problems in situations where employees were hired to prepare data and statistics.[52]

Table 8.2 shows that 18 percent of copyright cases dealt with the question of whether the copyright owner had lost enforcement rights because of abandonment, or because they failed to comply with the requirements of the statute. Here again, the judiciary held that patents were to be interpreted liberally, but copyrights were to be more strictly construed. Copyrights were overturned on seemingly inconsequential grounds: a painting had not been described in the registration; a copyright mark was omitted or its placing was inappropriate; the copyright notice had not been put on every copy of the work; the requisite number of copies had not been deposited in the time specified; the wording of the notice was not exactly the same as the statutes; the notice in the book included the author's trade name rather than her own name. In short, "authors take their rights under and subject to the law; and, when assailed, the burden is upon them to show literal compliance with each and every statutory requirement in the nature of conditions

---

[51] For an excellent and comprehensive legal treatment of the development of work for hire, see Catherine L. Fisk, "Authors at Work: The Origins of the Work-for-Hire Doctrine," 15 Yale J.L. & Human (Winter 2003):1–70.

[52] *Donaldson v. Wright*, 7 App. D.C. 45 (1895). In *Jones v. American Law Book Co.* (1908), the author of articles for an encyclopedia was not even allowed the right to have his name on the article; whether or not it was attributed to him was at the discretion of the defendants, who had hired him to write the articles.

precedent."[53] The issue of literal compliance for copyrights formed a persistent theme from the earliest years through much of the twentieth century.[54]

This insistence on compliance might seem to indicate a lack of appreciation for intellectual property rights. However, such an interpretation is unlikely to be valid for a number of reasons. First, many of the same individuals who reached these conclusions in copyright jurisprudence strongly favored patent grants. Their attitude can be discerned in rulings such as an 1862 decision declaring copyright law "should be liberally construed in favor of authors, and, leaving their comparative merits to be settled by critics, at the tribunal of public opinion, it should protect and encourage their labors. The fruits of their literary toils should be secured to them by the highest title, for they keep open the springs of thought which feed the intellectual life of the nation."[55] In the light of the number of cases overturning individual copyrights, this might seem to be an isolated opinion. However, in the realm of copyrights, compliance was part of the bargain with the public that courts had to enforce, regardless of their recognition of the merits of copyright holders, and was key to the utilitarian calculus.

Second, copyrights were not examined, merely registered, in return for the public grant of an exclusive right, which meant that a large variety of useless items would be granted such rights. One way of filtering out useless items was to increase the cost of filing by insisting on strict compliance with the statutes so that "casual filers" were penalized. Of relevance here is the landmark case *Wheaton v. Peters*, 33 U.S. 591 (1834), which ruled that the copyright statute extinguished any common-law rights that an author had in his unpublished work. Justice McLean, who also had delivered decisions in patent cases, explained the majority view regarding the necessity for strict compliance: "If any difference shall be made, as it respects a strict conformity to the law, it would seem to be more reasonable, to make the requirement of the author, rather than the inventor. The papers of the latter are examined in the department of state, and require the sanction of the attorney-general; but the author takes every step on his own responsibility, unchecked by the scrutiny of sanction of any public functionary." Third, copyright owners themselves benefited from the compliance rule, which ensured that markets in copyrights operated more efficiently. Registration lowered the transaction costs of identifying owners, and facilitated trade. The requirements of registration and copyright notices on articles that claimed statutory protection

---

[53] *Osgood v. A. S. Aloe Instrument Co.*, 83 F. 470 (1895).

[54] For instance, *Ewer v. Coxe*, 8 F. Cas. 917 (1824) held that technical compliance was required. For a contrary view, see *Nichols v. Ruggles*, 3 Day 145 (1808), stating that the statutory provisions were "merely directory."

[55] Judge Shipman of the Southern District of New York, in *Boucicault v. Fox et al.*, 3 F. Cas. 977 (1862).

also reduced the likelihood of unconscious infringement, and made litigation and enforcement more transparent. If the statute had not provided for these requirements, it is likely that private organizations would have had to replicate these services at far higher cost and lower effectiveness. Moreover, the deposit requirement significantly benefited authors and the public, in the form of a localized repository of national culture in the Library of Congress.

Questions about the scope of protection accounted for 10.8 percent of disputes between 1790 and 1879, and this figure almost doubled in subsequent years. Two issues in particular are worth noting in this regard: the expansion of derivative rights granted to copyright holders; and the extension of protection to works created by new technology. Derivative rights refer to original works that are based on a prior work, such as the translation of a novel, an abridgment or digest of legal reports, a recording of a musical composition, or a photograph of a painting.[56] The appendix shows that the statutory rights of the owner of the preexisting work were initially narrowly circumscribed, but the scope and depth of these rights increased over time. For instance, in 1853 Harriet Beecher Stowe was held not to possess the rights to translations of her own novel, but by 1910 the original copyright holder was granted the right to translate literary works into other languages; performance rights; and the rights to arrange and adapt musical works, among others.[57] In many respects, these rulings about derivative rights were decisions about the extent of the market that could be monopolized by the copyright holder. In the early period when such markets were thin, the danger of an extensive monopoly was likely to be much higher because of a lower availability of substitutes. However, as markets in substitutes developed and the danger of individual monopolies receded, copyright holders were granted access to more markets through extensions in the depth of copyrights.

The copyright system evolved to encompass technological innovations and changes in the marketplace. The burgeoning scope of copyright protection

---

[56] It is interesting to note in this regard that, as Table 8.2 shows, paintings only became an important subject of litigation toward the end of the nineteenth century, when the means of reproducing them multiplied.

[57] *Stowe v. Thomas*, 23 F. Cas. 201 (1853). *Kalem Co. v. Harper Bros.*, 222 U.S. 55 (1911) affirmed a lower court ruling that General Lew Wallace's novel *Ben Hur* was infringed by a movie based on the novel. *Daly v. Palmer et al.*, 6 F. Cas 1132 (1868) dealt with performance rights for a copyrighted play, which rights were conferred by the Copyright Act of 1856. According to William M. Landes and Richard A. Posner, "An economic analysis of copyright law," *Journal of Legal Studies*, vol. 18 (June) 1989: 325–363, derivative rights increase the incentives to create new works, reduce the incentive to gain lead-time in derivative markets by postponing the original work, and lower transactions costs of search and information by granting the copyright holder control over derivative rights.

that technological advances created raised numerous questions about the rights of authors and publishers relative to the public, and courts continually were confronted with the need to delineate the boundaries of private property in such a way as to guard the public domain. A number of the technological innovations of the nineteenth century were sufficiently different from existing technologies as to make judicial analogies to existing doctrines somewhat strained, and ultimately required accommodation by the legislature instead. As the Supreme Court pointed out: "From its beginning, the law of copyright has developed in response to significant changes in technology. Indeed, it was the invention of a new form of copying equipment – the printing press – that gave rise to the original need for copyright protection. Repeatedly, as new developments have occurred in this country, it has been the Congress that has fashioned the new rules that new technology made necessary."[58]

The 1831 copyright statute extended protection to musical compositions, at that time limited to sheet music. The creation of the player piano and the phonograph raised questions about the relevance of existing copyright rules, in part because the analogy between sheet music and the inputs to these machines appeared remote. *Stern v. Rosey*, 17 App. D.C. 562 (1901), dealt with the question of whether an injunction should issue against a manufacturer of phonograph records who had used copyrighted music. The court rejected the notion that copyright protection for music extended to such a different technological transformation. *Kennedy v. McTammany*, 33 F. 584 (1888), was brought by the copyright owner of a song entitled "Cradle's Empty, Baby's Gone." Judge Colt failed to accept the plaintiff's argument that McTammany's perforated piano rolls infringed on the copyright for the music, because he could "find no decided cases which, directly or by analogy, support the position of the plaintiffs." It was not evident that the perforated strips of paper – a mechanical invention – comprised illegal copies of sheet music. In 1908 the Supreme Court affirmed this position when it considered the claim brought by a music publishing company against the manufacturer of player piano rolls. The following year Congress responded by revising the copyright law to give composers the right to the first mechanical reproduction of their music. However, after the first recording, the statute permitted a compulsory license to issue for copyrighted musical compositions: that is to say, anyone could subsequently make their own recording of the composition on payment of a fee that was set by the statute at 2¢ per recording. In effect, the property right was transformed into a weaker liability rule that allowed users access without consensual exchange.

The advent of photography created a new form of "authorship," which was granted copyright protection in 1865. Photography also offered a ready means of copying books, paintings, and engravings that led to copyright infringement litigation. *Rossiter v. Hall*, 20 F. Cas. 1253 (1866), dealt with

---

[58] *Sony Corp. of America v. Universal City Studios, Inc.*, 464 U.S. 417 (1984).

photographic copies that had been taken of a copyrighted engraving of Washington's house that the statutes protected against unauthorized reprints. The defendant argued unsuccessfully that, as photography had not been invented at the time of the statute, it followed that this form of copying was not prohibited. Although the judiciary was reluctant to appropriate the task of Congress and create new policies, at times they were able to adjudicate cases relating to new technologies by stretching the existing analogy. This was apparent in the development of litigation surrounding movies not long after Edison obtained his 1896 patent for a kinetoscope. The lower court rejected Edison's copyright of moving pictures under the statutory category of photographs, but this decision was overturned by the appellate court: "To say that the continuous method by which this negative was secured was unknown when the act was passed, and therefore a photograph of it was not covered by the act, is to beg the question. Such construction is at variance with the object of the act, which was passed to further the constitutional grant of power to "promote the progress of science and useful arts." . . . [Congress] must have recognized there would be change and advance in making photographs, just as there has been in making books, printing chromos, and other subjects of copyright protection."[59]

Technological innovation created new cultural properties to be protected, but many of these also facilitated infringement through mechanical means of reproduction that lowered the costs of duplicating copyrighted works. Congress responded to the creation of new subject matter by expanding the scope of the copyright laws. The legislature also repeatedly lengthened the term of copyright, possibly in order to support the value of copyright protection in the face of falling costs of infringement. Nevertheless, it is worth repeating that the largely utilitarian rationale of the American statutes ("to promote learning") precluded perpetual grants, the term of copyright protection in United States was among the most abbreviated in the world, and the United States offered the most liberal opportunities for unauthorized use of copyrighted material.

Copyright cases drew a distinction between an author's expression (which could be monopolized) and the underlying ideas (which lay in the public domain and could not be monopolized through copyright protection.)[60] Thus, "the copyright book is sacred, but not the subject of which it treats."[61]

[59] *Edison v. Lubin*, 122 F. Cas. 240 (1903).
[60] Relevant cases include *Holmes v. Hurst*, 174 U.S. 86 (1899); *Baker v. Selden*, 101 U.S. 99 (1880); *Johnson v. Donaldson*, 3 Fed. Rep. 22 (1880); *Perris v. Hexamer*, 99 U.S. 674 (1878); *Bobbs-Merrill Co. v. Straus*, 210 U.S. 339 (1908). According to a recent decision, "the Framers intended copyright itself to be the engine of free expression. By establishing a marketable right to the use of one's expression, copyright supplies the economic incentive to create and disseminate ideas." *Harper & Row Publishers v. Nation Assoc.*, 105 S. Ct. 2218 (1985).
[61] *Griggs v. Perrin*, 49 F. 15 (1892).

In *Johnson v. Donaldson*, the court pointed out that copyright "does not rest upon any theory that the author has an exclusive property in his ideas, or in the words in which he has clothed them."[62] The distinction between idea and expression tried to achieve a difficult balance between the public welfare in the form of free access to primary facts and ideas, and the right that copyright holders had in their investments in expression.[63] The lack of protection for facts and ideas, as opposed to expression, lowered the production and search costs of users and creators, and avoided potential rent-seeking.[64]

This difficult quest for balance was also illustrated in other dimensions of litigation. Table 8.2 indicates that more than a quarter of all copyright disputes dealt with the issue of ownership and originality, or the question of whether the defendant had infringed the rights of the plaintiff. The answer to this question was complicated by the fact that copyright statutes permitted independent creation because originality under the law did not imply novelty.[65] Moreover, unlike patents, users were allowed unauthorized access to copyrighted works under certain conditions. The "fair use doctrine" had originated in England but, while it dwindled there, fair use became a lasting and integral element of copyright policy in the United States. Joseph Story outlined the American doctrine in *Gray v. Russell* (1839) and the more frequently cited *Folsom v. Marsh* (1841). This doctrine allowed that it was permitted to use some portion of a copyrighted work, although exactly how much copying was permissible was (and remains today) "one of the most difficult points that can well arise for judicial discussion." Story offered a number of guidelines that are enshrined in modern statute: "We must often, in deciding questions of this sort, look to the nature and objects of the selections made, the quantity and value of the materials used, and the degree in which the use may prejudice the sale, or diminish the profits, or supersede the objects, of the original work."[66]

---

[62] *Johnson v. Donaldson*, 3 F. 22 (1880). The plaintiff claimed rights in chromos that both plaintiff and defendants had copied from a picture in a foreign publication. The case was dismissed, because both parties had independently created the chromos.

[63] See *Harper & Row v. Nation Enterprises*, 723 F.2d 195 (1983) referring to "a definitional balance between the First Amendment and the Copyright Act by permitting free communication of facts while still protecting an author's expression."

[64] See William M. Landes and Richard A. Posner, "An economic analysis of copyright law," *Journal of Legal Studies*, vol. 18 (June)1989: 325–363.

[65] See *Emerson v. Davies*, 8 F. Cas. 615 (1845): "In truth, in literature, in science and in art, there are, and can be, few, if any, things, which, in an abstract sense, are strictly new and original throughout. Every book in literature, science and art, borrows, and must necessarily borrow, and use much which was well known and used before." Landes and Posner (1989) point out that the lack of emphasis on novelty in copyright makes economic sense because it would be difficult to verify, given the huge number of works that would have to be surveyed, and the harm of accidental duplication is minimal.

[66] *Gray et al. v. Russell et al.*, 10 F. Cas. 1035 (1839); *Folsom et al. v. Marsh et al.*, 9 F. Cas. 342 (1841). Ironically, one of Justice Story's books was the subject of a fair use dispute brought

One of the striking features of the fair use doctrine was the extent to which property rights were defined in terms of market valuations, or the impact on sales and profits, as opposed to a clear holding of the exclusivity of property. Indeed, the fair use doctrine can be interpreted as the most stringent form of compulsory licensing in existence, because it transforms a property rule (free market exchange based on consent) for copyright material into a liability rule (nonconsensual exchange) with zero compensation. Such liability rules are regarded as socially efficient when transaction costs are so high that they inhibit the free functioning of market exchanges.[67] The fair use doctrine minimized the "takings" problem by allowing unauthorized use only when the property owner's loss was estimated to be small; property rules and consensual exchange applied when the copyright owner stood to forfeit significant profits. Fair use doctrine thus illustrated the extent to which policy makers weighed the benefits of diffusion against the costs of exclusion. If copyrights were as strictly construed as patents, it would serve to reduce scholarship, prohibit public access for noncommercial purposes, increase transaction costs for potential users, and inhibit learning which the statutes were meant to promote.

Transaction costs undoubtedly play some part in justifying fair use, as some law and economics scholars have argued. However, fair use in the United States was not formulated simply as a function of transaction costs, nor was it limited because courts recognized the (moral or other) rights of authors. It is important to understand this point, because the development of technologies that influenced the ability to monitor use and exclude unauthorized users without judicial intervention posed a danger to the central role of fair use in the American copyright system. Indeed, if monitoring costs were zero, and all use could be traced by the author, fair use doctrines would be all the more relevant to attaining the ultimate function of property rights in cultural products. Without fair use, copyright would be transmuted into an exclusive monopoly right that would restrict public access and violate the Constitution's mandate to promote the progress of science. The economic

by his heirs against an abridger, in *Story v. Holcombe et al.*, 23 F. Cas. 171 (1847). The Copyright Act of 1976 (17 U.S.C. § 107 (1976)) codified the fair use doctrine:

"The fair use of a copyrighted work ... for purposes such as criticism, comment, news reporting, teaching ... scholarship, or research, is not an infringement of copyright. In determining whether the use made of a work in any particular case is a fair use the factors to be considered shall include – (1) the purpose and character of the use, including whether such use is of a commercial nature or is for nonprofit educational purposes; (2) the nature of the copyrighted work; (3) the amount and substantiality of the portion used in relation to the copyrighted work as a whole; and (4) the effect of the use upon the potential market for or value of the copyrighted work."

[67] The distinction between property rules and liability rules is discussed in G. Calabresi and A. D. Melamed, "Property Rules and Liability Rules, and Inalienability: One View of the Cathedral," *Harvard Law Review, vol. 85* (1972): 1089–1128.

history of intellectual property in the United States reveals that fair use by the public was not regarded as an exception to the grant of copyright; instead, the grant of copyright comprised a limited exception to the primacy of the public domain.

The fair use doctrine also addressed the importance of freedom of speech, in cases that justified unauthorized use of copyrighted material for purposes of parody and criticism. In *Bloom & Hamlin v. Nixon et al.*, 125 F. 977 (1903), Fay Templeton (an actress with "unusual powers of mimicry") imitated on stage another actress who performed the plaintiffs' copyrighted song in a musical production of *The Wizard of Oz*. When the plaintiffs sued for infringement, the court ruled that "a parody would not infringe the copyright of the work parodied" and was not prohibited by either the letter or spirit of the copyright statutes.[68] Thus, the court in effect ruled that the benefit of a more exclusive property right to the plaintiffs fell short of the potential benefit to society of access to that property. In a true parody, exclusion would not significantly enhance the benefits to the copyright holder, as the market for the original and the profits of the copyright holder were unlikely to be affected by a parody that did not substitute for the original performance. In short, private property rights, held to be "sacred" in nineteenth-century patent jurisprudence, were noticeably impinged on in the allowance of unauthorized use of copyrighted materials.

The increasingly polarized scholarly debate about the scope of copyrights tends to overlook the importance of allied rights that are available through other forms of the law such as contract and unfair competition. A noticeable feature of the nineteenth-century judiciary was their willingness to extend protection to noncopyrighted works under alternative doctrines in the common law, although the judicial mind in 1915 balked at the thought of extending free speech protections to commercial productions such as movies. More than 10 percent of "copyright" cases were decided using concepts of unfair competition, where the court rejected copyright claims but still protected the work against unauthorized users by applying fair trade doctrines. Some 7.7 percent dealt with contracts, which raised questions such as ownership of photographs in cases of "work for hire." A further 12 percent encompassed issues of trade secrets, misappropriation, and the right to privacy. Many of these cases included instances in which the courts could not in all conscience grant protection under copyright statutes, but nevertheless protected the plaintiff from unauthorized or unfair use of property.[69] For

---

[68] See also *Green et al. v. Minzensheimer*, 177 F. 286 (1909), which followed the same ruling. According to the reasoning in some modern legal decisions, parody is permitted on the grounds of free speech and because a license is unlikely to be given for a work that makes the copyright the subject of ridicule.

[69] *Baker v. Selden*, 101 U.S. 99 (1880) argued that copyrights (unlike patents) made no examination for novelty, so to give the author an exclusive right in a method that was described

instance, in *Keene v. Wheatley et al.*, 14 F. Cas. 180 (1860), the plaintiff did not have a statutory copyright in the play that was infringed.[70] However, she was awarded damages on the basis of her proprietary common-law right in an unpublished work, and because the defendants had taken advantage of a breach of confidence by one of her former employees. Similarly, the courts offered protection against misappropriation of information, such as occurred when the defendants in *Chamber of Commerce of Minneapolis v. Wells et al.*, 111 N.W. 157 (1907) surreptitiously obtained stock market information by peering in windows, eavesdropping, and spying.

Several other examples relate to the more traditional copyright subject of the book trade. E. P. Dutton & Company published a series of Christmas books that another publisher photographed, and offered as a series with similar appearance and style but at lower prices. Dutton claimed to have been injured by a loss of profits and a loss of reputation as a maker of fine books. The firm did not have copyright in the series, but they essentially claimed a right in the "look and feel" of the books. The court agreed: "The decisive fact is that the defendants are unfairly and fraudulently attempting to trade upon the reputation which plaintiff has built up for its books. The right to injunctive relief in such a case is too firmly established to require the citation of authorities."[71] In a case that will resonate with academics, a surgery professor at the University of Pennsylvania was held to have a common law property right in the lectures he presented, and a student could not publish them without his permission.[72] Titles could not be copyrighted, but were protected as trademarks and under unfair competition doctrines.[73] In this way, in numerous lawsuits G. C. Merriam & Co, the original publishers of

in a book "would be a surprise and a fraud upon the public." To some extent, the application of unfair competition rulings to these species of property is a natural extension of the differences between patents and copyrights. As I pointed out in the chapter on patent litigation, courts argued that the patent right did not involve monopoly rights, because the patentee created something new (novelty) and dedicated it to the public welfare whereas the monopolist made private what had previously belonged to the public. As *Baker v. Selden*, 101 U.S. 99, 102 (1880) emphasized, "novelty of the art or thing described has nothing to do with the validity of the copyright." Copyright, by granting exclusion without novelty, approximates restraint of trade practices more closely than patents. It therefore seems a natural extension of this logic to grant protection for matter that falls outside the range of the copyright statutes through laws regarding unfair competition.

[70] Similarly, in *Crowe v. Aiken*, 6 F. Cas. 904 (1869), the unauthorized performance of a play was enjoined even though the play was not covered by copyright protection.

[71] *E. P. Dutton & Company v. Victor W. Cupples & Arthur T. Leon*, 117 App. Div. 172 (1907).

[72] Miller's Appeal, 15 Wkly. Notes Cas. 27 (1884).

[73] For instance, a perpetual injunction was issued against a play entitled "Sherlock Holmes, Detective" not because it was felt to unfairly infringe on any property rights that the plaintiff had in the name of his play "Sherlock Holmes," but because it was likely to deceive the public.

Webster's Dictionary, restrained the actions of competitors who published the dictionary once the copyrights had expired.[74]

The development of the right to privacy is especially interesting, because it illustrates the creation of a new legal concept at common law to compensate for the potential of new technologies to infringe on third-party rights. Samuel Warren and Louis Brandeis, in what has been touted as the most effective law review piece of all time, argued that "modern enterprise and invention" subjected the ordinary individual to unwarranted suffering that could not be effectively alleviated through existing copyright or tort laws. Instant photographs and "numerous mechanical devices" led to the "evil of invasion of privacy."[75] The concept of a legal right to privacy immediately entered into litigated arguments, and the New York Supreme Court, in *Schuyler v. Curtis et al.*, 15 N.Y.S. 787 (1891), quoted directly from the 1890 law review, but distinguished between private individuals and public figures who by implication ceded the right to privacy. In a Massachusetts case three years later the wife of the great inventor George H. Corliss tried to enjoin the publication of a photograph of her late husband. The court rejected the plea because her husband was "among the first of American inventors, and he sought recognition as such," permitting thousands of his photographs to be distributed at the Centennial Exposition in Philadelphia. In 1903, the New York legislature passed a right to privacy statute that levied criminal and civil liability for the unauthorized use of the "name, portrait or picture of any living person" for "advertising purposes, or for the purposes of trade," and several other states did the same.

CONCLUSION

The intellectual property clause of the United States Constitution was one of the few to be passed without debate and with unanimous approval. This clause, which appears in the very first article of a document that distilled democratic precepts, authorized the protection of the works of authors and inventors. Congress did so by approving separate statutes to create the patent system and the copyright system. Despite their common Constitutional heritage, and despite the fact that many of the same individuals who influenced patent policies also shaped copyright policies, the two developed in distinctly different directions. According to a legal scholar in the nineteenth century, "...the mistake should not be made of supposing that patents and copyrights stand on the same basis as to natural exclusive right, for they do not;

---

[74] Some of these cases include *Merriam v. Ogilvie*, 159 F. 638 (1908), and *Merriam v. Texas Siftings Pub. Co.*, 49 F. 944 (1892).

[75] Samuel Warren and Louis. D. Brandeis, "The Right to Privacy," *Harvard Law Review*, 4 (5) 1890:193–220.

the difference between them, in this regard, is radical."[76] American policies toward literary property were not only different from patent policies, they also diverged dramatically from European precedent.

These features of the American system naturally raise the question of why it happened to evolve in this direction. In this chapter, I argued that the unique structure of the U.S. intellectual property system can be explained in terms of the characteristics of American democracy, and the belief that decentralized markets, economic democracy, and political freedoms are linked. To a much greater degree than in Europe, the American system melded democratic values of equal access, the diffusion of information, and the importance of an informed and literate electorate with the decentralized means of achieving those goals through a market economy that was supported by an efficient legal system and an enlightened judiciary. These beliefs and objectives were more compatible in the case of patent grants; in the case of copyrights the social tradeoffs were more evident. Patent rights responded more strongly to market incentives and induced novel additions to public welfare, whereas the public costs of limited access were more at the forefront of deliberations on copyright. Unlike property rights in patents, the character of copyright was more akin to a monopoly grant. Thus, it is not surprising that the creators of the U.S. intellectual property clause should have endorsed two such different systems for protecting the works of authors and inventors.

The divergence between patents and copyrights was most apparent in the international arena. John Ruggles, who had recommended the changes in the patent system in 1836, also was a member of an 1838 Committee to reform international copyrights. He argued that "American ingenuity in the arts and practical sciences, would derive at least as much benefit from international patent laws, as that of foreigners. Not so with authorship and book-making. The difference is too obvious to admit of controversy."[77] By 1836, the only difference between foreign and American patentees lay in the fees charged, and after the Civil War foreigners and Americans alike received identical treatment in the Patent Office. However, in the realm of copyright, Americans were unabashed pirates of European culture, and resisted numerous attempts to alter their "obnoxious laws" for fully one century. According to Ruggles, an international copyright law was ill-advised, because "The answer is found in the significant inquiry of the British reviewer – 'Who ever reads an American book?'...the profits of trade and manufacture, and all the benefits...would be, for us, on the wrong side of the leger [sic]." The following chapter examines this episode more closely and analyzes the effect of copyright piracy on American society.

---

[76] W. E. Simonds, "International Copyright," in G. H. Putnam, *The Question of Copyright*, New York: G. P. Putnam's Sons, 1896: 77–130.

[77] Report to accompany Senate Bill No. 32, 25th Congress, 2nd Session, June 25 (1838).

APPENDIX: COPYRIGHT STATUTES IN THE UNITED
STATES, 1790–1910

January 1783   Connecticut passes first statute "for the encouragement of
literature and genius."
Preamble: "Whereas it is perfectly agreeable to the
principles of natural equity and justice, that every author
should be secured in receiving the profits that may arise
from the sale of his works, and such security may
encourage men of learning and genius to publish their
writings; which may do honor to their country, and
service to mankind."

May 31, 1790   First Copyright Act of the United States [1 Stat. 124.]
Preamble: "An Act for the encouragement of learning, by
securing copies of maps, charts, and books, to authors
and proprietors of such copies, during the times therein
mentioned."

• protected copyright in books, charts, and maps
• granted right to creators and their assignees
• right to print, reprint, publish, and sell
• term of fourteen years from registration, with possibility of renewal for
  further fourteen years if the author were still alive
• copyright limited to citizens and residents of the United States
• penalties for infringement: confiscation of copies and fine of 50 cents per
  page to be recovered in a court of record (half of the damages accruing
  to the author and half to the United States)
• requirements: deposit copy of work in district court, payment of fee of
  60 cents and publication of record in at least one newspaper for four
  weeks.

April 29, 1802

• Extended Act of 1790 to include protection of "the arts of designing,
  engraving, and etching historical and other prints"
• Notice of copyright should be included on protected work
• fine for infringement of prints of one dollar per print to be recovered in
  court (half for the owner and half for the use of the United States)
• False notice of copyright fined one hundred dollars

February 15, 1819

• circuit courts granted jurisdiction over patents and copyrights in equity
  and at law
• ability to grant injunctions in equity
• right of appeal to Supreme Court

February 3, 1831

- Copyrightable matter includes books, maps, charts, musical compositions, prints, woodcuts, and engravings
- Term of twenty-eight years from time of registration with right of renewal for 14 more years
- "That nothing in this act shall be construed to extend to prohibit the importation or vending, printing, or publishing," of the works of foreign residents

June 30, 1834

- Assignments must be recorded within sixty days to be legally valid

August 10, 1846

- One copy of copyrighted material to be deposited at the Smithsonian, and one to the Library of Congress within three months of publication

August 18, 1856

- Copyright extended to dramatic compositions, and includes performance rights

March 3, 1865

- Copyright extended to photographs and negatives
- Copy of work must be deposited with Library of Congress

July 8, 1870

- Library of Congress declared new registry of copyrights
- Copyright extended to statues, chromos, models, and designs that are "works of fine arts"

March 3, 1891
"International Copyright Act"

- Copyright protection granted to residents of foreign countries that accord similar protection to American residents.

1909 Major revision in copyright statutes

- Scope of protection broadened to all works of authorship
- Term: twenty-eight years plus extension of further twenty-eight years

Source: Copyright Enactments of the United States, 1783–1906, compiled by Thorvald Solberg, Washington, D.C.: G.P.O., 1906; Library of Congress, *United States Copyright Office: a Brief History and Overview* [www.loc.gov/copyright/cpypub/circ1a.html].

9

American Copyright Piracy

"The country will deprive itself of the honour and service of letters, and the improvement of science, unless sufficient laws are made to prevent depredation on literary property."

– Thomas Paine (1782)

Previous chapters highlighted the liberality of American policies toward foreign inventors, and it is not surprising that the United States led the movement for harmonization of patent laws. A nation of artificers and innovators, both as consumers and producers, American citizens were confident of their global competitiveness in technology, and accordingly took an active role in international patent conventions. Although they excelled at pragmatic contrivances, Americans were advisedly less sanguine about their efforts in the realm of music, art, literature, and drama. As a developing country, the United States was initially a net debtor in flows of material culture from Europe. England and other European countries entered into multilateral treaties to establish reciprocity in the recognition of foreign copyrights. The United States, in contrast, engaged in protectionist policies that benefited its residents at the expense of authors and artists in other countries, most notably England. The most notorious aspect of its strategy in the international sphere was the official support for American "piracy" of foreign copyright products. The first copyright statute in 1790 authorized Americans to take free advantage of the cultural output of other countries and the legal system continued to encourage international copyright piracy for a century.

In view of the strong protections for inventors under the U.S. patent system, overseas observers found its copyright policies to be all the more reprehensible. A plaintive note of bewilderment is evident in the Report of the 1878 British Commission on Copyrights: "The original works published in America are, as yet, less numerous than those published in Great Britain. This naturally affords a temptation to the Americans to take advantage of the works of the older country.... Were there in American law no recognition of the rights of authors, no copyright legislation, the position of the United States would be logical. But they have copyright laws; they afford protection to citizen or resident authors, while they exclude all others from the benefit of that protection. The position of the American people in this respect is the more striking, from the circumstance that, with regard to the analogous right of patents for invention, they have entered into a treaty with this country for the reciprocal protection of inventors."[1]

U.S. copyright laws toward other countries in the nineteenth century comprised one of the most colorful episodes in the history of intellectual property. According to Ainsworth Spofford, Librarian of Congress between 1864 and 1897, "a group of publishing houses in the United States, which made a specialty of cheap books, vied with each other in the business of appropriating English and Continental trash, and printed this under villainous covers, in type ugly enough to risk a serious increase of ophthalmia among American readers."[2] Unlikely coalitions formed to lobby for recognition of the rights of overseas authors, including American writers with international reputations such as Henry Wadsworth Longfellow and Louisa May Alcott; educational institutions such as Longfellow's alma mater, Bowdoin College, and the newly founded University of California; miscellaneous groups such as the American Medical Association and the citizens of Portland, Maine. An array of Europeans decried copyright piracy, ranging from Charles Dickens, Edmund Burke, and Harriet Martineau, to Gilbert and Sullivan. Equally vociferous were groups in the United States that lobbied to maintain the existing policies, including concerned citizens from Richmond, Virginia, to Bellow Falls, Vermont; paper producers in Boston, Newark, and Pennsylvania; and typographical unions, Toledo printers, New York publishers, and Hartford bookbinders.

The rationale for supporting the worldwide expansion of patent laws while simultaneously engaging in copyright piracy was quite pragmatic. John Ruggles, who had overseen the reform of the patent laws, in 1838 declared that American self-interest favored international patent laws, but in copyright issues its self-interest was aligned with piracy.[3] Recognition of international copyright would harm U.S. employment and manufacturing prospects, and lead to few reciprocal benefits as there was little demand for American culture in nineteenth-century Europe. The demand for books in the United States was sufficiently elastic that cheaper prices actually increased revenues for publishers, and network effects led to expanding markets. He pointed out that copyright piracy promoted "that general diffusion of knowledge and intelligence, on which depends so essentially the preservation and support of our free institutions." In short, the social accounting from an international copyright law was initially on the wrong side of the ledger as far as the United States was concerned. The tendency to freely reprint foreign works

---

[1] Reprint of Report in G H Putnam *The Question of Copyright*, New York: G. P. Putnam's Sons, 1890, pp. 269–70.

[2] Cited p. 170, *The Question of Copyright*, compiled by George Haven Putnam, New York: G. P. Putnam's Sons, 1896 (second edition.)

[3] Report to accompany Senate Bill No. 32, 25th Congress, 2nd Session, June 25, 1838.

was encouraged by the existence of tariffs on imported books that ranged as high as 25 percent.[4]

International copyright petitions (from supporters on both sides of the debate) were submitted on more than one hundred occasions in the congressional sessions through 1875.[5] Earlier attempts were defeated by publishers, printers, and representatives of the Democratic Party, and it was not until 1891 that the Chace Act granted copyright protection to selected foreign residents.[6] However, the Chace Act also included significant concessions to printers' unions in the form of "manufacturing clauses." First, a book had to be published in the United States before or at the same time as the publication date in its country of origin. Second, the work had to be printed here, or printed from type set in the United States or from plates made from type set in the United States. These clauses resulted in U.S. failure to qualify for admission to the Berne Convention until 1988, one hundred years after the initial convention.[7] It is thus more than a little ironic that today the United States is at the forefront of efforts to compel developing countries to forego "piracy" and to recognize foreign copyrights.[8]

This episode provides a convenient way of investigating the likely dynamic effects of ignoring international legal standards. What was the impact of this policy of international piracy on the "progress of science and useful arts" in the United States? The general consensus after the fact seems to be that U.S. international copyright piracy was an embarrassing aberration that had harmed social and economic progress. Commentators claimed that American publishing companies "indiscriminately reprinted books by

---

[4] See Donald Marquand Dozer, "The Tariff on Books," *Mississippi Valley Historical Review*, Vol. 36, No. 1. (Jun., 1949): 73–96.

[5] For instance, S. 223 (1837); H.R. 779 (1868), "A Bill For securing to authors in certain cases the benefit of international copyright, advancing the development of American literature, and promoting the interests of publishers and book-buyers in the United States;" H.R. 470 (1871); and S. 688 (1872), among others. On February 18th, 1853, Millard Fillmore, President of the United States, sent to the Senate "with the view to its ratification, a convention which was yesterday concluded between the United States and Great Britain for the establishment of international Copyright," but the Senate refused to comply with the request. See the Journal of the executive proceedings of the Senate of the United States of America, 1852–1855, February 24, 1853, p. 35.

[6] International Copyright Act of 1891, 26 Stat. 1106.

[7] Berne Convention for the Protection of Literary and Artistic Works, opened for signature Sept. 9, 1886, 828 U.N.T.S. 221, S. Treaty Doc. No. 99–27, 99th Cong. (1986) (revised at Paris, July 24, 1979.)

[8] The movement for international copyright is ostensibly under the aegis of GATT. The Uruguay Round of GATT established an Agreement on Trade-Related Aspects of Intellectual Property Rights (TRIPS) in 1994, to be administered by the World Trade Organization. TRIPS protects general copyright clauses, such as the grant of property in expression and it protects computer programs as literary works. See General Agreement on Tariffs and Trade: Agreement on Trade-Related Aspects of Intellectual Property Rights, Dec. 15, 1993, 33 I.L.M. 81 art. 9 (1994.)

foreign authors without even the pretence of acknowledgement," that foreigners "dumped" cultural goods on the domestic market, and American literature was retarded.[9] This chapter examines such assertions regarding international copyright piracy in the context of the publishing industry. It explores the economics of the book trade in order to assess the welfare effects of unauthorized copying on publishers, authors, and the public in general. I conclude that the United States was not harmed, and likely benefited from its refusal to recognize international copyrights. Toward the end of the nineteenth century, the growth of internationally competitive commercial literature and culture among U.S. "authors" created an incentive for this country to change its laws.

THE EFFECTS OF THE INTERNATIONAL COPYRIGHT ACT OF 1891

The first U.S. Copyright Act of 1790 specified that "nothing in this act shall be construed to extend to prohibit the importation or vending, reprinting or publishing within the United States, of any map, chart, book or books ... by any person not a citizen of the United States."[10] The law therefore explicitly permitted copying of foreign works without payment to their owners. Other countries retaliated by refusing to grant copyright protection for the literary and artistic products of American residents. After a century of lobbying by interested parties on both sides of the Atlantic, based on reasons that ranged from the economic to the moral, in 1891 copyright laws were finally changed to allow foreign artists and authors to obtain copyrights in this country. This section explores the relationship between international copyright piracy and the book trade in order to estimate whether the United States was harmed or benefited from its policy of flouting international laws.

A natural place to start such an inquiry might initially seem to be the patterns of copyright filings. Figure 9.1 shows the level and growth rate in copyrights filed in the United States before and after the 1891 reform. The number of copyright registrations does not reflect any significant alteration over the period and suggests that the movement to harmonization did not affect overall filings. The critical change in the laws to allow foreign authors to obtain American copyright protection was accompanied by an immediate increase in the growth rate of registrations from 4.4 percent to 14.3 percent in 1891 and 11.9 percent in the following year. However, marked changes

---

[9] See John Feather, *Publishing, piracy and politics : an historical study of copyright in Britain*, London ; New York, N.Y. : Mansell, 1994 , p. 154.

[10] Original Copyright Act, First Congress, Second Session, Chapter 15, May 31, 1790: "An Act for the encouragement of learning, by securing the copies of maps, charts, and books, to the authors and proprietors of such copies, during the times herein mentioned." See Library of Congress, *Copyright Enactments of the United States, 1783–1906*, compiled by Thorvald Solberg, Washington, D.C., 1906.

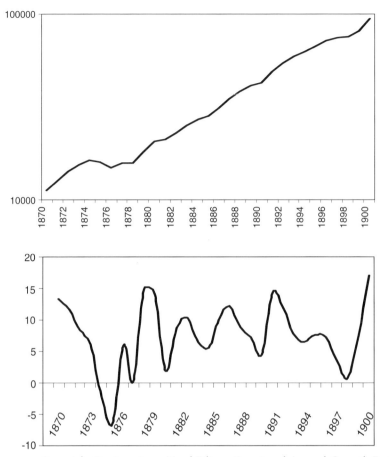

Figure 9.1. Copyright Registrations: Total Filings (Logs) and Annual Growth Rate,
1870–1900
Source: Historical Statistics of the United States, Series W 82–95.

in the growth rates had been a feature of the time series for the previous
two decades as well, so one cannot credibly attribute the pattern entirely to
statutory changes.

In 1900 the U.S. Senate authorized Carroll D. Wright, the Commissioner
of Labor, to investigate the effect of the reforms in the copyright system.
Wright was discouraged from any statistical analysis by the marked lack of
data on the publishing industry, and instead conducted a survey of printers and publishers, to find out whether the new law was viewed as "detrimental or beneficial." Table 9.1 classifies the written answers of respondents to the Wright survey. The survey assessed the impact of the reforms in
terms of four groups: publishers, authors, employees in the printing industry, and the book-reading public. Respondents felt that foreign authors were

Table 9.1. *Effects of Changes in Copyright Law*
Survey of Firms in the U.S. Book Trade, 1900

Effects on:		
American Authors		
Beneficial	44	74.6%
Harmful	2	6.8
Mixed effects	2	3.4
None	9	12.9
Foreign Authors		
Beneficial	54	93.0%
Harmful	0	0.0
Mixed effects	2	3.5
None	2	3.5
Publishers		
Beneficial	52	74.3%
Harmful	13	18.6
Mixed effects	3	4.3
None	2	2.9
Public		
Beneficial	35	59.3%
Harmful	15	25.4
Mixed effects	6	10.2
None	3	5.0
Prices of Books		
Increased	25	47.2%
Decreased	7	13.2
Mixed	6	9.4
No change	16	30.2
Effects of Piracy		
Beneficial	15	23.4%
Harmful	41	64.1
Mixed effects	6	9.4
No effect	2	3.1

Notes: The survey was conducted in accordance with a resolution of the U.S. Senate in 1900. Questions included: "Has the international copyright law been detrimental or beneficial to – a. publishers or book manufacturers; b. printers and their employees; c. American authors; d. foreign authors; e. the book-purchasing public?" "Has the effect of the law been to increase or reduce the selling price of books?" and "Was "piracy" as practiced prior to the enactment of the international copyright law, beneficial or injurious to printers or publishers?" The questions were answered by printers and publishers in Boston (11), Buffalo and New York (34), Chicago (5), Cleveland (3) and Philadelphia (17.) The respondents gave their opinions in essay form, which I have tabulated, not including in the totals instances in which the question was not addressed.

Source: Carroll D. Wright, *A Report on the Effect of the International Copyright Law in the United States*, Washington, D.C.: G.P.O., 1901.

unambiguously better off after the reforms. American authors were held to have benefited because the previous regime had exposed them to "unfair competition" in the form of popular uncopyrighted works, from Britain in particular, which had discouraged the development of domestic literature. Moreover, Americans would no longer face retaliation from other countries. Publishers who dealt in copyrighted books also were better off because they could now exclude unauthorized reprinters, whereas the latter class of publishers were quickly driven into bankruptcy. Printers' unions felt that the reforms had not caused any real change in the circumstances of their members. This was unsurprising, as the reforms included manufacturing provisions to protect domestic printers. As for the public, results were regarded as mixed: prices of copyrighted books had increased, fewer books of the "cheap and nasty sort" from the pens of foreign novelists were available, and the overall quality of available books had improved.

The survey concluded that piracy was not only immoral, it had been economically costly to the United States. In sum, the contributors to Wright's study argued that social welfare had increased as a result of the legal reforms. As a bonus, the move toward harmonization of copyright laws also meant that the United States was in better political standing with other countries. No one explained why the United States for one hundred years had stoutly resisted all invitations to change its international copyright law, despite these allegedly immoral and costly effects.[11] Carroll D. Wright, a sophisticated collator of economic data, was forced to rely on these subjective assessments because of the lack of statistical information on books and the publishing industry (an irony he surely must have appreciated.) Such data are still unavailable and not as complete as one would wish. However, today it is possible to conduct a more systematic analysis of the impact of international copyright laws on the book trade than Wright was able to provide. My analysis employs data on prices, books, the publishing industry, and biographical information about authors. These data are inadequate to precisely estimate the welfare effects of "piracy" in the nineteenth century but do allow us to assess the validity of several assertions that featured in the debate about the impact of copyrights in foreign books.

[11] Indeed, the passage of the Act was in doubt right to the end: "While a member of the Fifty-first Congress, an international copyright bill was reported by the Judiciary Committee, debated for two days, and failed of passage by a negative majority of about forty. Mr. Simonds then redrafted the bill, added its famous thirteenth section, and procured its favorable report to the House. On the third day of the short term he secured its passage through the House, after a vigorous fight, by a majority of about forty. By reason of parliamentary tactics and maneuvers, it had to pass the House, in one shape or another, three times subsequently, each time after a fight over it, the last passage being about two o'clock on the morning of March 4, 1891, the day on which Congress adjourned. For this service in connection with international copyright the government of France conferred upon him the Cross of the Legion of Honor." *Scientific American*, vol. 66 (n.s.), 18 June 1892, p 389.

BOOKS AND AUTHORS

According to Arthur Schlesinger, "So long as publishers . . . could reprint, or pirate, popular English authors without payment of royalty, and so long as readers could buy such volumes far cheaper than books written by Americans, native authorship remained at a marked disadvantage."[12] Piracy discouraged professional authorship among Americans because it was difficult to compete with established foreign authors such as Scott, Dickens, and Tennyson, and as a result "much of beauty, value and interest was lost to the world."[13] In G. H. Putnam's view, "An international copyright is the first step toward that long-awaited-for 'great American novel.'"[14] This argument is somewhat suspect on its face, for a number of reasons. First, it supposes that the highest valued product was deterred, rather than works at the margin. Second, it assumes that there was a high degree of substitutability between cheap reprints and domestic books. Third, if the claim were true, one would expect that domestic authors would respond to the competition by accepting lower royalties and less favorable contracts. Instead, one observes over time *higher* royalties and better terms being offered to American writers.[15]

Such observations do not disprove the counterfactual claim that, if the laws had protected foreign copyrights, even better terms would have prevailed for native writers. But because that possibility cannot be tested, it is more pragmatic to focus on hypotheses that can be assessed for consistency with the available evidence. Consider the claim that foreign books

---

[12] Arthur M. Schlesinger, *The Rise of the City, 1878–1898*, New York: Macmillan, 1933, p. 252.

[13] See Aubert J. Clark, *The Movement for International Copyright in Nineteenth Century America*, Washington, D.C.: Catholic University of America Press, 1960, p. 49: "Writing as a profession would never be attractive to native talent as long as the average author had to compete with the great masters of England whose works were appropriated without cost." Similarly, see *Federal Copyright Records, 1790–1800*, edited and with an introduction by James Gilreath; compiled by Elizabeth Carter Wills, Washington: Library of Congress, U.S. G.P.O., 1987, p. xxiii: "The grant of copyright protection only to American citizens pushed the publishing industry in a direction that injured those who sought to make a living by creative writing in America."

[14] "International Copyright," in *Publishers' Weekly*, Feb. 22 (371) 1879, p. 237.

[15] Many of the earlier books were published at author's risk, or on commission. "Half-profits" or profit-sharing between the publisher and author was also a way of sheltering publishers from risk that prevailed until the 1830s. In the 1840s, reputable authors received an average of 10 percent, and between 10 to 20 percent in royalties. However, there was wide variation in contracts for unknown authors. For instance, as discussed in *Bean v. Carleton et al.*, 12 NYS 519 (1890), Fanny Bean advanced $900 to publishers George W. Carleton & Co, to be repaid when two thousand copies of the book were sold, on the expectation of royalties on sales after the first two thousand copies. Until the 1890s, authors had few means of monitoring their publisher; the 1896 decision in *Savage v. Neely*, for the first time gave authors the right to inspect accounts of their publishers. The improvements in contractual terms could be due to sample selection, if lower quality authors were selected out of the market.

were dominant because they were sold at lower prices than books by American authors.[16] Proponents of the copyright reforms frequently referred to the cheap "Libraries," such as the Fireside and Franklin Square series that published English reprints at a retail price of 10¢, and argued that American authors were driven from the market by such prices.[17] This argument confuses cause and effect, because "dime novels" were a quintessentially American innovation, and reprinters of low-end foreign fiction priced their books to become competitive in this market. The first number of the Lakeside Library that reprinted the works of foreign authors appeared in 1875 *in response to* the success of cheap American fiction, and was followed by the Home, Seaside, and Franklin Square Libraries.

Moreover, one cannot compare the price of a gilt-edged volume of history bound in morocco with a detective story printed on cheap yellow paper. It is necessary to control for factors that might influence price, in addition to nationality, in order to determine whether books by American authors were indeed more expensive than those by foreign authors. Such factors as the literary quality of the book are difficult to quantify, especially since there is likely to be little agreement as to what constitutes a "good book." In order to control for differences across publishing firms, I consider within-firm variation in prices for books published by Ticknor and Fields between 1832 and 1858.[18] Ticknor and Fields (the precursor of Houghton Mifflin) was one of the leading publishers of this period, and was especially noted for its publication of foreign authors such as Dickens, Thackeray, Tennyson, Browning, Kingsley, Reade, and de Quincey. Moreover, the firm also published an impressive roster of well-known American writers including Hawthorne, Longfellow, Thoreau, and Lowell.[19] Other less eminent figures included Josiah Bumstead, the author of a set of best-selling children's readers, and Jacob Abbott, who

---

[16] John Tebbel, *A history of book publishing in the United States*, New York: R. R. Bowker Co., [1972–81], vol. 2, p. 23, cites an 1834 study that stated the average retail price for American authors was $1.20 and for foreign reprints, 75¢. This claim has been repeated in the literature. However, it is unclear how this price was arrived at, and to what it refers, much less what a price that averages across all books indicates.

[17] Indeed, according to a study of the fiction of this era (Quentin Reynolds, *The Fiction Factory*, New York: Random House, 1955), dime novels were initiated by Irwin P. Beadle and Co. in 1860 because they wished to publish American authors: "Its detractors could never deny the fact that this was a peculiarly American institution and not a pale replica of English tales," p. 72.

[18] The firm also published an extensive array of pamphlets, many on commission, which are not included in these data.

[19] According to the editors of the *Cost Book*, "Of the outstanding American writers of the period only three names are lacking from the Ticknor lists." These were Poe, Melville, and Whitman. See Warren S. Tryon and William Charvat (eds), *The cost books of Ticknor and Fields, and their predecessors, 1832–1858*, New York: Bibliographical Society of America, 1949, p. xviii, footnote 7.

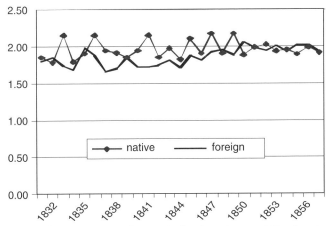

Figure 9.2. Log of Average Prices for Works by American and Foreign Authors
Source: Cost Books of Ticknor and Fields. See Text.

wrote the popular juvenile "Jonas" series. Figure 9.2 shows the pattern over time in the log of the average annual retail price of all books by American authors, relative to foreign authors. There is clearly a lot of noise in the data, especially for prices of American books, which is partly due to the unsettled state of the book trade in the 1830s and 1840s, and partly to heterogeneity among books and authors. However, by the 1850s the two series converge. We need to consider whether these patterns were associated with differences in nationality, holding other variables constant.

Table 9.2 presents the results from a multivariate regression, which examines the influence of variables such as time, gender, type of book, and nationality, on the log of nominal price. The unit of observation is an edition of an individual book published by the firm between 1832 and 1858. The regressions do not support the notion that American books were suffering from competition with cheaper foreign books. First editions are likely to be less predictable and thus more difficult to price than subsequent editions, but even here there is no significant difference between the price of a book by an American and a foreign reprint. Indeed, in the only instance in which the variable for American nationality is significant, the coefficient is negative. Variation in prices is mostly explained by average variable cost.[20] These results

[20] Average cost of publishing reflected strong economies of scale. Hence, independently of piracy, average cost in the United States was likely lower than in Britain because the market of readers was much more extensive in this country. Readers in urban centers in Britain were more likely to belong to commercial lending libraries or book clubs, which again would suggest a more narrow commercial market for an individual work.

Table 9.2. *Regressions of Prices for Books Published by Ticknor and Fields, 1832–1858*

Variable	All Editions			First Edition	
Intercept	4.14***	4.10***	1.34***	4.23***	2.08***
	(39.86)	(39.88)	(15.37)	(20.00)	(8.95)
Time Period					
1840–1844	−0.01	−0.02	0.17	0.29	0.15
	(0.06)	(0.14)	(1.56)	(1.07)	(0.76)
1845–1849	0.06	0.14	0.24***	0.02	0.12
	(0.50)	(0.12)	(3.66)	(0.08)	(0.74)
1850–1854	0.26**	0.15	0.18***	0.35	0.21
	(2.35)	(1.34)	(2.90)	(1.62)	(1.32)
1855–1858	0.31***	0.15	0.18***	0.37	0.21
	(2.85)	(1.37)	(2.90)	(1.68)	(1.30)
American	−0.06	−0.07	−0.12***	0.01	−0.01
Nationality	(1.39)	(1.65)	(4.97)	(0.09)	(0.12)
Gender	–	0.10	0.01	−0.06	−0.06
	–	(1.70)	(0.29)	(0.80)	(0.97)
Fiction	–	0.19***	0.02	−0.18**	−0.09
	–	(3.68)	(0.78)	(2.06)	(1.42)
Poetry	–	0.30***	0.10***	−0.24***	0.00
	–	(5.80)	(3.29)	(2.80)	(0.32)
Log(Average	–	–	0.84***	–	0.61***
cost)	–	–	(41.54)	–	(12.53)
	$R^2 = 0.04$	$R^2 = 0.08$	$R^2 = 0.72$	$R^2 = 0.10$	$R^2 = 0.52$
	$F = 6.06$***	$F = 8.12$***	$F = 216.15$***	$F = 2.48$***	$F = 21.51$***
	$N = 770$	$N = 756$	$N = 753$	$N = 190$	$N = 189$

Notes:
* Significant at 5 percent level
** between 1 and 5 percent
*** 1 percent level or below

Absolute value of t-statistics in parentheses. The observations refer to books published by Ticknor and Fields, and do not include annual publications that are not priced, such as the firm's catalogues. The dependent variable is the log of the stated retail price, unadjusted for inflation. The results for the nontrend variables are qualitatively the same when adjusted for inflation. Costs are variable costs, excluding expenses that the firm allocated to "overhead" (salaries, rent, advertising, insurance, interest, taxes, postage, and cost of travel.) Costs do not include fixed payments for early sheets made to foreign authors. They predominantly comprise production costs (paper, composition and printing, illustrations, binding) and royalties.

Source: Warren S. Tryon and William Charvat (eds), *The cost books of Ticknor and Fields, and their predecessors, 1832–1858*, New York: Bibliographical Society of America, 1949. See text.

suggest that, after controlling for the type of work, the cost of the work, and other objective factors, the prices of American books were *lower* than prices of foreign books. American book prices may have been lower to reflect lower perceived quality or other factors that caused imperfect substitutability between foreign and local products.[21] This is not surprising, as prices are not exogenously and arbitrarily fixed, but vary in accordance with a publisher's estimation of market factors such as the degree of competition and the responsiveness of demand to determinants. As one of the respondents to the Wright survey remarked: "The book-purchasing public has not been seriously affected by the act, inasmuch as the ordinary law of supply and demand is sufficient to protect the general public against unfair prices."[22]

A second question is whether domestic authors were deterred by foreign competition. This would depend on the degree to which books by foreign authors were substitutable for books by American authors. It also would depend on the extent to which foreign works prevailed in the American market. According to one of the leading histories of publishing in this era, by 1850 the majority of books in this country were written by Americans.[23] However, this is not entirely true for all classes of publications. Early in American history the majority of books were reprints of foreign titles.[24] It is important to note that nonfiction titles written by foreigners were less likely to be substitutable for nonfiction written by Americans; consequently, the supply of nonfiction soon tended to be provided by native-born authors. From an early period, grammars, readers, and juvenile texts were written by Americans.[25] Geology, geography, history, and similar works had to be adapted or completely rewritten to be appropriate for an American market.[26]

---

[21] Demand might have been lower for a number of reasons, such as the claim that "The difficulties of early American authorship are often attributed to American prejudice against American literature," p. 42, William Charvat, *Literary Publishing in America, 1790–1850*, Philadelphia: University of Pennsylvania Press, 1959. One may ascribe such "prejudice" to the higher perceived quality of foreign literature.

[22] Carroll D. Wright, *A Report on the Effect of the International Copyright Law in the United States*, Washington, D.C.: G.P.O., 1901, p. 44.

[23] "In all fields of authorship, American books were supplanting the British works. Goodrich estimates that in 1820 American authors wrote 30 per cent of the books, while British authors wrote 70 per cent, but for 1850 his estimate is reversed," p. 124, Hellmut Lehmann-Haupt, *The Book in America*, 2nd edition, New York: Bowker, 1952. Another frequently cited statistic is the claim that, between 1830 and 1842, "nearly half the publications issued in the United States were reprints of English books," and that in 1853 there were 733 new titles, which included 278 English reprints and 35 translations; in 1854, 765 titles and 277 reprints; and in 1855, 1092 titles and 250 reprints. These figures were originally produced by a firm of London booksellers, and reproduced by the *Publisher's Circular and Literary Gazette*, September 13, 2(37), 1856, p. 552. However, the *Gazette* later expressed doubts about the accuracy of the information, especially as even a casual count from publishers' trade lists reveal that the fraction of reprints was manifestly higher.

[24] According to David Saunders, *Authorship and Copyright*, London and New York: Routledge, 1992, p. 156, "Harper's first catalogue contained 234 titles of which 90 percent were English reprints, the same pattern being true for Wiley and for Putnam."

[25] See Gilreath, *Federal Copyright Records*, p. xxii.

[26] Carey and Lea had this experience when they wished to "pirate" a British encyclopaedia to publish in the American market. See David Kaser, *Messrs. Carey & Lea of Philadelphia; a study in the history of the booktrade*, Philadelphia: University of Pennsylvania Press, 1957.

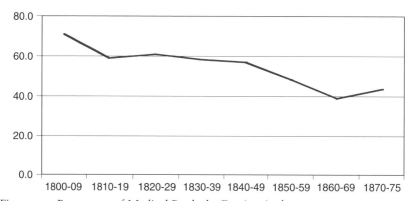

Figure 9.3. Percentage of Medical Books by Foreign Authors
Source: Francesco Cardasco, *Medical publishing in 19th century America: Lea of Philadelphia, William Wood & Company of New York City, and F. E. Boericke of Philadelphia,* Fairview, N.J.: Junius-Vaughn Press, 1990.

As early as 1835, American authors produced a third of poetry and drama publications, whereas 65 percent of science books, 82 percent of business-related texts, and 75 percent of law books were by Americans.[27] Figure 9.3 shows the fraction of medical books that were written by foreigners. Until the middle of the century, foreign authors were credited with about half of all medical books, but this figure fell to approximately 40 percent soon after. This was true even though the high fixed cost of production for medical volumes deterred rivalry among publishers of reprints, who feared predatory behavior would lead to large losses.[28] Thus, publishers of schoolbooks, legal treatises, medical volumes, and other nonfiction did not feel that the reforms of 1891 were relevant to their undertakings. According to a leading producer of educational texts, "The question of international copyright law is one which we have not considered very much, as it does not materially affect the schoolbook business. It has almost wholly to do with general literature. Each country has its own methods of teaching, and the school books of one country can not be pirated in another to advantage."[29]

By contrast, foreign authors dominated the field of fiction, so it is worth exploring whether there might be some validity to the idea that there was no Great American Novel in the nineteenth century because of the international

[27] See David Kaser, p. 71.
[28] William Wood Co., in the Wright survey, p. 88, testified that "Medical works of English authors have but a limited sale in the United States, and even when, with rare exceptions, a book of this class is found to prove unexpectedly popular, the cost of manufacturing such books is so great as to deter one publisher from reprinting on another, with it absolutely understood that the first party would reduce his price so as to make any competition ruinous."
[29] Ginn & Co. in the Wright survey, p. 74.

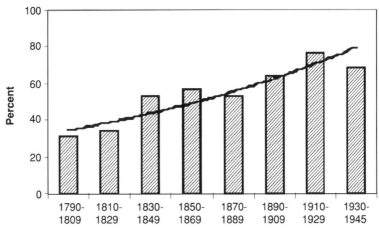

Figure 9.4. Best-Sellers by American Authors as a Percentage of Total, 1790–1945
Source: Frank Luther Mott, *Golden Multitudes: The Story of Best Sellers in the United States,*
New York: Macmillan, 1947, Appendix A. Best sellers are defined as books that had a total sale
of 1 percent of the population in the decade of publication. The list does not include bibles, hym-
nals, textbooks, almanacs, cookbooks, medical works, manuals, and reference books. Authors'
nationalities are determined by country of birth.

copyright laws. Although I agree that Americans did not produce any great
works of literature during this period (*Moby Dick* notwithstanding), I doubt
that the lacuna was because of the absence of copyright protection for foreign books. Figure 9.4 suggests a gradual decline over time in the role of
foreign authorship. In the first decade of the nineteenth century, 90 percent of titles comprised foreign reprints. In the period between 1790 and
1829, two thirds of all authors of fiction best-sellers were foreign.[30] A discrete change in the relative success of American writers occurred after the
1830s with the entrance of such literary figures as James Fenimore Cooper,
Henry Wadsworth Longfellow, Nathaniel Hawthorne, and R. H. Dana. By
the early twentieth century, Americans accounted for the majority of best-
selling authors in this country.[31] This gradual increase over time in the relative
importance of domestic authorship may have been because of a natural evolutionary process, or may have been caused by the change in the copyright
laws. Some have claimed that the cadre of professional American authors –
especially of novels – was small or nonexistent because of foreign competition. For instance, the biographer of Edgar Allan Poe contended that Poe

[30] See also Mott, *Golden Multitudes*, pp. 92–3.
[31] Alice P. Hackett and James Henry Burke, *Eighty Years of Best Sellers, 1895–1975*, New
York: Bowker, 1977, imply a more abrupt change, because they argue that in 1895 American
authors accounted for two of the top ten best-sellers, but by 1910 nine of the top ten were
written by Americans

switched to short story format because he was unable to profit from the market for novels.[32]

If it were indeed true that professional authorship was deterred by a lack of copyright protection, the reforms in 1891 should be associated with a rise in the number of Americans whose profession was writing, holding other factors constant. Some scholars define professional authors as individuals whose sole occupation or source of income was from writing. However, this definition is problematic as it is biased toward women writers, who were markedly less likely than men to engage in jobs outside the home.[33] It also to some extent equates professionalism with success, as one is less likely to depend on writing for one's income unless writing provides greater income than available alternatives. I define a professional author as a person who is listed in a biographical dictionary as an author, or had written more than ten books.

In order to investigate whether copyright reforms influenced the propensity for Americans to become professional authors, I constructed a random sample of 758 individuals from biographical dictionaries of authors. Table 9.3 describes the characteristics of the sample. Academic and religious books are less likely to be written for monetary returns, and their authors probably benefited from the wider circulation that lack of international copyright encouraged. However, the writers of these works declined in importance relative to writers of fiction, a category that grew from 6.4 percent before 1830 to 26.4 percent by the 1870s. The growth in fiction authors was associated with an increase in the number of books per author over the same period. Fifty-nine percent of the ninety-eight women writers in the sample published in the fiction-only category, but they did not account for more than 39 percent of all fiction authors. Improvements in transportation, and increases in the academic population probably played a large role in enabling individuals who lived outside the major publishing centers to become writers despite the distance.[34] As the market expanded, a larger fraction of writers could become professional authors.

The average age of a writer of nonfiction at time of first publication was approximately forty years, relative to fiction where age at first publication

---

[32] See Israfel Hervey Allen, *The Life and Times of Edgar Allan Poe*, New York: Farrar & Rinehart, 1934, p. 403. An alternative view (mine) is that even in the absence of any foreign competition Poe would have remained an indifferent novelist.

[33] See Lawrence Buell, *New England Literary Culture: From Revolution through Renaissance*, Cambridge: Cambridge University Press, 1986, pp. 375–92, who argues that women writers may have been the first professional writers, because they had few other sources of employment. Between 1820 and 1865 writing was the sole source of income for 34 percent of women authors, relative to 17 percent for men.

[34] For a discussion of the influence of transportation on book distribution, see Ronald Zboray, "The Transportation Revolution and Antebellum Book Distribution Reconsidered," *Journal of Southern History*, vol. 14 (3) 1948: 305–330.

Table 9.3. *Characteristics of Authorship by Birth Cohort*

	Before 1830	1830–49	1850–69	1870–89	Total Number
		Year of Birth of Author			**Total Number**
		Percent of Authors			
Type of Book					
Religion	21.1	9.8	5.1	6.2	98
Fiction and juvenile	6.4	11.0	20.5	26.4	104
Poetry and drama	1.6	4.9	11.0	9.1	41
Both fiction and nonfiction	7.7	8.0	14.6	12.5	75
Nonfiction	63.2	66.3	48.8	45.8	439
Total Number of Books Published					
1–2 books	35.8	37.4	21.9	10.3	218
3–5	37.1	33.7	36.5	34.5	271
6–10	21.1	19.6	27.7	38.6	192
More than 10	6.1	9.2	13.9	16.6	77
Profession					
Listed as Author	8.0	13.5	26.3	24.5	118
Professional Author	7.8	12.4	17.6	18.2	88
Region of Birth					
Mid-Atlantic	27.5	37.7	28.4	27.1	223
Midwest	1.9	9.9	18.7	33.3	95
New England	46.9	35.2	26.9	16.7	262
South	12.0	6.8	14.9	12.5	86
Foreign	11.7	10.5	11.2	10.4	83
Residence					
Mid-Atlantic	45.6	46.5	36.5	39.0	246
Midwest	5.6	12.7	16.2	27.1	62
New England	29.1	33.1	35.1	20.3	168
South	15.4	6.3	6.8	11.9	65
Foreign	4.2	1.4	5.4	1.7	19
Residence in Urban Center					
(Philadelphia, Boston, New York, Chicago)	33.2	31.9	17.5	0.7	181
Percentage Women	6.4	19.0	19.7	14.5	99
Average Age at First Publication (Years)					
Nonfiction	42.6	44.6	41.9	40.2	
Fiction	30.2	31.7	33.9	34.8	
Sample Size	313	163	137	145	754

Source: See text for sources.

was in the early thirties. Because the data are organized by birth cohort, this implies that authors of fiction who were born in the 1860s were the most likely to have been influenced in their choices by the change in the copyright laws. The regressions in Table 9.4 are directed toward the question of whether writers were discouraged from choosing authorship as a career by the lack of international copyright protection. The results do not seem to support this contention. The first set of regressions reports the coefficients from a linear probability model that estimates the factors that influenced whether an author was a professional author. The time dummies suggest a fairly steady increase over time in the likelihood of this occurrence, with the biggest increase in the cohort born in the 1880s, who would have become writers around 1910 or 1920.[35] For the critical category of fiction, the biggest increase occurs for the birth cohort between the 1840s and the 1850s, the members of which would have entered the market well before 1891.

Although these results do not support the hypothesis that the lack of copyright protection discouraged professional authors, this does not imply that piracy was of little economic significance. Marginal authors may have been discouraged, although the deterrence of yet another poor work of fiction probably did not constitute a great loss to social welfare. The inframarginal foreign writers were able to obtain returns through competition on the part of American publishers to gain their "authorization." Successful authors directed more effort to the production of complementary goods and services, such as readings and lecture tours. They were able to exploit network effects as piracy increased the scale of readership and encouraged the growth of a mass market in the United States, in some instances far in excess of the high-priced and restricted European markets. Charles Dickens, who publicly and in his writings launched bitter diatribes against "the continental Brigands" in the United States, was a major beneficiary of such bandwagon and network effects. He played publishers off against each other, and as many as four companies paid him large sums and had legitimate claims for considering themselves his sole American representative. Moreover, Dickens was able to parlay his popularity among readers into a heightened demand for complementary lectures. His U.S. reading tour of 1867–1868 comprised seventy-six appearances that earned the author the astonishing sum of $228,000 in total receipts.[36]

---

[35] The notion that cultural progress was evolutionary is reflected in the views of a German judge at the Philadelphia exhibition in 1876, who opined that "the United States of America already outstripped most of the older nations, except in matters of art, and as art required time, America would eventually not be behind other nations even in that," cited in "Arguments before the Committee on Patents of the Senate and House of Representatives," 45th Congress, 2nd Session, Mis. Doc. No. 50, Washington, D.C.:Government Printing Office, 1878: 445.

[36] See Andrew J. Kappel and Robert L. Patten, "Dickens' Second American Tour and his "Utterly Worthless and Profitless" American Rights," in *Dickens Studies Annual*, vol. 7 (1978): 1–33.

Table 9.4. *Factors Influencing Authorship in the Nineteenth Century*

Variable	Dependent Variable: Professional Authors		Dependent Variable: Fiction Authors	
Intercept	0.12***	0.09*	0.02	−0.00**
	(2.20)	(1.92)	(0.41)	(0.94)
Decade of Birth				
1810–1819	0.00	0.02	−0.02	0.01
	(0.04)	(0.36)	(0.27)	(0.13)
1820–1829	0.11*	0.09*	0.02	0.01
	(1.93)	(1.85)	(0.39)	(0.17)
1830–1839	0.11*	0.03	0.12**	0.06
	(1.95)	(0.63)	(2.22)	(1.29)
1840–1849	0.14***	0.08	0.08	0.03
	(2.49)	(1.49)	(1.58)	(0.56)
1850–1859	0.29***	0.15***	0.26***	0.19***
	(4.91)	(2.77)	(4.73)	(3.68)
1860–1869	0.28***	0.17***	0.23***	0.18***
	(4.45)	(2.97)	(4.03)	(3.38)
1870–1879	0.20***	0.09	0.24***	0.21***
	(3.30)	(1.74)	(4.34)	(4.16)
1880–1889	0.48***	0.30***	0.35***	0.29***
	(4.94)	(3.42)	(5.94)	(5.24)
Region of Birth				
Midwest	−0.12	−0.12*	−0.01	0.02
	(1.85)	(1.96)	(0.13)	(0.34)
South	−0.00	−0.04	0.10	0.10*
	(0.04)	(0.72)	(1.78)	(1.94)
Mid-Atlantic	−0.04	−0.06	0.06	0.06
	(0.81)	(1.24)	(1.28)	(1.42)
New England	0.01	−0.01	0.06	0.05
	(0.19)	(0.29)	(1.34)	(1.14)
	–	0.25***	–	0.41***
Gender				
	–	(5.47)	–	(10.57)
Fiction	–	0.39***	–	–
	–	(9.60)	–	–
	R² = 0.08	R² = 0.28	R² = 0.09	R² = 0.21
	F = 4.64***	F = 19.04***	F = 6.28***	F = 15.25***
	N = 699	N = 699	N = 754	N = 754

Notes:
* Significant at 5 percent level
** between 1 and 5 percent
*** 1 percent level or below

Absolute value of t-statistics in parentheses. The dependent variable in first two regressions has the value of 1 if the individual's primary occupation was listed as author OR if he or she had published more than 10 books. The dependent variable in the next two regressions takes on a value of 1 if the individual's primary occupation was listed as a fiction author. The excluded regional dummy represents authors who were born in other countries. Gender is 0 if male, 1 if female. Fiction is a dummy that has a value of 1 if the author published only in the area of fiction, poetry or drama. The results do not vary if a probit or logit model is used instead of the linear probability model.

Source: See text for sources.

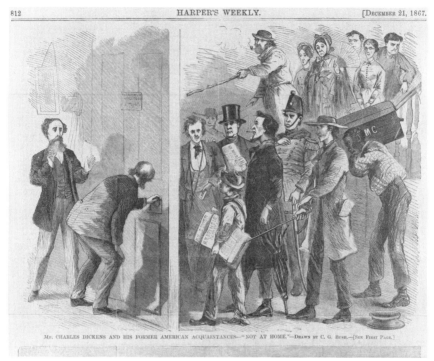

Mr. CHARLES DICKENS AND HIS FORMER AMERICAN ACQUAINTANCES—"NOT AT HOME."—DRAWN BY C. G. BUSH.—[SEE FIRST PAGE.]

Illustration 11. This 1867 drawing by C.G. Bush is entitled "Mr. Charles Dickens and his former American acquaintants – "not at home." The illustration shows a fearful Charles Dickens, who assailed U.S. copyright piracy, being confronted by irate American characters from his novel *Martin Chuzzlewit*. (*Source*: Library of Congress.)

The lack of foreign copyrights led to some misallocation of efforts, such as attempts to circumvent the rules. Authors changed their residence temporarily when books were about to be published in order to qualify for copyright protection. The English author Frederick Marryat claimed copyright protection because he was living in the United States in 1838 but failed because the court ruled that one also must have the intention to become a citizen. American authors visited Canada in order to satisfy the more lenient British regulations, which permitted copyright protection for books whose authors were within the borders of Britain or its colonies at time of publication. Others obtained copyright overseas by arranging to coauthor with a foreign citizen. T. H. Huxley adopted this strategy, arranging to coauthor with "a young Yankee friend. . . . Otherwise the thing would be pillaged at once."[37] An American publisher suggested that Rudyard Kipling should find "a hack writer, whose name would be of use simply on account of its carrying the copyright." Harriet Beecher Stowe proposed a partnership with

[37] From p. 70, Simon Nowell-Smith, *International Copyright Law and the Publisher in the Reign of Queen Victoria*, Oxford: Clarendon Press, 1968.

Elizabeth Gaskell, so they could "secure copyright mutually in our respective countries and divide the profits."[38]

Courts were somewhat sympathetic to these stratagems, as revealed in a lawsuit involving the *Encyclopaedia Britannica*. The British publishers, Adam & Charles Black, of Edinburgh, Scotland, engaged a number of American contributors for the volumes, and these individuals obtained copyright protection, which infringers of the *Encyclopaedia* challenged as a mere evasion of the law. The court ruled that "The acts of the Messrs. Black were for the purpose of making a use of the statutes which might assist them against pecuniary loss, and give them a more unobstructed field for their large commercial venture.... There was no impropriety in soliciting competent citizens of the United States to write upon its history, and I can perceive no unfairness or injustice toward the defendant company in the plaintiffs' use of the copyright laws for their pecuniary advantage, and as a weapon with which to repel a competition which is more enterprising than considerate."[39]

PUBLISHERS

The previous analysis related to authors, but it is widely acknowledged that copyrights in books during this period tended to be the concern of publishers rather than of authors (although the two are naturally not independent of each other.) Copyright in Europe was largely enforced to regulate the book

---

[38] Susan Coultrap-McQuin, *Doing Literary Business*, Chapel Hill: UNC Press, 1990, p. 89. Elizabeth Gaskell was not persuaded by the argument.

[39] See *Black v. Henry G. Allen Co.*, New York, 42 F. 618, June 26, 1890. "The Encyclopaedia Britannica, as a whole, was the production of aliens, who could obtain no copyright in this country, and is a work of great value to the whole people. The employment of citizens of the United States to write articles which were to be used in some of its volumes, and the purchase of an interest in the copyright of such articles, were an attempt to deprive the defendant, and other like-minded persons, of a privilege which they would have otherwise enjoyed, and were for the purpose of giving the foreign owners of the encyclopaedia an advantage in the sales of the work in this country. The attempt contained an element of unfairness, because the book, if written by foreigners, could be reproduced here, and the complainants have only a color of copyright interest, and therefore should not receive the sanction of the courts.... The acts of the Messrs. Black were for the purpose of making a use of the statutes which might assist them against pecuniary loss, and give them a more unobstructed field for their large commercial venture. The disputed point is whether there was an element of fraud or injustice in the scheme which would prevent a court from regarding it with favor." See also *Carte v. Evans*, Circuit Court, D. Massachusetts, 27 F. 861 (June 21, 1886) which related to a pianoforte arrangement for Gilbert and Sullivan's *Mikado*: "There is nothing in our copyright law to prevent one of our own citizens from taking out a copyright of an original work composed by him, even though the work of composition was performed at the procurement and in the employment of an alien; or from assigning his copyright to an alien under an agreement made either before or after the composing of the work. A nonresident foreigner is not within our copyright law, but he may take and hold by assignment a copyright granted to one of our own citizens. The proprietor as well as the author is entitled to enter the work for copyright."

trade and to ensure that publications were non-seditious. Early publishers obtained copyrights in the books they produced, and authors frequently sold the copyright to the publisher outright, thus transferring all risk in return for a potentially lower but more certain payment. Similarly, in the United States it was a common practice from the first decades of the copyright statute for the publisher to hold the copyright in a book. Even when authors retained the copyright, publishers were most at risk because they were required to make large fixed investments that might be lost if the sales of the book were low because of piracy.

As discussed in Chapter 8, publishers in this country were able to prevent unauthorized copying of books by American authors, and to enforce their property rights in the United States through the threat of legal action.[40] The growth in litigation was minimal until the 1880s, suggesting that infringement of domestic authors was within manageable proportions.[41] Many of the early copyright cases dealt with genuine questions regarding the boundaries of property rights in literary, dramatic, and artistic works, rather than blatant disregard of the claims of legitimate property owners. However, the situation was quite different for books by foreign authors in which no copyright protection existed. If all firms produced rival editions, competition was likely to drive prices down to marginal cost, in which case the high initial fixed investments would not be recovered.

Throughout the period, publishers attempted to avoid "ruinous competition" and engaged in numerous unsuccessful attempts to fix prices. In the early years of the nineteenth century, publishers engaged in copyright races in order to be the first in the market with popular foreign books such as the works of Sir Walter Scott.[42] A Waverley novel could be reprinted within twenty-four hours through a gang system in which the book was divided among as many as a dozen printers working at full capacity. Carey & Lea, a prominent Philadelphia firm, saturated the frontier markets before selling in New York, where rival printers stood ready to reprint at the first appearance of the book. If they judged the size of the market accurately, the winners of the race were able to sell all copies that they had printed, whereas the other firms lost their initial outlay.

---

[40] The landmark Supreme Court case, *Wheaton v. Peters*, 33 U.S. 591 (1834), did not recognize state common law rights for publications, in the interests of a national, uniform policy. Thus, the boundaries of property in patents and copyrights in this country are specified by federal statute and enforced by litigation in the federal courts. The Supreme Court found that no common-law copyright protection existed for published works, which were products of the existing statutes. Unpublished works, however, were protected under common law. The dissenting minority opinion argued that authors held an inherent right in their creations beyond their statutory right.

[41] See Kaser, p. 143, "the second quarter of the nineteenth century saw few copyright violations disturb the comparative quiet of the domestic publishing scene."

[42] The details about the firm of Carey & Lea are from Kaser, 1957.

One of the consequences of copyright races was a greater likelihood of mistakes or deliberate alterations in the attempt to be first and to reduce costs.[43] For instance, Carey & Lea paid Sir Walter Scott $1,475 for an early manuscript copy of his *Life of Napoleon*. Subsequently, readers were concerned that Scott had made changes after the proofs had been pulled, and these amendments were not incorporated in the American edition. Within one month of the initial American publication date, a small New York firm produced an abridged version, without the author's consent, which was advertised as preferable to the "voluminous" original. Complaints also were rife about Carey & Lea's edition of *The Pirate*, which had omitted an entire chapter. Robert Browning sent a list of errata to Ticknor and Fields, in the hope that the American edition would be updated, but the corrections were never made.[44] Other complaints included charges that the spelling in Macaulay's *History of England* was Americanized, that hack authors were sometimes put to the task of creating a version that was more likely to appeal to native tastes, or even that enterprising Yankees marketed their work under the guise of a more meritorious foreign author's name. These allegations might suggest that the lack of formal copyrights and the prevalence of copyright races led to lower concerns about quality in the literary market. However, if consumers cared about quality over price, this would have created an incentive for sorting among publishers, thus leading to appropriation through reputation and, indeed, the more "reputable" publishers were able to secure greater returns in part because they offered products that were more likely to be free of defects.

To the firm that won the race, the profitability of foreign books was likely to be higher than for American works. But this was partly because the market for writers such as Scott and Dickens was more predictable and certain, independent of copyright questions. By trading in on the established reputation of foreign authors, the publisher also avoided high advertising and marketing costs. For the winner of the race, foreign books entailed less risk at lower cost and higher margins. But competition and the probability of being the loser in the race decreased these advantages. As markets expanded and the probability of copying increased, the relative advantage to publishers of some means of exclusion became greater. Some publishers made advance payments for early proof sheets to get an advantage over others who waited until the first imprinting. Henry Carey paid an agent $250 per year to send English titles to his firm in Philadelphia, and was so concerned about the delay of several days at the New York customhouse that he hired

---

[43] "Speed was of the greatest importance in any reprinting venture; and speed bred carelessness. American editions became more and more sloppily printed and bound. Workmanship degenerated. Proofreaders corrected only the most obvious errors. Printed sheets and bindings were often not properly pressed," according to Kaser, p. 92.

[44] *Cost Books*, p. 338.

another agent in New York to expedite the process.[45] Ticknor and Fields paid foreign authors significant sums for early sheets in the form of royalties, or simple lump sums out of profits. For instance, the company offered £60 for the advance sheets of Robert Browning's *Men and Women* in 1855, and the following year paid £100 for the early sheets and engravings for Mayne Reid's juvenile fiction work, *The Bush Boys*. The firm also sent several unsolicited payments to Tennyson over the years out of their profits on his poetry reprints. A number of shrewd authors exploited the competitive nature of the industry by playing one publishing firm off against the others. Charles Dickens abandoned Harper Brothers in April 1867, and proclaimed that Ticknor and Fields had become the only "authorized" representatives in America of his books. Payments to foreigners ensured the coincidence of publishers' and authors' interests, and were recognized by reputable publishers as "copyrights."[46] However, they naturally did not confer property rights that could be enforced at law.[47]

These were problems that publishers in England had faced before, in the market for books that were out of copyright, such as Shakespeare and Fielding.[48] Their solution was to collude in the form of strictly regulated cartels or "printing congers." Cooperation resulted in risk sharing and a greater ability to cover fixed expenses. The congers created divisible property in books that they traded, such as a 160th share in Samuel Johnson's *Dictionary* that was sold for £23 in 1805. The unstable copyright races in the United States similarly settled down during the 1840s to collusive standards that were termed "trade custom" or "courtesy of the trade." Publishing houses were acknowledged to have the right to reprint specific authors. For instance, Harper Brothers were associated with Edward Bulwer-Lytton, whereas Frederick Marryat was customarily reprinted by Carey & Lea. In

---

[45] The distance between Philadelphia and New York translated into a significant disadvantage for publishers in Philadelphia, and may ultimately have granted New York its precedence in the publishing industry.

[46] See the exchange between Charles Reade and Ticknor and Fields, p. 372, *Cost Books*. Reade authorized the firm to reprint his work *It is Never Too Late to Mend*. When it seemed that the Appletons would publish another edition, he wrote to Ticknor and Fields that this was unlikely because Appleton would desist when they found out that they would have to publish with a one-month delay behind Ticknor: "They might do the wrong thing for the Tea, but they are too respectable to do it for the Tea leaves!"

[47] As late as 1902, this issue was brought before the courts. See *Fraser v. Yack et al.* 116 F. 285 May 6 (1902): "We are of opinion that the contract conferred no rights of proprietorship in the manuscript, but only the right of publication coincidently with or in advance of the publication of the work in England."

[48] See A. S. Collins, *Authorship in the Days of Johnson*, London: Robert Holden and Co., 1927. Aileen Fyfe, "Copyrights and competition: producing and protecting children's books in the nineteenth century," *Publishing History*, vol. 45 (1999): 35–59, argues that the British "share-book" system survived until the middle of the 19th century in the market for children's books. The system served as a means through which participants could spread and share risk, raise capital, and also control competition.

the case of newer authors, the first publisher to receive the item or the first to list the work in a trade publication was deemed to have the right to exclude other reprinters. Firms that violated these rules were punished or at least threatened with punishment.[49]

If publishers were harmed by the lack of legal copyright we would expect that this would be reflected in their profits, which would tend to be declining or negative as a result of the competition. Table 9.5 presents information on the profit margins for Ticknor and Fields, one of the leading reprinters in the United States during the nineteenth century. Ticknor and Fields was well known for the quality of its poetry publications, and the results show that they provided a source of profit for the firm relative to other types of books. The lack of statistical significance on the time dummies before 1860 in these regressions do not support the view that profits were declining as a result of competition. Profits were somewhat higher for foreign titles, as shown by the negative coefficient on the dummy variable representing American nationality, but the magnitude of the effect is not large, especially as the costs do not include early lump-sum payments to foreign authors. The publishing industry was able to secure excess returns because, in the two decades before the Civil War, competition among the major firms had settled into a relatively stable situation of tacit collusion. American firms, like their British counterparts in the previous century, were able to appropriate returns from synthetic copyrights that were created by publishers in the absence of legal protection.

The case of *Sheldon v. Houghton*, 21 F. Cas 1239 (1865), illustrates the role of synthetic copyrights in the publishing industry. The very fact that a firm would file a plea for the court to protect their claim indicates how vested a right it had become. The lawsuit emphasized that these rights were considered to be "very valuable, and is often made the subject of contracts, sales, and transfers, among booksellers and publishers."[50] Henry Houghton,

---

[49] According to Kaser, p. 150, "[Henry Carey] wrote almost weekly to the New York firm [Harpers] warning them, threatening them, advising them, not to challenge his firm to an all-out war."

[50] The full claim noted "that, by the custom of the trade of booksellers and publishers in the United States, when any person or firm engaged in that business has undertaken the printing, publication and sale of a book not the subject of statute copyright, and has actually printed, published, and offered an edition of such book to the public for sale, other persons and firms in the same trade, having respect to the trade priority so acquired in the publication and sale of such book, or the particular edition thereof, refrain from entering into competition with such publisher by publishing such book in a rival edition, and that thereby, and by reason and operation of the custom aforesaid, the publication of such book becomes a good will in the hands of the person or firm so first publishing the same, where such book is one for which there is an extensive popular demand, and especially in the case of foreign authors of established reputation, whose works are not the subject of statute copyright in this country, and that such good will is often very valuable, and is often made the subject of contracts, sales, and transfers, among booksellers and publishers."

Table 9.5. *Profit Margins for Ticknor and Fields, 1832–1858*
(Weighted by Number of Copies Published)

Variable	(1)	(2)
Intercept	−0.61***	−0.57***
	(15.99)	(15.34)
Time Period		
1840–1844	0.01	−0.01
	(0.14)	(0.12)
1845–1849	0.03	0.06
	(0.77)	(1.43)
1850–1854	−0.01	0.00
	(0.28)	(0.04)
1855–1858	0.01	0.01
	(0.28)	(0.26)
Gender	0.02	0.04***
	(1.12)	(2.37)
Fiction	0.02	0.01
	(1.39)	(0.63)
Poetry	0.08***	0.08***
	(4.95)	(5.24)
Edition	0.01***	0.01***
	(3.27)	(4.66)
American Nationality	–	−0.08***
	–	(6.71)
	$R^2 = 0.06$	$R^2 = 0.11$
	$F = 5.72***$	$F = 10.38***$
	$N = 750$	$N = 750$

Notes:
* Significant at 5 percent level
** between 1 and 5 percent
*** 1 percent level or below
Absolute value of t-statistics in parentheses. The observations refer to editions published by Ticknor and Fields, and do not include annual publications that are not priced such as the firm's catalogues. The dependent variable is the profit margin ([price − average cost]/price.) The data are unadjusted for inflation; the conclusions are unchanged when the data are adjusted for inflation. The dummy variable American has a value of 0 if foreign, 1 if American; Gender takes a value of 0 if male, 1 if female. Fiction includes drama and juvenile fiction. The regressions are weighted by the number of copies of each edition that was published. Because some copies may have been sold at a discount of the retail price, revenues are likely overestimated. The firm made fixed payments to foreign authors that were not always recorded in the cost books so costs for foreign works are underestimated. Costs refer to publishing costs, and exclude labor costs and certain fixed expenses such as advertising.
Source: Cost Books of Ticknor and Fields.

who purchased the initial synthetic right from O. W. Wight, had formed a partnership with Sheldon & Co. of New York to publish, print, and market the "Household Edition" of Charles Dickens's works. In 1865 Houghton decided to terminate the contract, which Sheldon contested in court because the market value of the publication right had increased under the partnership to some $30,000. The plaintiff wished the court to acknowledge and protect his claim in the disputed property because "such custom is a reasonable one, and tends to prevent injurious competition in business, and to the investment of capital in publishing enterprises that are of advantage to the reading public."

The court dismissed the case, because "if anything which can be called, in any legal sense, property, was transferred to this partnership, it must have been that incorporeal right to publishing this edition of Dickens . . . founded upon the custom of the trade to forbear competition." However, this form of trade courtesy was "very far from being a legal custom, furnishing a solid foundation upon which an inviolable title to property can rest, which courts can protect from invasion. . . . It may be an advantage to the party enjoying it for the time being, but its protection rests in the voluntary and unconstrained forbearance of the trade. I know of no way in which the publishers of this country can republish the works of a foreign author, and secure to themselves the exclusive right to such publication. . . . For this court to recognize any other literary property in the works of a foreign author, would contravene the settled policy of Congress." Thus, synthetic rights differed from copyrights in the degree of security that was offered by the enforcement power of the courts. Nevertheless, in the absence of legal property rights in foreign works, synthetic copyrights were able to transform a competitive environment into a quasi-monopolistic arena. These title-specific rights of exclusion decreased uncertainty, enabled publishers to recoup their fixed costs, and avoided the wasteful duplication of resources that would otherwise have occurred. In short, American publishers were able to achieve some degree of appropriation through industry structure rather than through government-mandated monopolies.[51]

CONCLUSIONS

The question of the appropriate role of intellectual property in development is complex. Few studies provide empirical assessments, especially from the point of view of developing countries. Theoretical models yield ambiguous answers to the question of whether "piracy" results in net welfare benefits or costs, and whether the interests of all parties coincide or conflict. For

---

[51] Foreign firms that attempted to use unfair competition laws to enforce their perceived rights were also unsuccessful at law. See *Black v. Ehrich*, 44 F. 793 (1891), one of the numerous cases brought by the British publishers of the *Encyclopaedia Britannica*.

instance, studies argue that profits fall when firms are unable to appropriate the benefits of copies of their products, but others point out that price discrimination of nonprivate goods can result in net welfare benefits for society and for the individual firm.[52] Economists have hypothesized that infringement can lead firms to adopt a strategy of producing higher-quality commodities if the cost of imitation increases with quality.[53] Static welfare gains to developing countries from infringement may exceed the costs to the owners of copyright in developed countries, but the dynamic consequences of ignoring intellectual property rights are difficult to estimate. Thus, some insights may be gleaned from a period when the United States was itself a developing country. The United States maintained very different policies toward authors and inventors. In the case of patents, the social good was seen as coincident with the award of secure and strong patent rights to individual inventors, regardless of their citizenship. However, the rationale for copyrights was held to be much weaker because of the lower incentive from their grant, and the higher social costs of restricted access.

This chapter investigated the welfare effects of "piracy" of foreign copyrighted material, and focused on the impact on authors, publishers, and the general public in the nineteenth century. Claims had been made that prices of foreign books were so low that Americans could not compete; that professional authors were deterred by foreign competition; that publishers' profits were driven down over time by the inability to exclude competitors; and that American society suffered from a lack of quality domestic literature as a result of copyright policies. I find little support for these contentions. Publishers appear to have priced in accordance with the dictates of the market. In cases where American literature was relatively cheaper, prices were for the most part determined by lower demand or lower perceived quality. According to conventional economic analysis, in the absence of legal protection the market prices of books would be competitively bid down to marginal cost, and publishers would be deterred by their inability to recover fixed costs. This was not the case, for, despite the lack of copyright protection, publishing houses were able to appropriate returns through cartels, price discrimination across firms, and the creation of synthetic copyrights. Moreover, the expansion in the size of the market generated incentives for technological improvements in publishing that gave American firms a long-run competitive advantage in developing capital-intensive copyrighted products. However, the lack of formal enforcement of property rights may have led to higher costs of exclusion for the industry, lower investments in quality, and a diversion of resources from production to rent-seeking.

---

[52] See Harold Demsetz, "The Private Production of Public Goods," *Journal of Law & Economics*, vol. 13 (2) 1970: 293–306.

[53] Lynne M. Pepall and Daniel J. Richards, "Innovation, Imitation, and Social Welfare," *Southern Economic Journal*, vol. 60 (3) 1994: 673–84.

After the copyright reforms in 1891, both English and American authors were disappointed to find that the change in the law did not lead to significant gains from foreign royalties.[54] This is consistent with the regression results, which suggested that professional American authorship seems to have developed through a natural evolutionary process. Foreign authors may even have benefited from the lack of copyright protection in the United States. Despite the cartelization of publishing, competition for these synthetic copyrights ensured that foreign authors were able to bid up payments that American firms made to secure the right to be first on the market. It also can be argued that foreign authors were able to reap higher total returns from the expansion of the market to incorporate the lower valuations of a developing country. In other words, the lack of copyright protection functioned as a form of price discrimination, where the product was sold at a higher price in the developed country, and at a lower price in the poorer country. Returns under such circumstances may be higher for goods that are associated with demand externalities or network effects, such as "best-sellers," where consumer valuation of the book increases with the size of the market.[55] Even if the American copyright rules created a public good in foreign cultural products, bundling of complementary goods and services such as lecture tours with an expanding demand for free music, art, and literature also could increase potential returns to foreign producers.

In general, the greater the responsiveness of authors to financial returns, the stronger the case for copyright protection, other things being equal. It is important to emphasize that the converse also is true, as social welfare will increase from weak or no copyright protection in cases in which authors are not primarily motivated by the prospect of pecuniary returns. For instance, financial incentives to authors tend to be relatively unimportant in case of nonfiction, where authors benefit more from diffusion (proselytizing and reputational effects), and we have noted the predominance of nonfiction titles in the earlier part of the century. Thus, copyright piracy most affected the commercial market for new American fiction but, from the point of view of many contemporary commentators, fiction was regarded as a marginal and discretionary good for much of the nineteenth century. The movement for international copyright was only able to gain impetus toward the end of the century because of the growth in the mass market for popular fiction.

[54] This section is based on "Results of the Copyright Law," in G. H. Putnam, *The Question of Copyright*, New York, G. P. Putnam's Sons, 1896: 162–74. After the change in the copyright law, publishers price discriminated across time rather than across region. They tended to bring out the higher priced, more elaborately bound volumes first, and the cheaper versions only after a year or two.

[55] See Lisa Takeyama, "The Welfare Implications of Unauthorized Reproduction of Intellectual Property in the Presence of Demand Network Externalities," *Journal of Industrial Economics*, vol. 42 (2) 1994: 155–66.

The reading public gained from the lack of copyright, which increased access to foreign works, especially fiction. After 1891, this "unnatural demand" for cheap fiction went unsatisfied in the case of new titles, but because the law was not retroactive formerly unprotected works were still in the public domain. Books were no longer printed on the "scramble system" and tended to be characterized by higher quality and accuracy. After the reforms, the prices of some books were higher, and the range of choices less extensive than would have been the case if the law remained unchanged. Still, the loss to consumers from the reforms may not have been significant, because the books and firms that had depended on the subsidy from lack of copyright prior to the 1890s were likely of marginal value. A number of cheap reprint establishments went bankrupt, although some observers attributed this not to the law but to the "cutthroat competition" that prevailed among fringe firms. Moreover, the copyright reforms coincided with the worst economic depression of the century, so many of these firms might have failed even in the absence of legal changes.

U.S. policy makers today show great concern about the consequences for corporate profits of both domestic and international "piracy." However, Congress in the nineteenth century repeatedly rejected proposals for reform of copyright laws because the emphasis in that era was on fulfilling the objectives of the Constitution in promoting the progress of social welfare. In a democratic society this was interpreted as a mandate for ensuring that the public had ready access to literature, information, education and other conduits for achieving equality of opportunity. Democratic values may have furthered the interests of those who were the subject of so-called piracy because, as discussed here, even in the absence of copyright protection, foreign authors directly or indirectly benefited from the larger fraction of literate consumers in the United States. U.S. publishers were not demonstrably harmed by the lack of formal protection because they were able to create parallel rights that were privately enforced, and evolved firm-level strategies such as variation in price and quality. This finding is borne out by the fact that the highest profit margins in book publishing today are derived from reprints of out-of-copyright "classics."

At present there is a narrow emphasis on state-created rights and less on private market-generated means of exclusion such as private contracts or monitoring. However, given that firms' strategies regarding appropriation are endogenous to the security of copyrights, strong measures by the state to counter "piracy" may lead to social overinvestment in property rights enforcement. Some scholars have expressed concern that technological methods of exclusion at the firm level have the capacity to unduly restrict public access in perpetuity, without the social balance of costs and benefits that underly welfare maximization. For others, the censure of both copyright "piracy" and price discrimination may rest on outmoded notions of competition; and in some contexts, copyright "piracy" may merely constitute fair use by another name. Some lessons may be derived from the period when the United States flourished as a "continental Brigand," and for a century successfully resisted international pressures to conform. It is worth emphasizing that, once the United States had developed its own native stock of literary capital, it voluntarily had an incentive to recognize international copyrights. In sum, the U.S. experience during the nineteenth century suggests that appropriate intellectual property institutions are not independent of the level of economic and social development.

10

Intellectual Property and Economic Development

"It is only by considering the trend of legal development that we can make sure of the direction in which efforts toward improvement can be guided most effectively."

– Brander Matthews (1890)

The nineteenth century stands out in terms of the diversity across nations in intellectual property institutions, but this period also saw the origins of the movement toward the "harmonization" of laws that at present dominates global debates. Among the now-developed countries, the United States stood out for its conviction that democratic intellectual property rules and standards were key to achieving economic development. Democratic ends required security in patent property to create incentives for invention and innovation, whereas copyrights had to be abridged to ensure that the public interest in learning was not precluded. This calculus was distinct from that of European countries, which tended to regard patent rights as monopolistic and restricted them to protect vested interests and existing jobs; but at the same time the Europeans were less concerned about enhancing mass literacy and public education, and viewed copyright owners as inherently meritorious and deserving of strong protection. European copyright regimes thus evolved in the direction of author's rights, whereas the United States lagged behind the rest of the world in terms of both domestic and foreign copyright protection. Americans not only refused to adhere to international copyright treaties long upheld by European countries, they continued to engage in copyright piracy of foreign cultural products even in the face of widespread protests and condemnation. In direct contrast, the United States actively urged other countries to strengthen their patent laws in line with American policies, in order to benefit globally competitive American patentees.

This chapter describes the rich variation in the historical experience of (then) follower countries, beginning with Switzerland during the period before and after adopting patent protection. Follower countries such as Germany and Japan were strongly influenced by the American system, but they tailored their institutions to fit the needs of their own particular circumstances. I trace the movement to harmonize patent and copyright laws to two separate sources that culminated in stipulations for a system of uniformly strong patents and strong copyrights regardless of the level of economic development. Such a system did not exist anywhere in the world during the period under review here, when countries enjoyed greater freedom to choose appropriate institutions. Thus, the intellectual property system to which today's developing countries are required to subscribe constitutes an historical and economic anomaly. Whereas intellectual property institutions stimulated early American economic growth because of their flexible responses to economic and social circumstances, the menu of choices is much more limited for today's developing countries. Intellectual property regimes in the twenty-first century are not wholly self-determined; their structure and agenda are constrained by the political economic parameters set by the advanced nations.

As for current U.S. patent and copyright institutions, their original clarity has been clouded by historical amnesia: publishers are now privileged above the public, and even the patent system has generated widespread criticism.[1] Just as in the nineteenth century, there is growing discontent with intellectual property systems. Concerns are rife that overuse or misuse of patents and copyrights has led to an "anticommons," and some disillusioned modern commentators echo advocates from over a century ago who called for the abolition of property rights in invention. Before such proposals are entertained, more information is needed regarding the sources of technological change and economic growth, the impact of different rules and standards and their evolution over time, the feedback mechanism between economic factors and institutions, and the degree of substitutability across and within institutions. These all suggest it is worthwhile to explore why and how far the current system has deviated from its origins.

DIVERSITY IN NATIONAL PATENT SYSTEMS

Very few developed countries would now seriously consider eliminating statutory protection for inventions, but in the second half of the nineteenth century the "patent controversy" in Europe pitted advocates of patent rights against an effective abolitionist movement. For a short period, the abolitionists were strong enough to obtain support for dismantling patent systems in a number of European countries. In 1863 the Congress of German Economists declared "patents of invention are injurious to common welfare;" and the movement achieved its greatest victory in Holland, which

---

[1] For two such monographs, see Adam B. Jaffe and Josh Lerner, *Innovation and Its Discontents: How Our Broken Patent System is Endangering Innovation and Progress, and What to Do About It*, Princeton, N.J.: Princeton University Press, 2004; and Lawrence Lessig, *The Future of Ideas: The Fate of the Commons in a Connected World*, New York: Random House, 2001.

repealed its patent legislation in 1869.[2] The Swiss cantons did not adopt patent protection until 1888, with an extension in the scope of coverage in 1907.[3] The abolitionists based their arguments on the benefits of free trade and competition, and viewed patents as part of an anticompetitive and protectionist strategy analogous to tariffs on imports. Instead of state-sponsored monopoly awards, they argued, inventors could be rewarded by alternative policies, such as stipends from the government, payments from private industry or associations formed for that purpose, or simply through the lead time that the first inventor acquired over competitors by virtue of his prior knowledge.

According to one authority, the Netherlands eventually reinstated its patent system in 1912 and Switzerland introduced patent laws in 1888 largely because of a keen sense of morality, national pride, and international pressure to do so.[4] The appeal to "morality" or national pride as an explanatory factor is incapable of accounting for the timing and nature of changes in strategies. It bears repeating that nineteenth-century institutions were not exogenous and their introduction or revisions generally reflected the outcome of a self-interested balancing of costs and benefits. The Netherlands and Switzerland were initially able to benefit from their ability to free-ride on the investments that other countries had made in technological advances. As for the cost of lower incentives for discoveries by domestic inventors, the Netherlands was never vaunted as a leader in technological innovation, and this is reflected in their low per capita patenting rates both before and after the period without patent laws. They recorded a total of only 4,561 patents from 1800 to 1869 and, even after adjusting for population, the Dutch patenting rate in 1869 was a mere 13.4 percent of the U.S. patenting rate. Moreover, between 1851 and 1865, 88.6 percent of patents in the Netherlands had been granted to foreigners.[5] Thus, the Netherlands had little reason to adopt patent protection, except for external political pressures and the possibility that some types of foreign investment might be deterred.

The case was somewhat different for Switzerland, which was noted for being innovative, but in a markedly narrow range of pursuits. Because the

---

[2] This discussion draws on Fritz Machlup and Edith Penrose, "The Patent Controversy in the Nineteenth Century," *Journal of Economic History*, vol. 10(1) 1959: 1–29.

[3] National treatment in international patent laws meant that the Swiss could take out patents in foreign countries on the same grounds as citizens of that country, even though no one could get patents in Switzerland. After the patent reforms, the Swiss Law of 1888 protected only inventions that were represented by mechanical models, and excluded chemicals, pharmaceuticals and dyeing, as well as process inventions. The statute of June 1907 removed the model requirement but did not allow patents for chemical substances (processes were patentable.)

[4] See Eric Schiff, *Industrialization without National Patents: The Netherlands, 1869–1912; Switzerland, 1850–1907*, Princeton, N.J.: Princeton University Press, 1971.

[5] After the patent laws were reintroduced in 1912, the major beneficiaries were again foreign inventors, who obtained 79.3 percent of the patents issued in the Netherlands.

scale of output and markets were quite limited, much of Swiss industry generated few incentives for invention.[6] A number of the industries in which the Swiss excelled, such as handmade watches, chocolates, and food products, were less susceptible to invention that warranted patent protection. For instance, despite the much larger consumer market in the United States, during the entire nineteenth century U.S. inventors filed fewer than three hundred patents related to chocolate composition or production. Improvements in pursuits such as watch-making could be readily protected by trade secrecy as long as the industry remained artisanal. However, increased mechanization and worker mobility reduced the effectiveness of secrecy, and meant that innovators were less able to appropriate returns without more formal means of exclusion. According to contemporary observers, the Swiss resolved to introduce patent legislation not because of a sudden newfound sense of morality or international pressure, but because they discovered that American manufacturers were surpassing them as a result of patented innovations in the mass production of products such as boots, shoes and watches.[7] Indeed, before 1890, American inventors obtained more than 2,060 patents on watches, and the U.S. watch-making industry benefited from mechanization and strong economies of scale that led to rapidly falling prices of output, making them more competitive internationally. The implications are that the rates of industrial and technical progress in the United States were more rapid, and technological change was rendering artisanal methods obsolete in products with mass markets.

What was the impact of the introduction of patent protection in Switzerland? Foreign inventors could obtain patents in the United States

---

[6] Schiff (1971) claims (without providing any data) that the chocolate industry benefited or was not harmed by the lack of patent rights, for "The great days of the industry began in the 1870s" (p. 110). By contrast, other evidence suggests that the growth of the industry started only after the patent laws were changed. A. Muriel Farrer, "The Swiss Chocolate Industry," *Economic Journal*, vol. 18(69) 1908: 110–14, points out that the chocolate industry was initially quite small. After the patent reforms, "chocolate has attained a position of surprising eminence in an unusually short period of time." In 1890, total exports of Swiss cocoa products earned only £85,331 but after this period "exports increased from year to year by leaps and bounds" and were about £434,600 in 1900. (In comparison, the English family firm of Cadburys alone sold over £1 million in 1905 and £2.3 million in 1914. Charles Dellheim, "The Creation of a Company Culture: Cadburys, 1861–1931," *American Historical Review*, Vol. 92, No. 1, Supplement to Volume 92 [February 1987], pp. 13–44.) This is not to suggest any causality between the introduction of the patent system and the subsequent prosperity of the Swiss chocolate industry; instead, the facts refute Schiff's implication that the industry was flourishing in the period without a patent system.

[7] See *Scientific American*, vol. 54(18), p. 280, May 1, 1886: "A few years ago a commission of Swiss manufacturers who visited this country returned home almost in despair of competing with us even in the manufacture of watches; and in their report they recommend, as of the utmost necessity, the creation of a patent system in Switzerland similar to our own. Sir William Thomson, President of the Mathematical and Physical Section of the British Association, has declared that "if Europe does not amend its patent laws, America will speedily become the nursery of useful inventions for the world.""

292 of The Democratization of Invention

regardless of their domestic legislation, so we can approach this question tangentially by examining the patterns of patenting in the United States by Swiss residents before and after the 1888 reforms.[8] Between 1836 and 1888, Swiss residents obtained a grand total of 585 patents in the United States. Fully a third of these patents were for watches and music boxes, and only six were for textiles or dyeing, industries in which Switzerland was regarded as competitive early on. Swiss patentees were more oriented to the international market, rather than the small and unprotected domestic market where they could not hope to gain as much from their inventions. For instance, in 1872 Jean-Jacques Mullerpack of Basel collaborated with Leon Jarossonl of Lille, France, to invent an improvement in dyeing black with aniline colours, which they assigned to William Morgan Brown of London, England.[9] Another Basel inventor, Alfred Kern, assigned his 1883 patent for violet aniline dyes to the Badische Anilin and Soda Fabrik of Mannheim, Germany.

After the patent reforms, the rate of Swiss patenting in the United States immediately increased. Swiss patentees obtained an annual average of 32.8 patents in the United States in the decade before the patent law was enacted in Switzerland.[10] After the Swiss allowed patenting, this figure increased to an average of 111 each year in the following six years, and in the period between 1895 to 1900 a total of 821 Swiss patents were filed in the United States. The decadal rate of patenting per million residents increased from 111.8 for the ten years up to the reforms, to 451 per million residents in the 1890s, 513 in the 1900s, 458 in the 1910s, and 684 in the 1920s. U.S. statutes required worldwide novelty, and patents could not be granted for discoveries that had been in prior use, so the increase was not because of a backlog of trade secrets that were now patented. It is possible, of course, that the sustained increase in patenting (and citations) after the laws were introduced in 1888 was merely coincidental or that the reforms were adopted because they anticipated such increases. Interpretations of these patterns may vary, but it is plausible that the higher rates of patenting reflected rates of inventive activity that were induced by patent protection. This conclusion is consistent with the results that Shih-Tse Lo reports after a meticulous analysis of patent reforms in Taiwan.[11]

---

[8] In 1888 the population of Switzerland approached three million residents. It should be noted that a sample of "international patents" represents the upper tails of the distribution of patented inventions in terms of commercial value, and may not reflect the patterns for average inventions.

[9] Specification forming part of Letters Patent No. 134,066, dated December 17, 1872.

[10] These figures, from the U.S. Patent Office Records, include Swiss patents that were either filed or cited in the U.S. Patent Office. The conclusions hold if only patent grants are included.

[11] Shih-Tse Lo, "Strengthening Intellectual Property Rights: the Experience of the 1986 Taiwanese Patents Reforms," Department of Economics Working Paper No. 4004, Concordia University (2005,) concludes that "The evidence on the number of patents

Moreover, the introduction of Swiss patent laws also affected the direction of inventions that Swiss residents patented in the United States. After the passage of the law, such patents covered a much broader range of inventions, including gas generators, textile machines, explosives, turbines, paints and dyes, and drawing instruments and lamps. The relative importance of watches and music boxes immediately fell from about a third before the reforms to 6.2 percent and 2.1 percent respectively in the 1890s, and even further to 3.8 percent and 0.3 percent between 1900 and 1909. Another indication that international patenting was not entirely unconnected to domestic Swiss inventions can be discerned from the fraction of Swiss patents (filed in the United States) that related to process innovations. Before 1888, 21 percent of the patent specifications mentioned a process. Between 1888 and 1907, the Swiss statutes included the requirement that patents should include mechanical models, which precluded patenting of pure processes. The fraction of specifications that mentioned a process fell during the period between 1888 and 1907, but returned to 22 percent when the restriction was modified in 1907.

In short, although the Swiss experience is often cited as proof of the redundancy of patent protection, the limitations of this special case should be taken into account. The domestic market was quite small and offered minimal opportunity or inducements for inventors to take advantage of economies of scale or cost-reducing innovations. Manufacturing tended to cluster in a few industries where innovation was largely irrelevant, such as premium chocolates, or in artisanal production that initially was susceptible to trade secrecy, such as watches and music boxes. In other areas, notably chemicals, dyes, and pharmaceuticals, Swiss industries were export-oriented, but even today their output tends to be quite specialized and high-valued rather than mass-produced. Export-oriented inventors were likely to have been more concerned about patent protection in the important overseas markets, rather than in the home market. Thus, between 1888 and 1907, although Swiss laws excluded patents for chemicals, pharmaceuticals, and dyes, 20.7 percent of the Swiss patents filed in the United States were for just these types of inventions. During this period, the 68.4-percent rate of patent assignment at issue for these industries was quite high compared to other countries, and remained unchanged in the following decade even after chemicals gained Swiss patent protection. The scanty evidence on Switzerland suggests that

awarded to Taiwanese inventors as well as that on R&D spending in Taiwan suggests that the reforms stimulated additional inventive activity, especially in industries where patent protection is generally regarded as an effective strategy for extracting returns, and in industries which are more R&D intensive. The reforms also seemed to induce additional foreign direct investment in Taiwan. On the other hand, for industries that chiefly use other mechanisms to extract returns from their innovations, such as secrecy, the strengthening of patent rights had little effect on their inventive activity." See also Lo's forthcoming dissertation on the same subject.

the introduction of patent rights was accompanied by changes in the rate and direction of inventive activity, although the direction of causation is open to question. In any event, both the Netherlands and Switzerland featured unique circumstances that hold few lessons for developing countries today.

Instead, the German experience is more instructive. The "patent controversy" of the nineteenth century had been especially contentious among the states that comprised the German alliance.[12] The German Empire was founded in 1871, and in the first six years each state adopted its own policies. Alsace-Lorraine favored a French-style system, whereas others such as Hamburg and Bremen did not offer patent protection. One of the concerns expressed was that patent protection would allow an influx of American patentees, to the detriment of the domestic market. However, after strong lobbying by supporters on both sides of the debate, Germany passed a unified national Patent Act in 1877. The German patent law followed the American system, but it also incorporated features unique to Germany and its perceived needs. The 1877 statute created a centralized administration for the grant of a federal patent for original inventions. The patent examination process required that the patent should be new, nonobvious, and also capable of producing greater efficiency. Applications were initially examined by consultants to the Patent Office who were expert in their field but, because of perceived conflicts of interest, in 1891 examiners became permanent employees of the Patent Office, as in the United States. During the eight weeks before the grant, patent applications were open to public scrutiny and an opposition could be filed denying the validity of the patent. German patent fees were deliberately set high, with a renewal system that required payment of 30 marks for the first year, 50 marks for the second year, 100 marks for the third, and 50 marks annually after the third year. Part of the reason for high fees was to raise revenues by taxing foreign inventors and to eliminate protection for trivial inventions. The initial term of fifteen years was extended in 1923 to eighteen years. Industrial entrepreneurs succeeded in their objective of creating a "first to file" system, so patents were granted to the first applicant rather than to the "first and true inventor," but in 1936 the National Socialists introduced a "first to invent" system.

German patent policies exempted certain key industries in order to promote diffusion and development goals. Patents could not be obtained for food products, pharmaceuticals, or chemical products, although the process through which such items were produced could be protected. It was felt that open access to use innovations and the incentives to patent around existing processes spurred productivity and diffusion in these industries. The

---

[12] The information on the German system was drawn from J. Vojacek, *A Survey of the Principal National Patent Systems*, New York: Prentice Hall, 1936. The German patent system later influenced legislation in a number of countries, including that of Argentina, Austria, Brazil, Denmark, Finland, Holland, Norway, Poland, Russia, and Sweden.

authorities further ensured the dissemination of information by publishing patent claims and specification before they were granted. The German patent system facilitated the use of inventions by firms, with the early application of a work for hire doctrine that allowed enterprises usufruct in the rights to inventions created by their employees. After 1891 a parallel and weaker version of patent protection could be obtained through a *Gebrauchsmuster* or utility patent (sometimes called a petty patent), which was granted through a registration system.[13] Protection as a utility patent was available for inventions that could be represented by drawings or models with only a slight degree of novelty, and for a limited term of three years (renewable once for a total life of six years.) Patent protection based on coexisting systems of registration and examination served distinct but complementary purposes. As in the United States, once patents were granted, the courts adopted an extremely liberal attitude toward inventors in interpreting and enforcing existing rights.

Although the German system was close to the American patent system, it varied in important respects. The German regime resulted in patent grants that were lower in number, but likely higher in average value. Legal remedies for wilful infringement included not only fines, but also the possibility of imprisonment. The most significant departure from U.S. policies was that German patents were subject to working requirements. The grant of a patent could be revoked after the first three years if the patent was not worked, if the owner refused to grant licenses for the use of an invention that was deemed in the public interest, or if the invention was primarily being exploited outside of Germany. However, in most cases, a compulsory license was regarded as adequate. The German patent system, to many other developing countries, comprised an appropriate amalgam of incentives to domestic patentees, concessions to the fact that the roster of patentees tended to include more foreigners than domestic inventors, and policy measures to attenuate the social costs of exclusion.

This realization that successful institutions should be tailored to individual circumstances was especially evident in the early Japanese patent system. Japan emerged from the Meiji era as a follower nation, which designed its institutions to emulate those of the most advanced industrial countries. The first effective national patent statute in Japan copied many features of the U.S. system, including the examination procedures. However, the overall system ultimately reflected Japanese priorities and the "wise eclecticism of Japanese legislators."[14] Patents initially were not granted to foreigners; protection could not be obtained for fashion, food products, or medicines; patents that were not worked within three years could be revoked; and severe

---

[13] Geoge von Gehr, "A Survey of the Principal National Patent Systems from the Historical and Comparative points of View," *John Marshall Law Quarterly*, 1936: 334–400.

[14] Vojacek, p. 160.

remedies were imposed for infringement, including penal servitude. After Japan became a signatory of the Paris Convention a new law was passed in 1899, which amended existing legislation to accord with the agreements of the Convention, and extended protection to foreigners. The influence of the German laws was evident in subsequent reforms. In 1909 petty or utility patents were protected. A 1921 statute removed protection from chemical products, and work for hire doctrines were adopted. Japan introduced opposition procedures at this time, possibly because it wished to economize on costly resource inputs into patent institutions. The Act of 1921 also included a provision that permitted the state to revoke a patent grant on payment of appropriate compensation if it were deemed to serve the public interest. Medicines, food, and chemical products could not be patented, but protection could be obtained for processes relating to their manufacture.

Britain and France initially had a disproportionate influence on international patent systems because of their numerous colonies. French patent laws were not only adopted in its own colonies but also diffused to other countries through its influence on Spain's system since the Spanish Decree of 1811.[15] As for Spain itself, during the nineteenth century this country experienced lower rates and levels of economic development than the early industrializers. Like its European neighbours, early Spanish rules and institutions were vested in privileges that had deleterious effects that could be detected even in the later period after the system was reformed. The per capita rate of patenting in Spain was lower than in other major European countries, and foreigners filed a significant fraction of patented inventions. Between 1759 and 1878, roughly one half of all grants went to citizens of other countries, notably France and (to a lesser extent) Britain. Thus, the transfer of foreign technology was a major concern in the political economy of Spain.

This dependence on foreign technologies was reflected in the structure of the Spanish patent system, which permitted patents of introduction as well as patents for invention.[16] Patents of introduction were granted to entrepreneurs who wished to produce foreign technologies that were new to Spain, with no requirement that applicants should be the true inventor. Thus, the sole objective of these instruments was to enhance innovation and production in Spain. The owners of introduction patents could not prevent third parties from importing similar machines from abroad, and they

---

[15] "[P]ractically all European and most of the Latin American patent laws issued at this period were more or less modeled on the French law" Jan Vojacek, p. 135. The description of the Spanish system is drawn from Patricio Sáiz González's excellent study, *Invención, Patentes e Innovación en la España Contemporánea*, Oficina Española de Patentes y Marcas, Madrid, 1999.

[16] Thus, the "foreign content" of Spanish technology could be viewed as the sum of inventions patented by foreigners, and patents of introduction obtained by Spaniards for foreign inventions. This implied that roughly two thirds of Spanish patents were drawn from overseas sources.

also had incentives to maintain reasonable pricing structures. Introduction patents had a term of only five years, with a cost of 3,000 reales, whereas the fees of patents for invention varied from 1,000 reales for five years, 3,000 reales for ten years, and 6,000 reales for a term of fifteen years.[17] Patentees were required to work the patent within one year, and about a quarter of patents granted between 1826 and 1878 were actually implemented.[18] The brief term of patents of introduction encouraged the production of items with high expected profits and a quick payback period, after which the monopoly rights expired, and the country hoped to benefit from greater diffusion.

Developing countries throughout the world faced similar concerns as Japan, Spain, and Germany did in constructing intellectual property systems. Most of these countries were net importers of foreign technologies and investment, with minimal stocks of domestic inventions and innovations. Their policy goals were primarily to raise revenues and moderate the impact of foreign technologies, rather than to create incentives for domestic invention. They lacked the effective legal and antitrust institutions that reinforced and monitored the grant of property rights in patents in the United States and other developed countries. Working requirements and compulsory licenses were introduced although not always enforced. The majority of today's developing countries assessed very high fees when they established patent systems (especially given their low per capita incomes.)[19] Many of the societies in Central and South America, regardless of their colonial origins, levied the highest fees in the world for patent protection. The high costs might have owed to a number of factors, including the wish to raise revenues, a conviction that patent rights would be sought more by foreigners than by natives, and a desire to limit exclusive rights to valuable inventions. Nevertheless, the net impact of high fees was to restrict access to a privileged few, insulate businessmen with considerable resources from competition, and to perpetuate inequalities in wealth and enterprise.[20]

---

[17] See Patricio Sáiz González, *Invención, Patentes e Innovación*, p. 133. These fees were set in 1826, and maintained through 1878. During this period, the average annual salary for an official was 4,275 reales, and that of an agricultural worker was about 1,050 reales. Between 1759 and 1878, some 77.5 percent of patents were for inventions, and the rest for introductions. Seventy three percent of patents by Spaniards were for inventions, relative to some 80 percent of the patents obtained by French citizens.

[18] Only 16.5 percent of foreign patents were implemented, relative to 34.7 percent of Spanish patents, and 12.6 percent of patents obtained by nonresidents. See Patricio Sáiz González, "Patents, International Technology Transfer and Spanish Industrial Dependence (1759–1878)," p. 11, mimeo, 1999.

[19] B. Zorina Khan and Kenneth L. Sokoloff, "The Innovation of Patent Systems in the Nineteenth Century: A Comparative Perspective," Unpublished manuscript (2003).

[20] It should be noted that the influence of colonial heritage is not nearly so powerful as some have claimed. The general imperial policy of Britain toward its colonies allowed for original legislation in the constituent colonies in accordance with local conditions. There was, for

HARMONIZATION OF PATENT AND COPYRIGHT LAWS

The nineteenth century was a time of intense debate about the value of intellectual property regimes, and follower countries selected policies that fulfilled their national objectives, even to the extent of removing patent protection. The decisive victory of the patent proponents in the "patent controversy" shifted the focus of interest to the other extreme. The United States was the most prolific patenting nation in the world, many of the major American enterprises owed their success to patents and were rapidly expanding into international markets, and the U.S. patent system was recognized as the most successful. It is therefore not surprising that the United States was in the vanguard of efforts to attain uniformity in patent laws, and its experience played a key role in ensuring that the international debate was resolved in favor of well-enforced patent rights. Part of the impetus for change occurred because the costs of discordant national rules became disproportionately burdensome as the volume of international trade grew over time. Americans also were concerned about the lack of protection accorded to their exhibits in the increasingly prominent World's Fairs. Indeed, the first international patent convention was held in Austria in 1873 at the suggestion of U.S. policy makers, who wanted to be certain that their inventors would be adequately protected at the International Exposition in Vienna that year. The convention also yielded an opportunity to protest provisions in Austrian law that discriminated against foreigners, including a requirement that patents had to be worked within one year or risk invalidation. Patent harmonization over the next century would lead to convergence toward the American model, despite resistance from nations that feared that their domestic industry would be overwhelmed by American patents.[21]

In the international sphere, the preferences and interests of the United States were directed toward replicating its domestic policies toward patent holders, which have always been the most liberal in the world. However, from the very beginning of the movement to international harmonization, deep divisions existed regarding the extent to which restrictions should be placed on the rights of patentees. Most members continued to support measure that they deemed to be in the public interest, especially working requirements and compulsory licenses for patentees, which the United States strenuously

---

example, enormous diversity in the characteristics of the patent systems of the colonies that remained under British rule at this time.

[21] One commentator pointed to "the extremely liberal propositions of the United States, which one could only recognize as approaching the ideal of the future." Cited in Edith Penrose, *Economics of the International Patent System*, Baltimore: Johns Hopkins University Press, 1951, p. 81.

DESIGN.

A. BARTHOLDI.
Statue.

No. 11,023.   Patented Feb. 18, 1879.

LIBERTY ENLIGHTENING THE WORLD.

Auguste Bartholdi

Illustration 12. Design Patent No. 11,023, "*Liberty Enlightening the World*," issued in 1879 to Frédéric Auguste Bartholdi of Paris, France. Bartholdi's statue is commonly known as "the Statue of Liberty" and was presented to the United States by the people of France on July 4, 1884. (*Source*: U.S. Patent Office.)

opposed.[22] Such policy instruments had been widely used by other developed countries since the earliest years of the Venetian patent grants. France incorporated working requirements in its 1844 statutes; Germany stipulated both working requirements and compulsory licenses; and so did Britain in the early twentieth century. During the colonial period, such statutory exceptions to patent and copyrights also were prevalent among the American states. In the antebellum period, the United States itself briefly incorporated working requirements for alien patentees in its 1832 and 1836 patent statutes. Moreover, in the second half of the twentieth century consent decrees in U.S. antitrust actions led to large-scale infractions of patent rights that involved not only compulsory licenses but also the forced transfer of trade secrets and know-how.[23] The United States was therefore as zealous in its application of compulsory licensing in the context of competition policy within its own borders, as in its efforts to prohibit other nations from using such patent restrictions to promote their own interests.

International conventions proliferated in subsequent years, and their tenor tended to reflect the opinions of the convenors.[24] Their objective was not to reach compromise solutions that would reflect the needs and wishes of all participants but, rather, to promote the preconceived ideas of the most advanced nations. It became clear that the goal of complete uniformity was not practicable, given the different objectives, ideologies, and economic circumstances of participants. Nevertheless, in 1884 the International Union

---

[22] American opposition gradually had an effect, as seen in the history of revisions to the Paris Convention since 1883. At that time "parallel imports" were permitted and members were allowed to stipulate that the patent should be exploited. In 1911 patent rights could be revoked only after three years and only if the patentee was unable to justify why the patent was idle. At present, trade-related intellectual property rights agreements contain a weak provision that "members may provide limited exceptions to the exclusive rights conferred by a patent, provided that such exceptions do not unreasonably conflict with a normal exploitation of the patent and do not unreasonably prejudice the legitimate interests of the patent owner, taking account of the legitimate interests of third parties." TRIPS Agreement, Article 30: Exceptions to Rights Conferred.

[23] See B. Zorina Khan, "Federal Antitrust Agencies and Public Policy toward Patents and Innovation," *Cornell Journal of Law and Public Policy*, vol. 9 (Fall) 1999: 133–69; B. Zorina Khan, "The Calculus of Enforcement: Legal and Economic Issues in Antitrust and Innovation," *Advances in the Study of Entrepreneurship, Innovation, and Economic Growth*, vol. 12 (1999): 61–106. U.S. copyright policies also allowed for compulsory licenses in certain industries.

[24] See Edith Penrose, *Economics of the International Patent System*, Baltimore: Johns Hopkins University Press, 1951. These included conferences in 1878, 1880 and 1883. Participants of the 1880 conference were drawn from Argentina, Austria-Hungary, Belgium, Brazil, France, Britain, Guatemala, Italy, Luxembourg, Netherlands, Portugal, Russia, San Salvador, Sweden, Norway, Switzerland, Turkey, the United States, Uruguay, and Venezuela. There were additional meetings in Rome (1886), Madrid (1890–91), Brussels (1897–1900), Washington (1911), The Hague (1925), and London (1934.)

for the Protection of Industrial Property was formed.[25] The United States pressed for the adoption of reciprocity (which would ensure that American patentees were treated as favorably abroad as in the United States) but this principle was rejected in favor of "national treatment" (American patentees were to be granted the same rights as nationals of the foreign country.) Ironically, because its patent laws were the most liberal toward patentees, the United States found itself with weaker bargaining abilities than nations who could make concessions by changing their provisions. This likely influenced the U.S. tendency to use bilateral trade sanctions rather than multilateral conventions to obtain the reforms it desired in international patent policies. It was commonplace in the nineteenth century to rationalize and advocate close links between trade policies, protection, and international laws regarding intellectual property. Indeed, French commentators proclaimed that "the laws on industrial property ... will be truly disastrous if they do not have a counterweight in tariff legislation."[26] The movement to harmonize intellectual property laws showed that institutions did not exist in a vacuum, but were part of a bundle of rights that were affected by other laws and policies.

In view of the strong protection of inventors under the U.S. patent system, to foreign observers its copyright policies appeared to be all the more reprehensible. The United States, the most liberal in its policies toward patentees, had led the movement for harmonization of patent laws. In marked contrast, as previous chapters showed, throughout the history of the U.S. system its copyright grants in general were more abridged than in almost all other countries. The term of copyright grants to American citizens was among the shortest in the world, the country applied the broadest and most liberal interpretation of fair use doctrines, and the validity of the copyright depended on strict compliance with statutory requirements. U.S. failure to recognize the rights of foreign authors also was unique among the major industrial nations. Throughout the nineteenth century, proposals to reform the law and to acknowledge foreign copyrights were repeatedly brought before Congress and rejected. Even the bill that finally recognized international copyrights almost failed, and its passage required protectionist exemptions in favor of American workers and printing enterprises.

---

[25] The first signatories were Belgium, Portugal, France, Guatemala, Italy, the Netherlands, San Salvador, Serbia, Spain, and Switzerland. The United States became a member in 1887, and a significant number of developing countries followed suit, including Brazil, Bulgaria, Cuba, the Dominican Republic, Ceylon, Mexico, Trinidad and Tobago, and Indonesia, among others. Recall that neither Switzerland nor the Netherlands at this time had a patent system in place. According to the terms of the Union, nationals of these countries could obtain patents in other countries on equal terms with the citizens of the patent-granting domain.

[26] Cited in Penrose, *Economics*, p. 77.

In a parallel fashion to the status of the United States in patent matters, France's influence was evident in the subsequent evolution of international copyright laws. Other countries had long recognized the rights of foreign authors in national laws and bilateral treaties, but France stood out in its favorable treatment of domestic and foreign copyrights as "the foremost of all nations in the protection it accords to literary property."[27] This was especially true of its concessions to foreign authors and artists. For instance, France allowed copyrights to foreigners conditioned on manufacturing clauses in 1810, and granted foreign and domestic authors equal rights in 1852. In the following decade, France entered into almost two dozen bilateral treaties, prompting a movement toward multilateral negotiations, such as the Congress on Literary and Artistic Property in 1858. The International Literary and Artistic Association, which the French novelist Victor Hugo helped to establish, conceived of and organized the Convention that first met in Berne in 1883.

The Berne Convention included a number of countries that wished to establish an "International Union for the Protection of Literary and Artistic Works."[28] The preamble declared their intent to "protect effectively, and in as uniform a manner as possible, the rights of authors over their literary and artistic works." The actual Articles were more modest in scope, requiring national treatment of authors belonging to the Union and minimum protection for translation and public performance rights. The Convention authorized the establishment of a physical office in Switzerland, whose official language would be French. The rules were revised in 1908 to extend the duration of copyright and to make adjustments for modern technologies. Perhaps the most significant aspect of the convention was not its specific provisions, but the underlying property rights philosophy which was decidedly from the natural rights school. Berne abolished compliance with formalities as a prerequisite for copyright protection since the "creative act" itself was regarded as the source of the property right. In 1928 the Berne Convention followed the French precedent and acknowledged the moral rights of authors and artists.

Unlike its leadership in patent conventions, the United States declined an invitation to the pivotal copyright conference in Berne in 1883; it attended but refused to sign the 1886 agreement of the Berne Convention. Instead, the United States pursued international copyright policies in the context of the weaker Universal Copyright Convention (UCC), which was adopted in 1952 and formalized in 1955 as a complementary agreement to the Berne Convention. The UCC membership included many developing countries that

---

[27] Brander Matthews, "The Evolution of Copyright," in George H. Putnam, *The Question of Copyright*, New York: G. P. Putnam's Sons, 1896, p. 336.

[28] France, Belgium, Britain, Germany, Spain, Haiti, Italy, Switzerland, and Tunisia ratified the 1886 agreement.

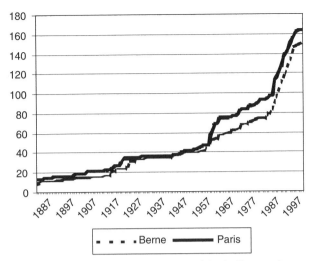

Figure 10.1. Cumulative Membership in Berne and Paris Conventions, 1887–2000
Notes and Sources: The figure shows the cumulative number of nations adhering to the Paris Convention for the Protection of Industrial Property (1883), and the Berne Convention for the Protection of Literary and Artistic Works (1886). The data are from the World Intellectual Property Organization (http://www.wipo.org/treaties), accessed February 20, 2003.

did not wish to comply with the Berne Convention because they viewed its provisions as overly favorable to the developed world.[29] As Figure 10.1 shows, the United States was among the last wave of entrants into the Berne Convention when it finally joined in 1988. In order to do so it complied by removing prerequisites for copyright protection such as registration, and also lengthened the term of copyrights. However, it still has not introduced federal legislation in accordance with Article 6bis, which declares the moral rights of authors "independently of the author's economic rights, and even after the transfer of the said rights."[30] Similarly, individual countries continued

[29] The original adherents to the Universal Copyright Convention (UCC) were the German Republic, Andorra, Argentina, Australia, Austria, Brazil, Canada, Cuba, Denmark, El Salvador, United States, France, Guatemala, Haiti, Honduras, India, Ireland, Israel, Italy, Liberia, Luxembourg, Monaco, Nicaragua, Norway, Portugal, the United Kingdom, San Marino, the Holy See, Sweden, Switzerland, Uruguay, and Yugoslavia. The four stipulations of the UCC were that member nations could not grant preferential treatment for domestic works relative to foreign works; formal copyright notice must appear in all copies of a work; the term of copyright protection must exceed the life of the author plus an additional twenty-five years; and members were required to grant an exclusive right of translation for a seven-year period to authors from other member countries.

[30] Although a few states have enacted legislation in the spirit of Berne, federal law relates only to artistic works in limited editions, according to 17 USC section 106A (Visual Artists Rights Act) 1990.

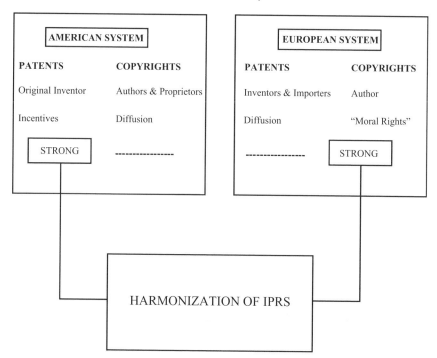

Figure 10.2. Harmonization of Intellectual Property Laws.

to differ in the extent to which multilateral provisions governed domestic legislation and practices.[31]

The quest for harmonization of intellectual property laws resulted in a "race to the top," directed by the efforts and self interest of the countries which had the strongest property rights (Figure 10.2). As outlined here, the movement to harmonize patents was driven by American efforts to ensure that its extraordinary patenting activity was remunerated beyond as well as within its borders. At the same time, the United States ignored international conventions to unify copyright legislation. Nevertheless, the harmonization of copyright laws proceeded, promoted by France and other civil law regimes which urged stronger protection for authors based on their "natural rights" while at the same time many of these countries abridged the rights of foreign inventors. The net result was that international pressure was applied to developing countries in the twentieth century to establish strong patents *and* strong copyrights, although no individual developed country had adhered

---

[31] Only a few countries complied with the letter of the law regarding formalities, and most kept stipulations such as deposit requirements through other types of legislation or regulations. In 1990 the majority of countries in the world still had a legal deposit system, even if deposits were not included in their copyright legislation.

to both concepts simultaneously during their own early growth phase. This occurred even though theoretical models did not offer persuasive support for intellectual property harmonization, and indeed suggested that uniform policies might be detrimental even to some developed countries and to overall global welfare.[32]

HISTORICAL PERSPECTIVES

The plethora of research on both the theoretical and empirical aspects of institutions, inventions and economic activity leaves many questions open.[33] This is in part due to a lack of attention to their historical dimensions. For instance, despite ample evidence from history that such institutions might be endogenous, some empirical studies still treat legal regimes, patent systems and other forms of property rights as exogenous determinants of economic performance. A number of economists have been persuaded by the superior theoretical properties of such alternative instruments as state-sponsored awards, buyouts and prizes. Instead, the abundant evidence from France and England illustrates the inefficiencies, corruption, and the lack of incentives for reforms that could arise from a nonmarket orientation. Our understanding of the so-called "explosion" in patenting and patent litigation in the last few decades of the twentieth century would surely benefit from a comparison to the much higher propensity to patent and litigate that prevailed among domestic patentees during much of the period before World War II. Moreover, unproductive rules and standards from the past can survive efforts at reform and hinder the potential for growth, implying the need to integrate economic analyses within an historical context. It is critical to examine the evolution of institutions if we are to improve on our knowledge of the nature and direction of long-run economic growth.

[32] Elhanan Helpman, "Innovation, imitation and intellectual property rights," *Econometrica*, vol. 61 (6) 1993: 1247–80; I. Diwan and D. Rodrik, "Patents, Appropriate Technology and North-South Trade," *Journal of International Economics*, vol. 30, 1991: 27–47. Developed countries such as Canada that have net inflows of intellectual property also may be harmed by stronger international intellectual property rights. See also Alan V. Deardorff, "Welfare Effects of Global Patent Protection," *Economica*, New Series, vol. 59, no. 233 (February 1992): 35–51. Deardorff attempted to assess the welfare implications of extending patent regimes from a country of innovation producers to a country of innovation consumers. He found that the welfare of the producer increased unambiguously, but the welfare of the consumer country fell, and it was possible for the net effects on global welfare to be negative overall.

[33] For an excellent overview of the debate on institutions and economic growth, see Stanley L. Engerman and Kenneth L. Sokoloff, "Institutional and Non-Institutional Explanations of Economic Differences," NBER Working Paper No. 9989, Sept. 2003; and Douglass North, *Institutions, Institutional Change and Economic Performance*, Cambridge: Cambridge University Press, 1990.

This monograph focused on the evolution of intellectual property institutions in the United States in the light of the European experience. The U.S. Constitution authorized an intellectual property system that has had a disproportionate impact on the course of global economic history. In the realm of patent policies, the stipulations of the U.S. Constitution can still be recognized both here and in other countries. The framers of the world's first modern patent institution paid close attention to the provision of broad access to, and strict enforcement of, property rights in new inventions. The early patent regime was extremely effective at stimulating the growth of a market for technology and promoting technical change. Another reason for its success, however, has been its flexibility and its utilitarian nature. Intellectual property institutions were from the outset in a state of continual evolution, and have undergone a number of fundamental modifications. Much of the change came through formal legislation or judicial initiatives and reinterpretation inspired by changing circumstances, but also important were innovations in the structure of the market for patented technologies and in copyrighted materials made directly by private agents responding to economic opportunities. That such adjustments so often proved to be constructive owed partly to the virtues of having the market as a central feature of the intellectual property system, and partly to the democratic nature of economic and political institutions.

By design, American statutes differentiated between patents and copyrights in ways that seemed warranted if the objective were to increase social welfare.[34] The patent system early on discriminated between nonresident and domestic inventors, but within a few decades changed to protect the right of any inventor who filed for an American patent regardless of nationality. The copyright statutes, in contrast, openly encouraged piracy of foreign goods on an astonishing scale for one hundred years, in defiance of the recriminations and pressures exerted by other countries. The American patent system required an initial search and examination that ensured the patentee was the "first and true" creator of the invention in the world, whereas copyrights were granted through mere registration. Patents were based on the assumption of novelty and held invalid if this assumption was violated, whereas essentially similar but independent creation was copyrightable. Copyright holders were granted the right to derivative works, whereas the patent holder was not. Unauthorized use of patented inventions was prohibited, whereas "fair use" of copyrighted material was permissible if certain conditions were met.[35] Patented inventions involved greater initial

---

[34] "Not with standing this allusion to patents, the mistake should not be made of supposing that patents and copyrights stand on the same basis as to natural exclusive right, for they do not; the difference between them, in this regard, is radical." W. E. Simonds, "International Copyright," in George H. Putnam, *The Question of Copyright*, New York: G. P. Putnam's Sons, 1896: 77–130.

[35] See Chapter 8.

investments, effort, and novelty than copyrighted products and tended to be more responsive to material incentives; whereas in many cases cultural goods would still be produced or only slightly reduced in the absence of such incentives.[36] Fair use apart from experimentation was not allowed in the case of patents, arguably because the disincentive effect was likely to be higher, whereas the costs of negotiation between the patentee and the more narrow market of potential users would generally be lower. If copyrights were as strongly enforced as patents it would benefit publishers and a small literary elite at the prohibitive cost of social investments in learning and education.

The experience in Europe and America underlined the importance of ensuring access to property rights and the return from individual efforts to all members of society, however humble their social background. Both the British and French patent systems reflected their origins in royal privilege. The British quite consciously limited property rights to groups who had greater wealth or better access to private information and capital, and favored inventors of more capital-intensive devices as opposed to smaller incremental inventions in labor-intensive industries. Their system fostered interest groups who had an incentive to protect their rents by blocking proposals for reform. Despite a series of changes in the laws, these patterns characterized patenting and trade in technological information in Britain until late in the nineteenth century and even beyond. In contrast, the United States was concerned with fashioning a system that induced enterprise from all members of society regardless of their social class or income. Consequently, when markets expanded, relatively ordinary individuals responded to these increases in profit opportunities. The remarkable advances in early American technology were associated with a process of democratization among both the creators of incremental inventions and the so-called "great inventors." Moreover, even among the relatively disadvantaged class of women inventors, a far greater number in the United States were able to obtain patents and profit from their ideas than was the case in England.

These individuals were responding in part to policies that successfully encouraged widespread diffusion, participation and inventiveness. A set of Patent Office reports from the nineteenth century still remains among the possessions of the last of the Shakers, a religious sect in rural New Gloucester, Maine, that created innovative artisanal technologies. Institutional features that improved on access included low fees, protection of the rights of the first and true inventor, a centralized examination system, and secure property rights that were enforced by a legal system that also tried to protect social welfare. The examination system played an important part in ensuring that inventors who did not have the resources to conduct searches were able to

---

[36] *Alfred Bell & Co. v. Catalda Fine Arts, Inc.*, 191 F. 2d 99 (1951): "we have often distinguished between the limited protection accorded a copyright owner and the extensive protection granted a patent owner."

secure the services of a trained cadre of professional examiners at minimal cost. Private parties could always, as they did under the registration systems prevailing in Europe, expend the resources needed to make the same determination as the examiners, but there was a distributional impact, as well as scale economies and positive externalities, associated with a centralized examination system. This process ensured that inventive ideas could be transformed into tradeable assets, and the securitization of invention encouraged markets in patents. Such markets disproportionately benefited relatively poor inventors who did not own the resources to exploit their patented inventions but could still gain returns by selling or licensing their rights. The spread of markets allowed them to specialize through a division of labor in invention and commercialization.

The commitment to the most pragmatic aspects of provision and diffusion played a significant role in the democratization of innovation in America. The American system stood out in its insistence on a rationalized recordkeeping system, prompt publication of information, free distribution to libraries and patent offices, and the adherence to predictable rules and procedures. American statutes made careful provisions for the dissemination of information about the patent system itself, as well as investments in publishing patent specifications and expiration dates, and made them readily available. Among the detailed instructions to achieve these ends, an 1837 section of the federal patent laws authorized the appointment of special agents in locations throughout the country to receive and forward to the Patent Office technical models and specimens of compositions and manufactures. The French had indulged in high-flown rhetoric about the rights of man, but failed to follow through on the more mundane provisions. For instance, in its early laws, France stipulated that patent descriptions were to be made available to the public but, as no specific procedure for their publication was introduced, the effect was to limit diffusion. Similarly, England administered patents in such a convoluted fashion that it was prohibitively expensive to obtain information. The wealthy could maneuver around these barriers, but the attendant costs acted as a regressive tax that was disproportionately felt by the less well endowed.

Another critical distinction between the United States and other countries lay in effective enforcement and in relatively low-cost access to courts. The legal system comprised an important aspect of an intellectual property regime, because the value of any property right to its owner depended on his ability to enforce his claims. U.S. rules were so transparent and institutions were so accessible that some patentees (including a woman inventor) even successfully prosecuted their causes in court without formal representation by counsel. England possessed a judicial and legal system that extended back for centuries, and its common law influenced the progress of numerous countries in the world. Nevertheless, the intellectual property laws were interpreted by judges in a manner that reinforced the existing oligarchic class system and hindered market transactions. Their legal system was notorious for its inconsistency, arbitrary decisions, uncertainty and cost. The United States from the very beginning was fortunate to possess a remarkable cohort of judges and legal practitioners who adopted the maxim "*salus populi suprema lex est*" [social welfare is the ultimate law] and interpreted the law in ways that furthered economic development.[37] Early jurisprudence favored the security of property and contracts, and enhanced private incentives for invention and markets in invention. However, courts were very much aware of the needs of the community as well, and tempered their interpretations of property rights to ensure that a balance would be maintained between private welfare and social welfare. Cases at equity allowed decisions that incorporated delicate adjustments to the rights of all parties concerned. This calculus ensured that the legal system reinforced the rights of intellectual property holders while reducing the social costs of exclusion.

Such policies helped to propel the United States to the first rank among developed nations, and U.S. owners of patent property expanded into international markets. They found that foreign institutions were less favorable to the creation and dissemination of cultural goods and inventions. A number of countries recognized their comparative disadvantage and they reformed their rules to create a two-tiered system that offered incentives for domestic ingenuity as well as the diffusion of foreign inventions. Germany distinguished between the high-value/high-cost grant of a full patent awarded mainly to foreign multinationals, and a lower-value/low-cost petty patent grant. The cost of administration of petty patents was low because, unlike regular patents, they were not subject to an initial examination. In both Germany and Japan, they proved to be an effective way of allowing residents to participate in the patent system and created an incentive for the commercialization of follow-on inventions. As early as 1900 the majority of German patents were issued to its own citizens. Like any other right of exclusion they were subject to abuse, but the potential harm was lower than in the case of full patents because of their short life. Spanish patents similarly distinguished between full and truncated patent rights in order to promote the transfer of technology and commercialization.

It was widely recognized in the nineteenth century that developing countries required appropriate institutions that might differ from those of more advanced economies. At present, criticisms have been leveled against developing countries such as India (which did not offer patent protection for drugs, chemicals and alloys, optical glass, or semiconductors), Thailand (which did not allow patents for chemicals, drugs, food and beverages, and agricultural

---

37 Oliver Wendell Holmes even declared that "every lawyer ought to seek an understanding of economics," in an 1897 speech reprinted as "The Path of the Law After One Hundred Years," 110 Harv. L. Rev. 991, March (1997.)

machinery), and Brazil (chemicals, drugs, and foodstuffs were not protected before the 1990s) for not offering universal patent protection.[38] Historically, the majority of developed countries other than the United States exempted particular industries from protection. The French statute of 1791 exempted medicines from patent grants. England countered Continental supremacy in chemicals by not offering patent protection for such products, and until recently issued compulsory licenses for pharmaceuticals and food products. Similarly, Germany (emulated by Japan) did not issue patents for food products, pharmaceuticals, or chemical products, although firms could obtain protection for innovations in the manufacturing processes. Consequently, there was ample historical precedent for a policy of discretionary grants across sectors or products in order to enhance the public interest in diffusion and access.

The need to tailor institutions to the prevailing level of economic and social development was particularly evident in the realm of copyright protection for cultural goods. Early American policy makers were aware of the European history of copyrights as a form of censorship and as monopoly rights that imposed restraints on widespread literacy and learning. The European system of privileges and its attendant notion of an elite class that deserved supranormal returns were later enshrined in moral rights copyright regimes, because of the successful rhetoric of French and British publishers who appropriated the notion of author's rights for their own ends. In contrast, legislators in the United States pierced the distorting veil of authorship and rights of creativity and instead attended to the welfare of society in general. Appropriate policies toward copyright were complicated by larger considerations than economic questions of the role of material incentives to proprietors (more often publishers). Copyright grants had critical implications for freedom of speech, the diffusion of knowledge, democratic access to learning, and the opportunity for social mobility. A flexible and comprehensive fair use doctrine was viewed as a prerequisite for ensuring that copyrights were compatible with an effective social contract. Rather than a bona fide property right, American copyright often mimicked more limited legal mechanisms such as contract, trade restraint, or even liability rules.

A natural extension of this approach was support for the notion that in the early stages of economic growth copyright piracy simply comprised international fair use. The proponents of copyright piracy were not unaware of

---

[38] See Edwin Mansfield, "Intellectual Property Protection, FDI and Technology Transfer," IFC Discussion Paper No. 19, World Bank, 1994. Mansfield surveyed American multinational corporations and found that, from their point of view as well, IPRs protection "plays a somewhat different role in each of these industries" (Edwin Mansfield, "Unauthorized Use of Intellectual Property: Effects of Investment, Technology Transfer, and Innovation," p. 121, in Mitchel B. Wallerstein, Mary Ellen Mogee, Roberta A. Schoen (eds.), *Global Dimensions of Intellectual Property Rights in Science and Technology*, Washington, D.C., National Academy Press, 1993.

the potential for disincentives to foreign investment, deleterious effects on local industries, the misallocation of resources to counterfeiting, and a fall in quality. However, U.S. policies prior to the recognition of international copyrights in 1891 were based on the argument that publishing constituted an "infant industry." Far from being deterred by the reprinting of foreign intellectual assets, their ready availability promoted domestic output and technological change in the publishing industry to the extent that, by the turn of the century, the balance of trade shifted in favor of the United States. At this point, self-interest dictated reforms in the copyright laws, although the provisions still included clauses to ensure the protection of U.S. manufacturers and printers. The policies of Britain toward its colonies also were quite similar, in their recognition of the benefits of discriminating across countries based on their level of economic development. During the nineteenth century, the British administered a two-tiered international copyright system that attempted to address the needs of its colonies. The Foreign Reprints Act of 1847 allowed colonies to import the works of British authors without copyright protection, introduced taxes as a substitute for copyright, and also provided for legal price discrimination with significantly lower prices for overseas editions.

Markets proved to be particularly adaptive in the context of cultural goods that did not benefit from copyright protection. During the nineteenth century, American publishers of unprotected reprints were able to appropriate returns from a variety of strategies, including privately created tradeable rights of exclusion ("synthetic copyrights"), lead time, or first mover advantages, and through cartelization. The more "reputable" publishers were able to secure greater returns because they offered products that were more likely to be free of defects, thus leading to appropriation through reputation.[39] Price discrimination was a strategy that likely increased both private and social returns relative to state-enforced property rights. Legal decisions evolved in the direction of private law, with formalized protection of trade secrets and well-developed common law doctrines of unfair competition that attained similar ends. Adjustments in copyright regimes undoubtedly redistributed returns, but they did not bring about the dramatic changes that copyright owners predicted. In short, the comparative historical evidence reveals considerable substitutability both within and across institutions.

Some of the changes in the American and European intellectual property regimes this study assessed, such as the introduction of the examination of

---

[39] The reputational effect may partly explain why foreign pharmaceutical firms in Brazil increased their share of the domestic market even in the absence of patent protection. See C. R. Frischtak, "The Protection of Intellectual Property Rights and Industrial Technology Development in Brazil," in F. W. Rushing and C. G. Brown (eds.), *Intellectual Property Rights in Science, Technology, and Economic Performance*, Boulder: Westview Press, 1990.

patent applications or expansion in the subject matter of copyrights, implemented what might be thought of as technical improvements. However, others such as the extension of copyrights to foreign nationals, the gradual strengthening of copyright protection, product exemptions, and the use of compulsory licenses, involved adaptations that seem related to the stage of economic development. This analysis of the evolution of intellectual property regimes in Europe and the United States raises questions about the desirability of applying the same system to all places and circumstances at all times. Indeed, the major lesson that one derives from this economic history of the United States in the European mirror is that intellectual property rights best promoted the progress of science and useful arts when they evolved in tandem with other institutions and in accordance with the needs and interests of social and economic development.[40] Thus, the historical record suggests that appropriate intellectual property systems were not independent of the level of development nor of the overall institutional environment.

Many observers today pay willing attention to historical accounts of patents and copyrights, but still persist in regarding their implications as curiosities that are irrelevant to the new economy and "digital dilemmas" of the twenty-first century. It should be noted that observers in the past were just as impressed with the "new era" associated with the advent of railroads, telegraphs, electricity, radio, telephones, and manned flight, which dramatically increased productivity and transformed social practices. They were perhaps even more concerned than we might be about the consequences of technological change. They found that U.S. institutions were sufficiently flexible to deal with even the most radical new technologies.[41] Instead, they realized that a far more serious consideration was how to maintain the central principle of public access and democratization in the face of these invasive innovations. The checks and balances of interest group lobbies, the legislature and the judiciary worked effectively as long as each institution was relatively well matched in terms of size and influence. By the early twentieth century, technological change and market expansion led to the concentration of corporate interests, increases in the number of uncoordinated users

---

[40] See the *Roundtable on Intellectual Property and Indigenous Peoples*, World Intellectual Property Organization (July 23 and 24, 1998), referring to some of the problems of ensuring that IPRs do not operate to the disadvantage of community norms that regard new ideas and inventions as part of the public domain. William P. Alford, *To Steal a Book is an Elegant Offense*, Palo Alto: Stanford University Press, 1995, argues that Chinese IPR policy is explicated by such community values. Copying or "plagiarism" are not held to be reprehensible because they are consistent with principles that revere the ancestral past and ancient customs. Such practices were prevalent in classical Chinese literary and artistic works. Alford argues that, unlike China, Taiwan has succeeded in changing its political institutions and privatizing its culture and this commercial market orientation helps to explain its greater success in intellectual property reforms.

[41] B. Zorina Khan, *Technological Innovations and Endogenous Changes in U.S. Legal Institutions, 1790–1920*, NBER Working Paper w10346 (March 2004.)

over whom the social costs were spread, and international harmonization of laws. Such changes created a tendency toward the over-enforcement of intellectual property rights at both the private and public levels.

This was especially true in the area of copyrights. Eighteenth-century publishers, who had struggled and were repeatedly defeated in their efforts to attain copyrights in perpetuity, would have been envious of the ease with which publishers achieved their objectives some two hundred years later. Their modern counterparts could legally employ technological methods to extend their claims beyond the first sale of the item, and could even monitor and control what users were permitted to do in the privacy of their homes.[42] Copyright legislation in the twentieth century was produced through an inbred process of negotiation among industry representatives, with few to defend the public domain. In an era of costly access to courts, the holders of intellectual property rights were better able to employ litigation as a strategy to extend their rights rather than simply to protect them. The resulting uncertainty in the intellectual property arena led risk-averse individuals and educational institutions to err on the side of abandoning their right to free access rather than hazard expensive challenges and damaging litigation. Copyright had finally been transformed from a species of trade restraint that merely protected domestic publishers against illegal copying by rivals, into a virtually perpetual right against all of society, not just in the country of origin but globally.

The interpretation of copyright as pervasive monopoly became even more apt after international harmonization with European doctrines introduced major changes in American copyright provisions. One of the most significant of these changes also was one of the least debated: compliance with the precepts of the Berne Convention accorded automatic copyright protection to all creations on their fixation in tangible form. This measure had far-reaching consequences, because it reversed the relationship between copyright and the public domain that the U.S. Constitution stipulated. According to original U.S. copyright doctrines, the public domain was the default, and copyright merely comprised a limited exemption to the public domain; after the alignment with Berne, copyright became the default, and the rights of the public merely comprised a limited exception

---

[42] Copyright scholars have been concerned that modern technologies such as digital music have disturbed this balance by reducing existing consumer rights and facilitating enforcement that infringes on the public domain and on social welfare. See, for instance, Jessica Litman, *Digital Copyright*, New York: Prometheus Books, 2001, p. 14, who argues that "copyright is now seen as a tool for copyright owners to extract all potential commercial value from works of authorship, even if that means that uses that have long been deemed legal are now brought within the copyright owner's control." It should be noted that many of the features these scholars find objectionable – such as the ability of digital copyright owners to control use after the first sale of the item – are perfectly in keeping with the Berne Convention in which the United States is now a member.

to the primacy of copyright. Additions to the public domain from this point on would have to be achieved through affirmative actions and by means of specific exemptions. European notions of "creativity" implied that academic and government output of information in the public domain could be used to subsidize the copyrighted duplications of commercial sellers who infused the good with the minimal amount of "creativity" that the law required. As Joseph Story had long ago pointed out in the context of patent laws, such "metaphysical" precepts did not accord well with the utilitarian intent to promote the progress of science and useful arts. A number of commentators were equally concerned about other dimensions of the globalization of intellectual property rights, such as the movement to emulate European grants of property rights in databases and their potential to inhibit diffusion and learning. Still, the reaction to these fundamental reverses was muted, perhaps in part because of the feeling that democratization was an outdated objective in the corporate and high technology environment of the twenty-first century.

As in the nineteenth century, today there is no shortage of proposals among economists regarding the features of intellectual property institutions that might best promote social and economic development. Some scholars have highlighted the role that enforceable patent rights might play in encouraging trade, inflows of investment, and technology transfer. Others regard the patent system as "broken," and those in the "open source movement" see little virtue in formal copyright rules regardless of the level of development. Most, if not all, would agree that is simplistic to suppose that the institutions and policies from any one country can be transferred in their entirety, across time or space, with any degree of success. Still, this does not imply that historical analysis yields no lessons, although the precise nature of those lessons will differ according to individual circumstance. This study drew attention to the democratizing role of intellectual property in an era when copyright was secondary to the public interest in science and learning, and well-enforced patents fostered markets in invention. Secure property rights in patented inventions helped to create tradeable assets, and this securitization disproportionately benefited creative individuals who were relatively disadvantaged. This central feature of U.S. intellectual property right doctrines seems especially relevant to the countries that are now attempting to extract the mass of their population from poverty.[43]

---

[43] Two recent studies sponsored by the World Bank come to the same conclusion. They propose that creative approaches to intellectual property rights have the potential to promote increases in social welfare in underdeveloped economies. See J. Michael Finger and Philip Schuler (eds), *Poor People's Knowledge: Promoting Intellectual Property in Developing Countries*, World Bank and Oxford University Press, 2004; Carsten Fink and Keith E. Maskus (eds), *Intellectual Property and Economic Development: Lessons from Recent Economic Research*, Washington, D.C.: World Bank and Oxford University Press, 2005.

Knowledge, information, literacy and learning are all key determinants of prospects for growth, so it is crucial for copyright doctrines to facilitate this process through liberal use of fair use exceptions and expansions of the public domain. Economic development also requires decentralized strategies that extend to the informal economy and to rural communities that tend to be untouched by the large-scale urban projects that incorporate foreign technologies. Some might question the extent to which untutored peasants or women in the household can contribute to invention or productivity gains, or whether they respond to market incentives. Similar doubts were equally present among nineteenth-century European commentators on the American experiment. The American response was based on the conviction that increased efforts and self-determination could be induced through a system that offered access to a broad spectrum of the population and allowed them to participate in expansions in the market. Today, appropriate intellectual property institutions would enhance the incentives for residents in developing countries to create and market incremental inventions that have the potential to dramatically improve their standards of living. Although the specific rules and standards that are needed might differ in part from their historical precursors, the principle of the democratization of invention is still vital to achieving advances in global welfare.

Index